PRENTICE-HALL ELECTRICAL ENGINEERING SERIES

William L. Everitt, *editor*

INFORMATION THEORY SERIES

Thomas Kailath, *editor*

Di Franco & Rubin
Radar Detection

Downing
Modulation Systems and Noise

Dubes
The Theory of Applied Probability

Franks
Signal Theory

Golomb, Baumert, Easterling, Stiffler, Viterbi
Digital Communications with Space Applications

Raemer
Statistical Communication Theory and Applications

SIGNAL THEORY

PRENTICE-HALL INTERNATIONAL, INC. *London*
PRENTICE-HALL OF AUSTRALIA, PTY. LTD. *Sydney*
PRENTICE-HALL OF CANADA, LTD. *Toronto*
PRENTICE-HALL OF INDIA PRIVATE LTD. *New Delhi*
PRENTICE-HALL OF JAPAN, INC. *Tokyo*

L. E. Franks

Bell Telephone Laboratories
North Andover, Massachusetts

SIGNAL THEORY

Prentice-Hall, Inc., Englewood Cliffs, N.J.

Library of Congress Catalog Card Number 75-77901

Printed in the United States of America

13-810077-2

Current Printing (last digit):
10 9 8 7 6 5 4 3 2 1

PREFACE

Anyone concerned with a study of the observable attributes of a physical system faces the task of finding suitable means for representing and classifying signals. In considering signals as entities by themselves, more or less separately from the systems that produce them, one is confronted with an immense variety of possibilities for representation and classification. The suitability of the various techniques depends largely upon how the observer intends to utilize the information carried by the signals. To a large extent, a generalized and unified treatment of these techniques is provided by the mathematical concepts embodied in the subject of functional analysis. In this book, I have tried to present a subset of these concepts, particularly those dealing with linear spaces, in order to tie together many familiar techniques of signal analysis and to provide the framework for a more generalized and effective application of these techniques to engineering problems. A major part of the advantage gained by familiarity with these ideas is the insight afforded by geometrical visualization of signals and signal processing operations. By means of several examples in each chapter, I have emphasized physical interpretation of the concepts rather than mathematical rigor. Because of my own background, the examples have a strong communication systems flavor, but hopefully the broader applications of the ideas, e.g., in automatic control, biophysical, and geophysical systems will be evident. Also, the examples tend to be idealized and are usually capable of solution by less general methods. This choice of familiar, often classical, examples is intentional in order to better serve the purpose of clarifying the physical interpretation of individual mathematical concepts.

Regarding content and organization of the book, the first five chapters present a step-by-step development of the signal space concept, beginning with sets and equivalence relations, proceeding to metric spaces, and finally to finite-dimensional and infinite-dimensional linear spaces with an inner product. Chapter 5, on linear operators, can be considered an introduction to the more abstract aspects of signal theory. Much of the material in this chapter is not essential to an understanding of the remaining topics in the book. For this first half of the book, the reader may quite possibly feel the need for a companion text of a more basic mathematical nature. For this purpose I can highly recommend the very readable text by G. F. Simmons, *Introduction to Topology and Modern Analysis*, New York, McGraw-Hill Book Company, 1963. The remaining five chapters represent the payoff in terms of applications to practical engineering problems. Several classes of problems, with well-established practical importance, are re-examined from the signal space viewpoint. Beginning with Chapter 7, problems involving random signal processes in addition to deterministic signals and system elements are introduced. As a general reference which is particularly appropriate, I would suggest the text by A. Papoulis, *Probability*, *Random Variables*, *and Stochastic Processes*, New York, McGraw-Hill Book Company, 1965. Additional references are listed at the end of each chapter. It is obvious that I have made no attempt to provide a complete bibliography of relevant papers and texts. Instead, I have listed a few references which I feel would be especially helpful for background material and for a starting point for further investigation of the various topics mentioned. The exercises, for the most part, are not merely routine manipulation of equations. They have been inserted throughout the text as a convenient means for introducing related concepts and providing additional examples.

For a formal course offering, the level of the text is appropriate for first-year engineering graduate students who have taken a course in linear systems (stressing applications of the Fourier transform) and an introduction to probability and random variables. Although the text might be used as supplementary material in an advanced communication theory course, it is primarily intended for a separate, more basic course in signal theory. There is ample evidence that such a course would be a desirable curriculum addition as preparation for more advanced system theory courses such as detection and estimation theory, state-variable techniques, and optimal control. The topics of generalized functions, digital signal processing, Kalman filtering, stochastic approximation, and information theory, omitted from this book, are certainly relevant to the subject; yet they can reasonably be relegated to subsequent courses of study.

The material in its present form has evolved from lecture notes prepared for various versions of a course given for the past several years in the Communications Technology Program at Bell Telephone Laboratories and for a graduate course given at Columbia University during 1965. My interaction with students participating in these courses has strongly influenced the selection and organization of the topics, and I gratefully acknowledge the benefits of these discussions. Other guidance has come from many sources. In particular, I would like to acknowledge the contribution of some of my colleagues at Bell Telephone Laboratories—Drs. Allen Gersho, Francis S. Hill, Jr., and Robert E. Maurer—for their efforts in reviewing the manuscript and their many valuable suggestions for improving the presentation.

L. E. FRANKS

CONTENTS

3

DISCRETE SIGNAL REPRESENTATIONS, 44

4

INTEGRAL TRANSFORMS FOR SIGNAL REPRESENTATION, 66

5

REPRESENTATION OF LINEAR OPERATORS, 98

6

CHARACTERIZATION OF SIGNAL PROPERTIES, 135

SIGNAL THEORY

INTRODUCTION

1

1.1 Signal Processing Systems

It is commonly understood that what is meant by a *signal* is a quantity which, in some manner, conveys information about the state of a physical system. In this context, it is natural to think of a signal as the result of a measurement on the physical system under observation. The measuring apparatus functions as a signal processor, taking "raw" data and transforming it into a recognizable form for the benefit of the observer. This apparatus may form a highly complex system which, for convenience, is best separated into smaller subsystems, each performing a specialized signal processing operation. A typical decomposition into subsystems is illustrated in Figure 1.1 which provides a model sufficiently general for our subsequent studies. The blocks shown represent a rather arbitrary division of the overall processing operation; however, most systems of interest contain these components, even if they appear in a very elementary form. The first *transducer* acts as a "probe" which converts the (mechanical, electrical, optical, thermal, chemical, etc.) physical quantities, denoted by x_1, into another set of physical quantities x_2, which are more amenable to subsequent processing. The type of transducer to be employed is very much dependent on the "state of the art" of the various technologies involved. For example, it is clear that, with present-day technology, television broadcasting based on an electro-optical system is much more feasible than a purely optical system. Having converted the physical quantities into, say, electrical signals, we can proceed to alter the format of the signals to emphasize the more significant attributes and de-emphasize or delete unwanted information concerning the state of the system under observation. This, in general terms, is the function of the *coding* apparatus. The function of the *modulator* is to provide a signal x_4 which is suitably matched to the characteristics of the *transmission channel* which will be employed in the case of remote observations. For instance, if a waveguide is to be used, the signal x_3 would normally modulate the amplitude or phase of a sinusoidal carrier at microwave frequencies. The demodulator and decoder serve to "undo" the special processing inserted to accommodate the characteristics of the transmission channel, and they provide an electrical

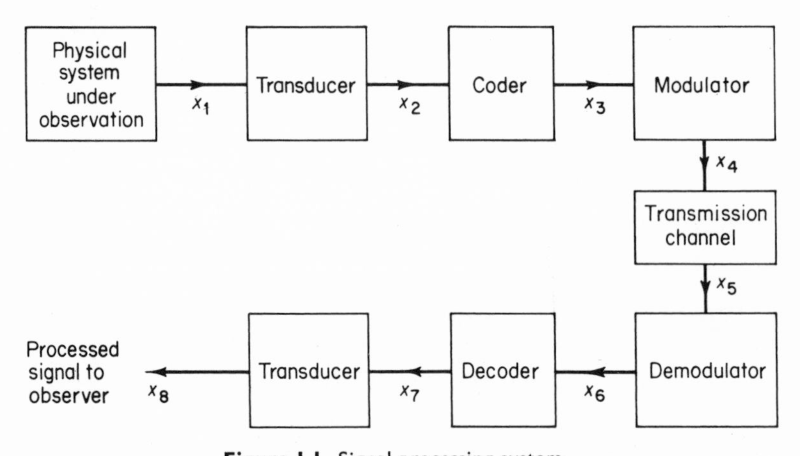

Figure 1.1. Signal processing system.

signal which, after passing through a transducer, is in the desired form for recognition by the observer. The reader will recognize the wide variety of forms possible in the implementation of these various functions by recalling design features of systems with which he is familiar. To enumerate a few obvious system examples, we have voice-to-voice communication, data telemetering, radar, spacecraft guidance, manufacturing process control, television, telegraphy, medical diagnostics, automatic character and pattern recognition, and atomic particle detection. It should also be noted that the processor modeled in Figure 1.1 could represent a subsystem of a larger system; e.g., it could form the feedback path delivering the control signals in an automatically controlled plant.

The purpose of the foregoing discussion is simply to point out the wide variety of signal formats encountered, even within a particular processing system. A signal theory must be sufficiently general to accommodate this variety of formats. With this in mind, we can state that the scope of signal theory includes the study of methods for the analytical representation of signals, the numerical characterization of significant signal properties, and the characterization of the signal transforming properties of various processing devices. Within this scope, we shall examine some of the more mathematically tractable aspects of these problems in this text.

In the examples mentioned above, most of the signals are quantities which fluctuate with time. It is natural and convenient for us to think of signals and time functions interchangeably, even if this occasionally means an artificial introduction of a time dependence. An optical image, for example, would be described by a function of spatial coordinates. However, the methods used for dealing with time functions are usually adaptable for use with functions of other variables.

At this point we are concerned with the problem of assigning a numerical representation to a time function [denoted abstractly by $x(t)$] which will

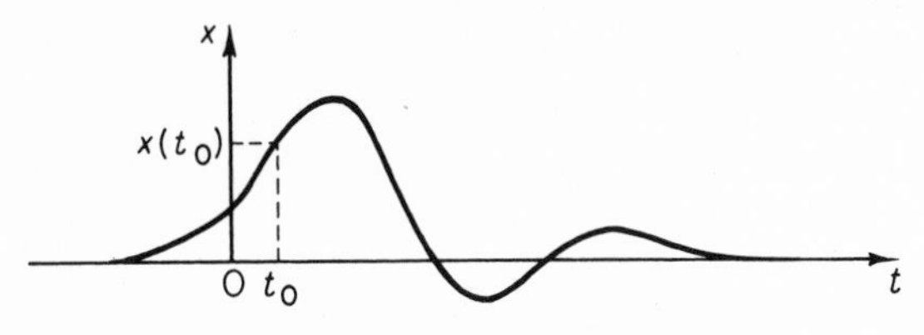

Figure 1.2. Graph of a signal.

serve to identify the function and to distinguish it from other functions which are substantially different from it. A familiar representation, in fact one that is deeply ingrained by usage, is the *graph* of a function. The graph is the collection of all the ordered pairs of numbers $\{t, x(t)\}$ most frequently displayed in rectangular coordinates, as shown in Figure 1.2.

We comment here that there is a long-standing ambiguity in the meaning of $x(t)$. Strictly speaking, $x(t)$ is simply a number which is the value of the function at the time instant t. Popular usage, however, allows us to call the function, which actually is the rule by which values of t and x are paired, simply by $x(t)$. We shall use x and $x(t)$ interchangeably to denote the function when this can be done without confusion. Some authors prefer to use $x(\cdot)$ to distinguish the function from the number $x(t)$.

The human observer, in a technological environment, has become very accustomed to graphically displayed signals and has developed a wide variety of oscillographic instruments for this purpose. A person can be trained to be highly perceptive in extracting information from radar displays, seismographs, cardiograms, polygraphs, etc., but his method of signal processing is apt to be rather mysterious and not readily adaptable to numerical analysis or to mechanized processing. From the standpoint of the system designer, the graphical representation is unmanageable simply because it consists of too many individual points. The simple expedient of representing a signal by a subset of its graph, e.g., the values of x at a set of instants equally spaced in time, is but one example of the general techniques for signal representation to be discussed in the following chapters.

1.2 Sets of Signals

In contrast to graphical representation where signals are represented by a complicated collection of points in a simple setting, i.e., a two-dimensional space, we shall instead establish more highly structured settings, *signal spaces*, whereby a signal can be considered as a single entity or point in the space.[1] As a first step in this direction, we consider a signal as an element of a set S. The set itself is described in terms of a property P which is a statement that must be satisfied by every element belonging to the set. This is indicated symbolically by $S = \{x; P\}$, i.e., the set of all x such that P is true. To introduce additional notation, we can write $P \Rightarrow x \in S$, which is read as "P is true implies that x is contained in S." The defining property P for a set creates a subset of all possible signals. If this property is sufficiently restrictive, the subset S is apt to be more easily manageable than an unrestricted set. Of course, if the property is too restrictive, then the subset will exclude too many signals of interest. The choice of P must necessarily be tailored to the problem at hand. A few examples of signal sets which are frequently encountered in signal analysis problems are given below.

Sinusoidal signals. Let S_C denote the set of all sinusoidal signals; i.e.,

$$S_C = \{x; x(t) = \text{Re}\,[e^{\alpha+j(\theta+2\pi ft)}], -\infty < t < \infty, \quad \alpha, \theta, f \in R\} \tag{1.1}$$

The statement, $\alpha, \theta, f \in R$, in (1.1) means that these parameters can be arbitrarily selected from the set of all real numbers R. This implies that S_C contains sinusoids having all possible values of amplitude, phase, and frequency. Often the defining property for a particular set can be expressed in an alternative form; e.g.,

$$S_C = \left\{x; \frac{d^2x(t)}{dt^2} + \lambda^2 x(t) = 0, -\infty < t < \infty, \lambda \in R\right\} \tag{1.2}$$

Periodic signals. Let $S_R(T)$ be the set of signals which are periodic, with period equal to T, then

$$S_R(T) = \{x; x(t+T) = x(t), -\infty < t < \infty\} \tag{1.3}$$

Bounded signals. The set of signals whose instantaneous values are bounded in magnitude by the real, positive number K is denoted by

$$S_M(K) = \{x; |x(t)| \leqslant K, -\infty < t < \infty\} \tag{1.4}$$

It is clear that

$$x \in S_M(K_1) \Rightarrow x \in S_M(K_2) \qquad \text{if } K_2 \geqslant K_1$$

Energy-limited signals. The signals in

$$S_E(K) = \left\{x; \int_{-\infty}^{\infty} x^2(t)\, dt \leqslant K\right\} \tag{1.5}$$

are said to have energy limited to K, where K is a real, positive number. It has become common practice to physically interpret the integral in (1.5) as the energy content of a signal. This is merely a convenient normalization, for if $x(t)$ is a voltage appearing across a 1-ohm resistive load, then the time integral of the square of this voltage is the total energy dissipated in the load.

Duration-limited signals. Let $S_D(T)$ be the set of signals which vanish outside a specified time interval $-T \leqslant t \leqslant T$.

$$S_D(T) = \{x; x(t) = 0 \text{ for all } |t| > T\} \tag{1.6}$$

Note that

$$x \in S_D(T_1) \Rightarrow x \in S_D(T_2) \qquad \text{if } T_2 \geqslant T_1$$

Band-limited signals. Let $S_B(W)$ be the set of signals band-limited to frequencies less than some specified frequency W, i.e.,

$$S_B(W) = \left\{x; X(f) = \int_{-\infty}^{\infty} x(t)e^{-j2\pi f t}\, dt = 0 \quad \text{for all } |f| > W\right\} \tag{1.7}$$

where $X(f)$ is the Fourier transform[1] of the time function $x(t)$.[2]

Set Operations

In dealing with signal sets, it is helpful to employ two of the elementary set operations: (1) the *union*, defined by

$$S_1 \cup S_2 = \{x; x \in S_1 \text{ or } x \in S_2\} \tag{1.8}$$

and (2) the *intersection*, defined by

$$S_1 \cap S_2 = \{x; x \in S_1 \text{ and } x \in S_2\} \tag{1.9}$$

The sets created by these operations are shown pictorially in Figure 1.3.

[1] We shall follow the conventional practice of assigning the corresponding capital letter to denote the Fourier transform of a time function.

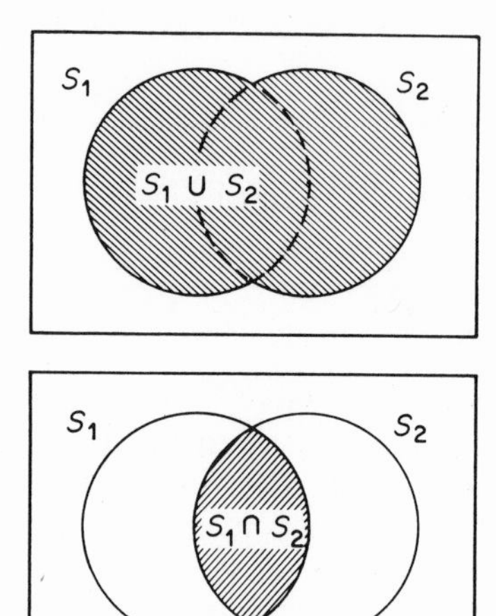

Figure 1.3. Pictorial representation of the union and intersection of two sets.

Example 1.1. It is a well-known fact to communication engineers that a signal cannot be both duration-limited and band-limited. This is because the integral

$$\int_{-T}^{T} x(t)e^{-j2\pi f t}\,dt$$

can be zero only for isolated values of f unless $x(t) = 0$ for all $|t| \leqslant T$. Hence we have

$$S_D(T) \cap S_B(W) = \{0\} = \{x;\, x(t) = 0 \text{ for all } t\} \qquad (1.10)$$

It may seem a trivial point, but it is important to distinguish $\{0\}$ from the *empty*, or *null*, set $\varnothing$ which contains no elements, whereas $\{0\}$ contains an element.

Exercise 1.1. Consider the countable set of signals

$$S_A = \{x_n(t);\, n = 1, 2, 3, \ldots\}$$

where

$$x_n(t) = ne^{-nt} \qquad \text{for } t \geqslant 0$$
$$= 0 \qquad \text{for } t < 0$$

Enumerate the elements in $S = S_A \cap S_M(10) \cap S_E(4)$.

Exercise 1.2. Describe the set which is the intersection of $S_R(T)$ and $S_E(K)$.

Partitions and Equivalence Relations

The set operations $\cup$ and $\cap$ can be used to describe the *partition*[3],[4] of a set into a sequence of disjoint subsets, as shown pictorially in Figure 1.4. We say that the set of sets $\{S_1, S_2, S_3, \ldots\}$ forms a partition of S if

$$S = S_1 \cup S_2 \cup S_3 \cup \cdots$$

and (1.11)

$$S_i \cap S_j = \varnothing \qquad \text{for } i \neq j$$

The motivation for partitioning signal sets is simply to create sets which are more manageable. It is possible, for example, to partition sets containing an uncountable number of elements into a finite or countable number of subsets, as some of the following examples illustrate. A partition can be generated by an *equivalence relation* and this is often the most convenient way to define a particular partition. We say that two elements are equivalent, $x \sim y$, if the relation $\sim$, defined for all pairs of elements, satisfies the following

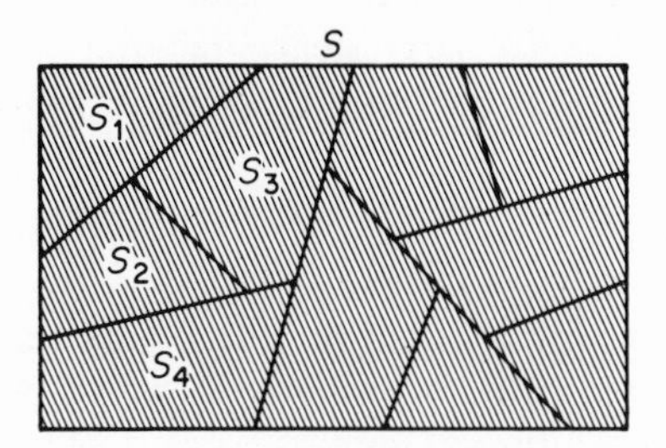

Figure 1.4. Partition of a set into disjoint subsets.

properties:[3],[4]

$$\begin{aligned}
&\text{a.} \quad x \sim x \qquad \text{for all } x \quad \text{(reflexivity)} \\
&\text{b.} \quad x \sim y \Rightarrow y \sim x \quad \text{(symmetry)} \\
&\text{c.} \quad x \sim y \quad \text{and} \quad y \sim z \Rightarrow x \sim z \quad \text{(transitivity)}
\end{aligned} \tag{1.12}$$

The resulting partition is simply the set of distinct subsets, called *equivalence sets*, with a typical equivalence set (containing the element x) given by

$$S_x = \{y; y \sim x\} \tag{1.13}$$

It is not difficult to show that every partition generates an equivalence relation; hence these two concepts really amount to the same thing, namely the grouping together into disjoint subsets all elements which are equivalent in some sense to each other.

Exercise 1.3. Show that an arbitrary partition (1.11) generates an equivalence relation; i.e., "$x \sim y$ if, and only if, x and y are contained in the same subset" satisfies (1.12) and hence is a valid equivalence relation. Conversely, show that an arbitrary equivalence relation (1.12) generates a partition; i.e., the distinct subsets defined by $S_x = \{y; y \sim x\}$ are disjoint and their union is the entire set as required by condition (1.11).

Example 1.2. Equality is an equivalence relation, but the equivalence sets contain only the individual elements. ▮

Example 1.3. Taking a familiar example from number theory, we consider the partition of the set of integers $\{n; n = 0, \pm 1, \pm 2, \ldots\}$ into a finite number m of equivalence sets

$$S_i = \{n; n = pm + i\} \qquad i = 0, 1, 2, \ldots, m-1 \tag{1.14}$$

where p is any integer. The corresponding equivalence relation $n_1 \sim n_2 \Rightarrow n_1 - n_2 = pm \Rightarrow n_1 = n_2 (\mathrm{mod}\, m)$ is called a *modulo m congruence*. For example, the mod 2 congruence partitions the integers into the sets of *even* and *odd* integers. ▮

Example 1.4. If we exclude from consideration the subset of signals $S_0 = \{x; x(t_0) = 0\}$, then the equivalence relation $x \sim y \Leftarrow x(t_0)y(t_0) > 0$ partitions the remaining signals into the two equivalence sets [2]

$$\begin{aligned} S_+ &= \{x; x(t_0) > 0\} \\ S_- &= \{x; x(t_0) < 0\} \end{aligned} \tag{1.15}$$

This partitioning is extensively used in binary data transmission where one binary digit is associated with all signals in S_+ and the other with all signals in S_-. Figure 1.5 shows a typical implementation. Although the transmitted set of signals may contain only two elements, the received signals may form a

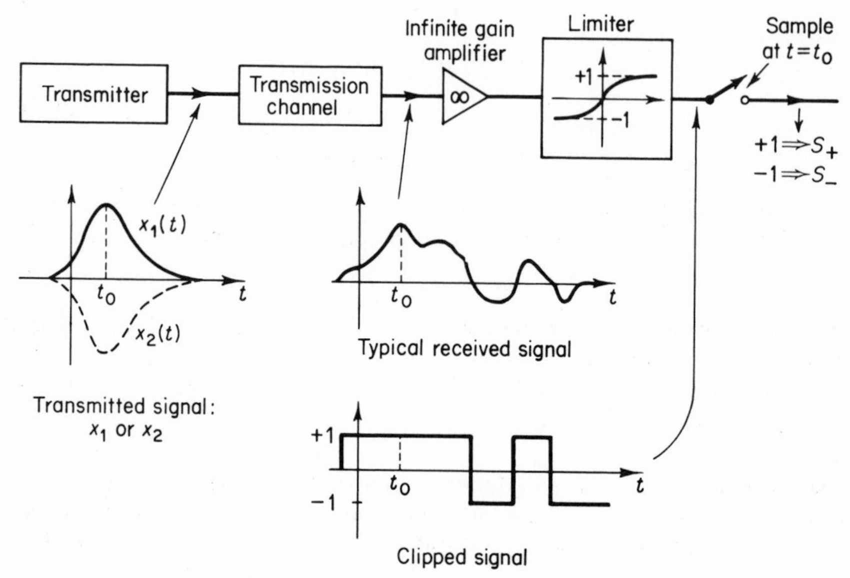

Figure 1.5. Binary data communication.

[2] $\Leftarrow$ may be read *is implied by*.

very large set because of noise or other disturbances introduced by the transmission channel. The observer decides which signal was transmitted according to the partition (1.15) which is indicated by the output of the threshold and sampling circuitry. As a practical matter, the set S_0 can be appended to either S_+ or S_- with no effect on performance because of the negligible chance that a received signal will fall in this set. ▮

Example 1.5. Another type of receiver for binary data transmission, which turns out to be more immune to channel disturbances, is one which

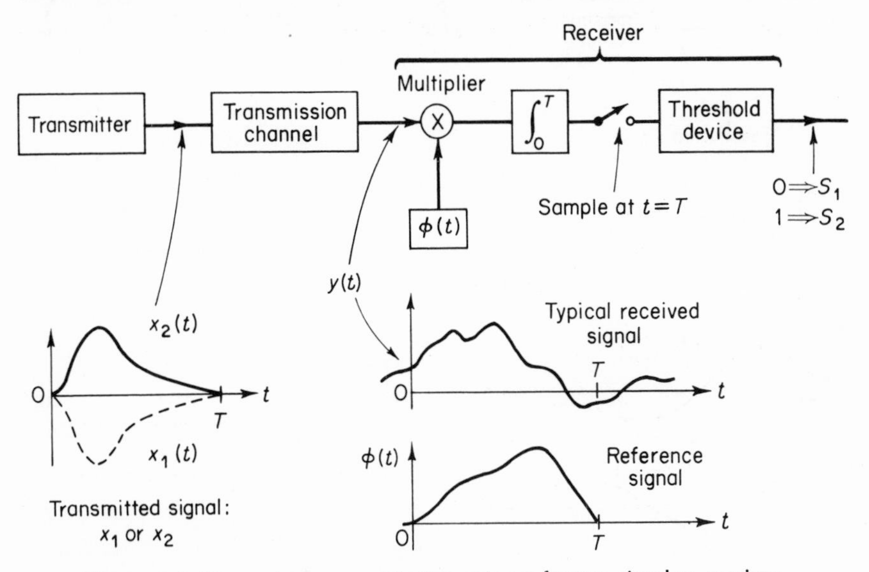

Figure 1.6. Binary data communication using reference signal at receiver.

employs a reference signal φ to partition the received signals into two sets. The partition into sets S_1 and S_2, corresponding to decisions as to whether x_1 or x_2 was transmitted, is performed on received signals y according to

$$\begin{aligned} S_1 &= \left\{ y; \int_0^T y(t)\varphi(t)\,dt < r \right\} \\ S_2 &= \left\{ y; \int_0^T y(t)\varphi(t)\,dt > r \right\} \end{aligned} \tag{1.16}$$

where r is some predetermined threshold value. In order to implement this receiver, a multiplier, integrator, sampler, and threshold device is used, as shown in Figure 1.6. Techniques for optimal selection of the reference signal and the threshold value are discussed in some detail in Chapter 10. ▮

Example 1.6. Another possibility for distinguishing signals is to count the number of zero crossings of the time functions in some selected time interval. We define the partition

$$S_n = \{x;\, x(t) \text{ has } n \text{ distinct zeros in the time interval}\} \qquad \text{where } n = 0, 1, 2, \ldots \tag{1.17}$$

We could make this a finite partition,

$$S = S_0 \cup S_1 \cup S_2 \cup \cdots S_{N-1} \cup S_{N^+}$$

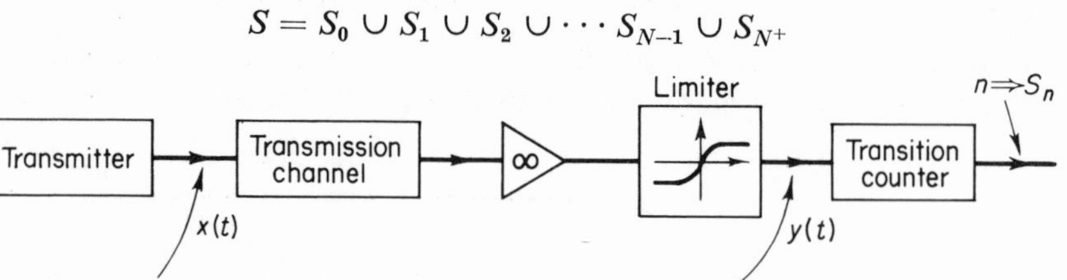

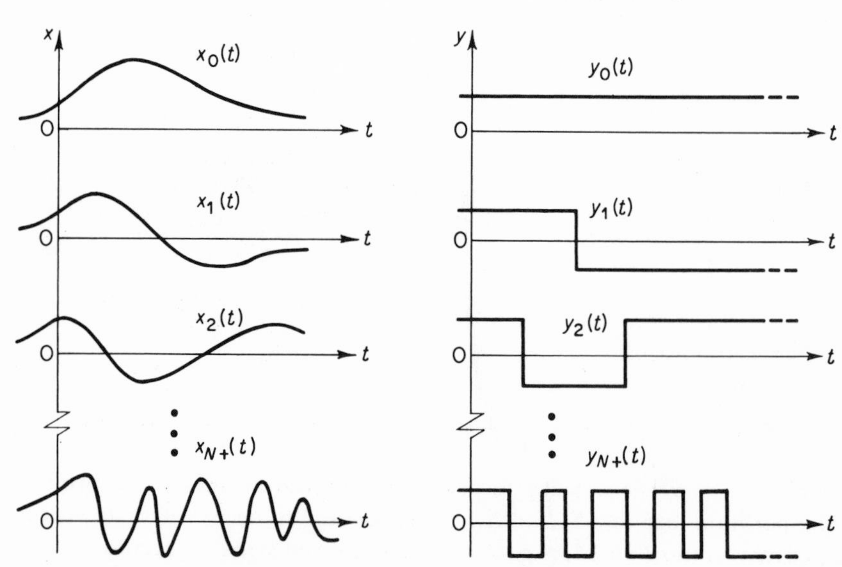

Figure 1.7. Zero-crossing signaling system.

provided S_{N^+} is defined as the set of all signals having N or more zero crossings in the interval. The implementation of an $(N + 1)$-ary alphabet data transmission system based on this partitioning is indicated in Figure 1.7. ∎

Example 1.7. Given a set of time functions $\{\varphi_i;\, i = 1, 2, \ldots, n\}$, we can define an equivalence relation by

$$x \sim y \Rightarrow \int_{-\infty}^{\infty} x(t)\varphi_i(t)\, dt = \int_{-\infty}^{\infty} y(t)\varphi_i(t)\, dt \qquad \text{for all } i = 1, 2, \ldots, n \tag{1.18}$$

This equivalence relation is actually a generalized form of congruence (Example 1.3) where

$$x \sim y \Rightarrow x = y(\text{mod } M) \Rightarrow x - y \in M$$

and M is the subset of functions defined by

$$M = \left\{z; \int_{-\infty}^{\infty} z(t)\varphi_i(t)\, dt = 0 \qquad i = 1, 2, \ldots, n\right\} \tag{1.19}$$

The resulting equivalence sets can each be characterized by a *representative element* $\hat{x}$ in the sense that

$$S_{\hat{x}} = \{x; x \sim \hat{x}\} = \{x; x = \hat{x} + z\}$$

where

$$z \in M \quad \text{and} \quad \hat{x}(t) = \sum_{k=1}^{n} a_k \varphi_k(t) \tag{1.20}$$

Under certain conditions on the set $\{\varphi_i; i = 1, 2, \ldots, n\}$, there is a one-to-one correspondence between the equivalence sets $S_{\hat{x}}$ and the ordered sequences of real numbers $\{a_1, a_2, \ldots, a_n\}$, called *n-tuples*. Hence, we see that the set of equivalence sets, as defined in this example, is a reasonably manageable set, i.e., the set of all n-tuples or, as we shall see later, an *n-dimensional vector space*. This example has a basic significance for our subsequent studies since it provides a generally applicable, as well as mathematically tractable, method for representation of signals. To be meaningful, it requires a basic understanding of the concepts of *metric spaces* and *linear spaces*, which are introduced in the following chapter. ▌

1.3 Mappings and Functionals

The previous section has introduced the concept, by means of equivalence relations, of providing alternative sets for the characterization of signals. Another useful concept, which is actually a more general form of relation between elements, is that of a *mapping* of elements from one set into elements of another set. A mapping is simply the rule by which elements in one set, say S_1, are assigned to elements in another set, say S_2. Symbolically, the mapping is denoted by f: $S_1 \rightarrow S_2$, which is a compact notation for

$$y = f(x); \qquad x \in S_1 \quad \text{and} \quad y \in S_2 \tag{1.21}$$

The element y in S_2 is called the *image* of x under the mapping f. The set S_1 is the *domain* of the mapping, and the set of all images of elements of S_1 (contained in S_2) is the *range* of the mapping. If the range of f is identical to S_2, then f is said to be a mapping of S_1 *onto* S_2. Otherwise, if there are elements in S_2 which are not images of elements in S_1, then the mapping is said to be *into* S_2. A mapping is always *single-valued* in the sense that there is a unique image for each element in S_1. If the images of distinct elements in S_1 are always distinct in S_2, however, then the mapping is *one-to-one*. If a mapping besides being one-to-one is onto, then we can speak of a mapping from S_2 onto S_1, $f^{-1}: S_2 \to S_1$, where f^{-1} is called the *inverse* mapping. In this situation, there results a *one-to-one correspondence* between elements in S_1 and elements in S_2.

It is often convenient to deal with a *composite* mapping which results from two or more consecutive mappings between sets. In Figure 1.8 is shown the

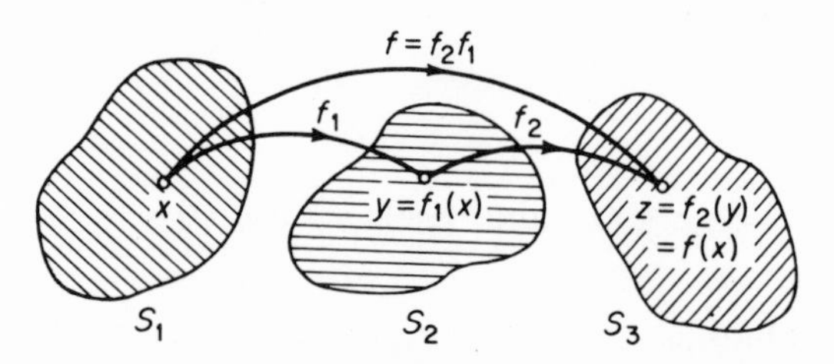

Figure 1.8. The composition of two set mappings.

mapping $f: S_1 \to S_3$ resulting from the two mappings $f_1: S_1 \to S_2$ and $f_2: S_2 \to S_3$. In this case we write $f = f_2 f_1$, which means that

$$f: S_1 \to S_3 \Rightarrow z = f_2(y) = f_2[f_1(x)] = f(x) \qquad \text{for all } x \in S_1 \quad (1.22)$$

To illustrate the idea of a composite mapping, the operation of the threshold and sampling circuitry in Example 1.4 can be decomposed into two mappings: (1) the mapping implied by the equivalence relation (1.15)

$$f_1: S \to \{S_+, S_-\}$$

where

$$S = \{x; x(t_0) \neq 0\}$$

and

$$\begin{aligned} f_1(x) &= S_+ \qquad \text{if } x(t_0) > 0 \\ &= S_- \qquad \text{if } x(t_0) < 0 \end{aligned} \qquad (1.23)$$

and (2) the mapping from the equivalence sets into numerical values

$$\begin{gathered} f_2: \{S_+, S_-\} \to \{+1, -1\} \\ f_2(S_+) = +1; \qquad f_2(S_-) = -1 \end{gathered} \qquad (1.24)$$

The composite mapping is simply

$$\begin{aligned} f(x) &= +1 \qquad \text{for } x(t_0) > 0 \\ &= -1 \qquad \text{for } x(t_0) < 0 \end{aligned} \qquad (1.25)$$

In this example we have used the fact that an equivalence relation (1.12) can be interpreted as a mapping, generally not one-to-one, of elements into their equivalence sets (1.13). In other words, any equivalence relation can be expressed by $f_\sim$, where

$$f_\sim : \{x\} \rightarrow \{S_x\} \Rightarrow f_\sim(x) = S_x \tag{1.26}$$

Perhaps more interesting is the fact that any mapping gives rise to an equivalence relation. For an arbitrary mapping f: $S_1 \rightarrow S_2$, the relation

$$x_1 \sim x_2 \Leftarrow f(x_1) = f(x_2) \tag{1.27}$$

between any pair of elements x_1 and x_2 in S_1 is a valid equivalence relation for the set S_1. For example, let f be the mapping described by

$$f(x) = \int_{-\infty}^{\infty} x^2(t)\, dt$$

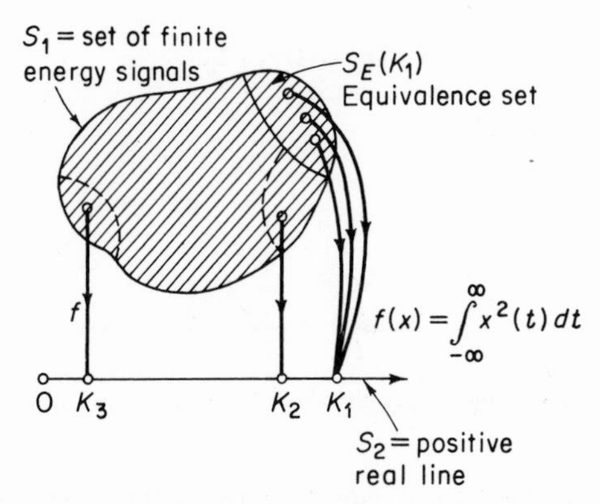

Figure 1.9. A mapping of signals into the real line.

Then we have a mapping from the set of signals into the real line according to the energy content of the signals as shown in Figure 1.9. The equivalence relation corresponding to f partitions S_1 into sets of signals of equal energy.

Fourier Transforms

A mapping extensively used in signal analysis is the Fourier transform. If S_1 is the set of bounded energy signals

$$S_1 = \left\{x; \int_{-\infty}^{\infty} x^2(t)\, dt < \infty\right\}$$

then the Fourier transform $\mathscr{F}: S_1 \rightarrow S_2$ is a mapping into another set of square-integrable functions[3]

$$S_2 = \left\{X; \int_{-\infty}^{\infty} |X(f)|^2\, df < \infty\right\}$$

[3] Square-integrability of $x(t)$ is only a sufficient condition, not a necessary condition, for the existence of a Fourier transform $X(f)$.[2]

This mapping is described by

$$\mathscr{F}: S_1 \to S_2 \Rightarrow X(f) = \int_{-\infty}^{\infty} x(t)e^{-j2\pi ft}\,dt \tag{1.28}$$

Strictly speaking, this mapping is not one-to-one. It is possible to have two or more time functions, such as those shown in Figure 1.10, which have identical Fourier transforms.

From this standpoint, it is clear that $\mathscr{F}$ is a many-to-one mapping. The equivalence sets defined by $\mathscr{F}$ contain time functions which can differ at a countable number of points in any time interval. It is also clear that such discontinuous behavior of signals is of no practical significance and we are perfectly willing to consider each equivalence set as a single signal. This type

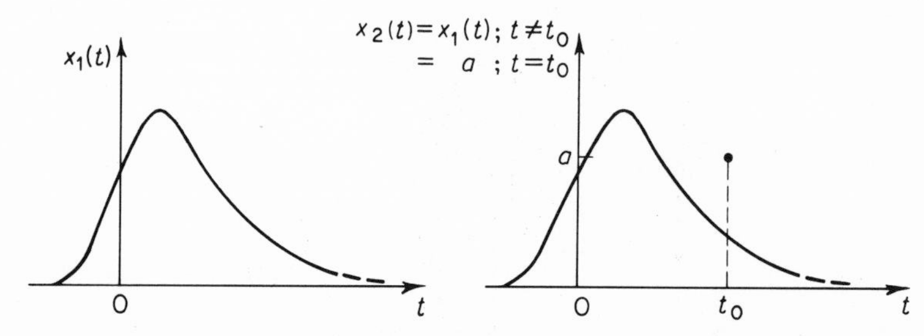

Figure 1.10. Two time functions which have the same Fourier transform.

of equivalence is referred to as *equal almost everywhere* and henceforth we shall not bother to distinguish between signals and their equivalence sets corresponding to the equal almost everywhere relation. Using this relation in both S_1 and S_2, we can now interpret $\mathscr{F}$ as a one-to-one and onto mapping. The inverse mapping is described by

$$\mathscr{F}^{-1}: S_2 \to S_1 \Rightarrow x(t) = \int_{-\infty}^{\infty} X(f)e^{j2\pi ft}\,df \tag{1.29}$$

The relations (1.28) and (1.29) taken together are called a *Fourier transform pair*.

Exercise 1.4. Show that, for an arbitrary mapping f, (1.27) describes a valid equivalence relation with equivalence sets given by

$$S_x = \{y; f(y) = f(x)\}$$

Exercise 1.5. Consider the equivalence sets corresponding to the mapping F described by $F(x) = |X(f)|^2$. Show by particular examples that, in contrast to the mapping $\mathscr{F}$ (1.28), elements in the same equivalence set can be substantially different from each other. *Hint:* $X(f)$ and $X(f)e^{j\theta(f)}$, where $\theta(f)$ is an arbitrary phase function, are Fourier transforms of elements in the same equivalence set.

Exercise 1.6. For an arbitrary finite-energy signal x, show that

$$\int_{-\infty}^{\infty} x^2(t)\,dt = \int_{-\infty}^{\infty} |X(f)|^2\,df$$

Functionals

Mappings from the more abstract signal sets into numerical values have a special significance since we would expect that any physical measurement on a signal should produce a particular numerical value. A mapping from an arbitrary set into a set of numerical values is commonly called a *function.*[4] In our applications, however, the domain will usually be a set of functions in the ordinary sense (mappings from one set of numerical values into another set of numerical values, e.g., time functions, frequency functions, etc.). To avoid confusion, it has become common practice to refer to a mapping from a domain of ordinary functions into numerical values as a *functional.* Thus, a functional is often thought of as a "function of a function."

At this point we should specify what we mean by numerical values. A reasonable choice would be the set R of real numbers; however, there are compelling analytical reasons for extending the set to include the complex numbers C, even though this seems to take us a step removed from physical measurements. "Reality" is restored by noting that each complex number can be associated with an ordered pair of real numbers. With this in mind we can enumerate some typical functionals.[5]

$$\begin{aligned}
f_1(x) &= \int_{-\infty}^{\infty} x(t)\varphi(t)\,dt \\
f_2(x) &= \int_{-\infty}^{\infty} w(t)x^2(t)\,dt \\
f_3(x) &= \int_{-\infty}^{\infty} x(t)e^{-j\Omega t}\,dt \triangleq X\left(\frac{\Omega}{2\pi}\right) \\
f_4(x) &= \int_{-\infty}^{\infty} x(t)\,\delta(t-t_0)\,dt \triangleq x(t_0) \\
f_5(x) &= \int_{-\infty}^{\infty} x(t)\,\delta^{(n)}(t-t_0)\,dt \triangleq (-1)^n \left.\frac{d^n x}{dt^n}\right|_{t=t_0} \\
f_6(x) &= \max\{|x(t)|;\, t_1 \leqslant t \leqslant t_2\} = \lim_{n\to\infty}\left[\int_{t_1}^{t_2} |x(t)|^n\,dt\right]^{1/n}
\end{aligned} \tag{1.30}$$

It is not accidental that the functionals above are all expressed as some form of time integral; this form for a functional is extremely convenient and is

[4] Some authors use "mapping" and "function" synonymously but it is a well-established convention to reserve the name "function" for the mappings described above.

[5] The symbol $\triangleq$ can be read as *equal by definition to.*

used even when a special type of function, such as the δ-function in f_4 and f_5, has to be defined in order to make this type of expression valid.[2]

Series Representations

Anticipating what is to follow (in Chapter 3), our goal in signal representation will be to find approximate series representations, denoted by $\simeq$, which can be described by a countable sequence of functionals $\{f_k; k = 1, 2, \ldots\}$ in the following manner.[6]

$$x(t) \simeq \sum_k f_k(x)\varphi_k(t); \qquad t \in T \tag{1.31}$$

where $\{\varphi_k(t); k = 1, 2, \ldots\}$ is a given, basic set of signals chosen independently of the particular signal $x(t)$ being represented. As a familiar example, the *time-series representation* for a signal involves a basic set formed from time-translated versions of a single, pulse-like signal;

$$\{\varphi_k(t);\ \varphi_k(t) = \varphi(t - k\tau), k = 0, \pm 1, \pm 2, \ldots\}$$

The pulse $\varphi(t)$ is called an *interpolating pulse* if it has the property that

$$\varphi(0) = 1$$

and

$$\varphi(k\tau) = 0 \qquad \text{for } k \neq 0$$

as shown in Figure 1.11.

With this basic set, a sensible representation can be obtained by making the f_k in (1.31) correspond to a mapping from the signal into its *sample*

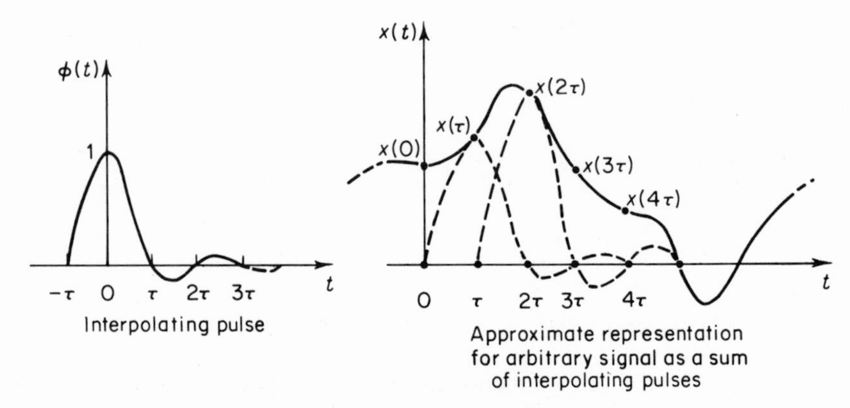

Figure 1.11. Time-series representation.

[6] As used in (1.31) and many of the equations that follow, T is not a number but rather a subset of the real line R denoting the time interval over which the series representation is applicable.

values; i.e.,

$$f_k(x) = x(k\tau); \qquad k = 0, \pm 1, \pm 2, \ldots \tag{1.32}$$

and

$$x(t) \cong \sum_k x(k\tau)\varphi(t - k\tau); \qquad -\infty < t < \infty \tag{1.33}$$

It is clear that this representation is exact at the particular instants $t = k\tau$, and if $x(t)$ does not fluctuate too rapidly (or if τ is sufficiently small), then with a suitably chosen interpolating pulse the error in the time-series representation (in the intersample region) can be made tolerably small. A much more definite statement can be made if $x(t)$ is restricted to lie in the set of band-limited signals. This is the well-known *sampling theorem*,[1],[2] which states that if $x \in S_B(W)$ [see (1.7)] then

$$x(t) = \sum_{k=-\infty}^{\infty} x\left(\frac{k}{2W}\right)\frac{\sin 2\pi W[t - (k/2W)]}{2\pi W[t - (k/2W)]}; \qquad -\infty < t < \infty \tag{1.34}$$

A band-limited signal with bandwidth W which is sampled every $\tau = 1/2W$ seconds, as in (1.34), is said to be sampled at the *Nyquist rate* and it has a unique and exact time-series representation using the interpolating pulse indicated in (1.34).

Another familiar example of series representation, applicable to duration-limited (or periodic) signals, is the *Fourier-series representation.* If $x \in S_D(T)$ [see (1.6)], then we have[7]

$$x(t) = \sum_{m=-\infty}^{\infty} c_m e^{j\frac{\pi m t}{T}}; \qquad |t| \leqslant T \tag{1.35}$$

where the series expansion coefficients c_m are given by the functionals

$$f_m(x) = c_m = \frac{1}{2T}\int_{-T}^{T} x(t)e^{-j\frac{\pi m t}{T}}\,dt; \qquad m = 0, \pm 1, \pm 2, \ldots \tag{1.36}$$

Time-Frequency Duality

As a final comment on mappings and functionals, we recall the one-to-one correspondence between the sets of square-integrable functions connected by the Fourier transform and we also note the essentially symmetric nature of the Fourier transform mapping and its inverse. Because of this, we find that for every relationship involving time functions there is a corresponding dual relationship involving Fourier transforms. This *time-frequency duality* property[5] exhibited by time functions and their Fourier transforms is frequently exploited in signal analysis problems and many examples of this duality will appear in the following chapters. For every signal analysis

[7] The representation in (1.35) is also appropriate for the periodic signals in $S_R(2T)$ [see (1.3)]. In this case, $-\infty < t < \infty$.

problem that we solve, we shall also have the solution of the dual problem which may, or may not, be of some practical interest. A simple illustration of results obtained by duality is shown by reconsidering the time-series and Fourier-series representations. From (1.35) and (1.36), we have

$$x \in S_D(T) \Rightarrow X(f) = \sum_{m=-\infty}^{\infty} X\left(\frac{m}{2T}\right)\frac{\sin 2\pi T[f-(m/2T)]}{2\pi T[f-(m/2T)]}; \qquad -\infty < f < \infty \tag{1.37}$$

whereas, from (1.34),

$$x \in S_B(W) \Rightarrow X(f) = \frac{1}{2W}\sum_{k=-\infty}^{\infty} x\left(\frac{k}{2W}\right)e^{-j\frac{\pi kf}{W}}; \qquad |f| \leqslant W$$

where

$$x\left(\frac{k}{2W}\right) = \int_{-W}^{W} X(f)e^{j\frac{\pi kf}{W}}\,df \tag{1.38}$$

Exercise 1.7. Discuss the various possibilities for interpreting "approximately equal" ($\simeq$) as an equivalence relation. As it is used in (1.31), is there an obvious equivalence relation implied? If so, describe the equivalence sets.

Exercise 1.8. Evaluate the Fourier transform of

$$x(t) = \frac{\sin 2\pi Wt}{2\pi Wt}; \qquad -\infty < t < \infty$$

Exercise 1.9. Using the sampling theorem (1.34), show that the set of band-limited, periodic signals can be represented exactly in terms of a finite set of functionals. In particular, if T is the period of $x(t)$ and if $x(t)$ has no frequency components above $W = N/T$, then

$$x(t) = \sum_{\ell=0}^{2N} x\left(\frac{\ell T}{2N+1}\right)\varphi\left(t - \frac{\ell T}{2N+1}\right); \qquad -\infty < t < \infty$$

where

$$\varphi(t) = \frac{\sin (2N+1)(\pi t/T)}{(2N+1)\sin (\pi t/T)}$$

Hint: Considering both the time-series and Fourier-series representations for x, we have

$$x(t) = \sum_{\ell=0}^{2N}\sum_{k=-\infty}^{\infty} x\left(\frac{\ell T}{2N+1} + kT\right)\frac{\sin (2N+1)(\pi/T)[t - kT - \ell T/(2N+1)]}{(2N+1)(\pi/T)[t - kT - \ell T/(2N+1)]}$$

and

$$x(t) = \frac{1}{T}\sum_{m=-N}^{N}\int_T x(s)e^{j\frac{2\pi m(t-s)}{T}}\,ds$$

Combining these expressions and performing further manipulations leads to the desired result.

Exercise 1.10. Prove the sampling theorem (1.34) for the case of finite-energy, band-limited signals. *Hint:* Consider a Fourier-series expansion of the Fourier transform of the signal.

Exercise 1.11. Consider the mapping f which maps real signals into an infinite sequence of real numbers according to

$$f(x) = \{x(k\tau); k = 0, \pm 1, \pm 2, \ldots\}$$

There is a very useful relationship involving Fourier transforms of any pair of elements in the same equivalence set as defined by this mapping. Show that if x and y are in the same equivalence set, i.e., if they have the same sample values, then

$$\sum_{m=-\infty}^{\infty} X\left(f - \frac{m}{\tau}\right) = \sum_{m=-\infty}^{\infty} Y\left(f - \frac{m}{\tau}\right); \qquad -\infty < f < \infty$$

From this result, show that if x has the sample values $x(0) = 1$ and $x(k\tau) = 0$ for $k \neq 0$, i.e., if x satisfies the requirements for an interpolating pulse, then

$$\sum_{m=-\infty}^{\infty} X\left(f - \frac{m}{\tau}\right) = \tau; \qquad -\infty < f < \infty$$

REFERENCES

1. C. E. Shannon, "Communication in the Presence of Noise," *Proc. IRE,* Vol. 37, No. 1, pp. 10–21 (January, 1949).
2. A. Papoulis, *The Fourier Integral and Its Applications*, McGraw-Hill, 1962.
3. G. F. Simmons, *Introduction to Topology and Modern Analysis*, McGraw-Hill, 1963.
4. A. N. Kolmogorov and S. V. Fomin, *Elements of the Theory of Functions and Functional Analysis, Vol.* 1, *Metric and Normed Spaces*, Graylock Press, 1957.
5. P. Bello, "Time-frequency Duality," *Trans. IEEE*, Vol. IT-10, No. 1, pp. 18–33, January, 1964.

SIGNAL SPACES

2

2.1 Metric Spaces

Having collected all signals exhibiting some common property into a set, our attention naturally turns to examining the distinctive properties of elements within the set. A particular signal is interesting only in relation to other signals in the set. We might ask, for example, of a particular signal relative to others, does it have more energy? does it last longer? does it fluctuate more rapidly? does it have more zero crossings? does it have a larger peak value? etc. A general approach, and one that is intuitively satisfying, for characterizing the difference between two elements of a set is to assign to each pair of elements a real, positive number. This number will be interpreted as the *distance* between the elements and the set itself begins to take on a geometric character. In fact, the set, with a suitably defined distance, will be referred to as a *signal space*. To define a distance, we need a functional which maps all pairs of elements from the set into the real line. Such a functional, $d: \{x, y\} \rightarrow R$, is called a *metric* if it possesses the following properties:

$$\begin{aligned} &\text{a.} \quad d(x, y) \geqslant 0 \quad \text{and} \quad d(x, y) = 0 \qquad \text{if, and only if, } x = y \\ &\text{b.} \quad d(x, y) = d(y, x) \quad \text{(symmetry)} \\ &\text{c.} \quad d(x, z) \leqslant d(x, y) + d(y, z) \quad \text{(triangle inequality)} \end{aligned} \tag{2.1}$$

These requirements are merely formalized statements reflecting the properties intuitively associated with a distance: (a) Distance is a non-negative number. (b) The distance from x to y is the same as the distance from y to x. (c) The length of one side of a triangle cannot exceed the sum of the lengths of the other two sides (here we are interpreting the elements x, y, and z geometrically as the vertices of a triangle).

A set of elements $\mathscr{X}$, together with a metric d, is called a *metric space* $(\mathscr{X}, d)$. It should be noted that two different metrics, defined on the same set of elements, form two different metric spaces.

Example 2.1. The real line R, comprising the set of all real numbers, is a metric space with a metric given by

$$d(x, y) = |x - y|; \qquad x, y \in R \tag{2.2}$$

This is known as the *usual metric* on R. It is helpful to think of many metric spaces as a generalization of this familiar example. ▮

Example 2.2. The set R^n of ordered sequences of n real numbers (real n-tuples) can be made into a metric space in various ways. If we let $x = \{\alpha_1, \alpha_2, \ldots, \alpha_n\}$ and $y = \{\beta_1, \beta_2, \ldots, \beta_n\}$, then the following functionals are examples of metrics that can be employed.

$$\begin{aligned} &\text{a. } d_1(x, y) = \sum_{i=1}^{n} |\alpha_i - \beta_i| \\ &\text{b. } d_2(x, y) = \left[\sum_{i=1}^{n} |\alpha_i - \beta_i|^2\right]^{\frac{1}{2}} \\ &\text{c. } d_3(x, y) = \max\{|\alpha_i - \beta_i|; i = 1, 2, \ldots, n\} \end{aligned} \tag{2.3}$$

These metrics can be applied directly to the set C^n of complex n-tuples (with α_i and β_i complex numbers) by expressing the magnitude of the complex number as the square root of the sum of the squares of the real and imaginary parts; i.e., if $\alpha = a + jb$, then $|\alpha| = (a^2 + b^2)^{\frac{1}{2}}$. The definitions can also be extended to apply to infinite sequences, thus providing metrics for R^∞ and C^∞. In this case, the metric (2.3c) is normally expressed as the *supremum*, or least upper bound, of the set $\{|\alpha_i - \beta_i|; i = 1, 2, \ldots\}$ and is written as

$$d_3(x, y) = \sup\{|\alpha_i - \beta_i|; i = 1, 2, \ldots\}$$

The metric (2.3b) coincides with the usual concept of distance in a three-dimensional space and is called the *Euclidean metric*. ▮

Example 2.3. In a communication system which transmits information in the form of a sequence of binary symbols (0 and 1), the message sequence may often be interpreted as a sequence of blocks or code words of fixed

	Information digits			Parity check digit
	α_1	α_2	α_3	α_4
x_1 :	0	0	0	0
x_2 :	0	0	1	1
x_3 :	0	1	0	1
x_4 :	0	1	1	0
x_5 :	1	0	0	1
x_6 :	1	0	1	0
x_7 :	1	1	0	0
x_8 :	1	1	1	1

$\alpha_4 = (\alpha_1 + \alpha_2 + \alpha_3) \bmod 2$

Figure 2.1. Code word set having minimum distance of 2.

length, say n symbols. The code words are n-tuples whose elements take on the values 0 or 1. The set of 2^n distinct words can be made into a metric space by defining the distance between any pair of words as the number of positions in the words where the symbols differ. This is equivalent to taking the ordinary sum, over all positions, of the modulo-2 sum of the symbols in

	Information digits				Check digits		
	α_1	α_2	α_3	α_4	α_5	α_6	α_7
x_1 :	0	0	0	0	0	0	0
x_2 :	0	0	0	1	1	1	1
x_3 :	0	0	1	0	0	1	1
x_4 :	0	0	1	1	1	0	0
x_5 :	0	1	0	0	1	0	1
•							
•							
•							
•							
x_{16} :	1	1	1	1	1	1	1

$$\alpha_5 = (\alpha_1 + \alpha_2 + \alpha_4) \bmod 2$$
$$\alpha_6 = (\alpha_1 + \alpha_3 + \alpha_4) \bmod 2$$
$$\alpha_7 = (\alpha_2 + \alpha_3 + \alpha_4) \bmod 2$$

Figure 2.2. Code word set having minimum distance of 3.

each position

$$d(x, y) = \sum_{i=1}^{n} [(\alpha_i + \beta_i) \bmod 2] \tag{2.4}$$

This metric is called the *Hamming distance* for binary code words and is useful for the study of error-detecting and error-correcting codes.[1],[2] An example of an error-detecting code is shown in Figure 2.1 where eight code words are selected from a set of sixteen in such a manner that the minimum distance between pairs of words is 2. This is accomplished by appending to the three information digits a *parity check digit* such that each word contains an even number of 1 s. Since the minimum distance between word pairs is 2, the occurrence of a single error in any four-digit word can be detected.

By adding additional parity check digits, we can create a code word set exhibiting a minimum distance of 3. In this case an error-correcting code results since a single error in transmission produces a code word closer to the correct word than any other word. An example of a seven-digit code having four information digits and three check digits is shown in Figure 2.2. ▌

Example 2.4. For an arbitrary set of real or complex time functions defined over a prescribed time interval $T = \{t; a \leqslant t \leqslant b\}$, metrics analogous to those in Example 2.2 can be defined.

$$\begin{aligned}
&\text{a. } d_1(x, y) = \int_T |x(t) - y(t)|\, dt \\
&\text{b. } d_2(x, y) = \left[\int_T |x(t) - y(t)|^2\, dt\right]^{\frac{1}{2}} \\
&\text{c. } d_3(x, y) = \sup\,\{|x(t) - y(t)|;\, t \in T\}
\end{aligned} \tag{2.5}$$

For the metrics d_1 and d_2 in (2.5) there is an apparent difficulty in that x and y might differ at a single point, say $x(t_0) \neq y(t_0)$; $t_0 \in T$; yet $d(x, y) = 0$ (see Figure 1.10). We overcome this difficulty by identifying functions which differ at most at a countable number of points in the interval T as occupying the same point in the metric space. In this situation we say that x and y are *equal almost everywhere.* ▌

Example 2.5. For an arbitrary set $\mathscr{X}$, a metric can be defined by the function d such that

$$\begin{aligned}
d(x, y) &= 0 \qquad \text{for } x = y \\
&= 1 \qquad \text{for } x \neq y
\end{aligned} \tag{2.6}$$

Although this is a rather trivial metric, it is sometimes useful in proving general theorems and providing counterexamples since it can be used with any set. ▌

Exercise 2.1. If the defining relations (2.1) for a metric are relaxed somewhat so that (a) $d(x, y) \geqslant 0$ and $d(x, y) = 0$ if $x = y$, (b) $d(x, y) = d(y, x)$, and (c) $d(x, z) \leqslant d(x, y) + d(y, z)$, then the functional $d(x, y)$ defined on the set $\mathscr{X}$ is called a *pseudo-metric*.[3] A pseudo-metric differs from a metric only in that it may vanish for $x \neq y$. Show that

$$x \sim y \Leftarrow d(x, y) = 0$$

is a valid equivalence relation in $\mathscr{X}$. Show how the set of equivalence sets generated by this equivalence relation can be made into a metric space. Interpret the "equal almost everywhere" relation in this sense.

Exercise 2.2. Let $f\colon \mathscr{X} \to R$ be an arbitrary functional defined on $\mathscr{X}$. Show that

$$d(x, y) = |f(x) - f(y)|$$

is pseudo-metric.

Exercise 2.3. Let $(\mathscr{X}, d)$ be a metric space and let

$$\tilde{d}(x, y) = \frac{d(x, y)}{1 + d(x, y)}; \qquad x, y \in \mathscr{X}$$

Show that $(\mathscr{X}, \tilde{d})$ is a metric space. What significant feature does this metric space possess?

2.2 Convergence and Continuity

In analysis problems we are often concerned with infinite sequences of elements $\{x_1, x_2, x_3, \ldots\}$ selected from some set $\mathscr{X}$. The distance concept of a metric space allows us to examine such sequences for the fundamental property called *convergence*.

We say that a sequence $\{x_n; x_n \in \mathscr{X}, n = 1, 2, \ldots\}$ is convergent if there exists an $x \in \mathscr{X}$ such that for any $\varepsilon > 0$, there exists a positive integer n_0 such that

$$n \geqslant n_0 \Rightarrow d(x_n, x) < \varepsilon$$

This is often expressed as

$$\lim_{n \to \infty} x_n = x$$

It is intuitively apparent that successive points in a convergent sequence must eventually tend to become closer and closer together as n increases. Any sequence with this property is called a *Cauchy sequence*. Specifically, if for any $\varepsilon > 0$ there exists a positive integer n_0, such that $m, n \geqslant n_0 \Rightarrow d(x_m, x_n) < \varepsilon$, then the sequence is a Cauchy sequence. It is clear from the triangle inequality $d(x_n, x_m) \leqslant d(x_n, x) + d(x, x_m)$ that a convergent sequence must be a Cauchy sequence. On the other hand, a Cauchy sequence may not be convergent simply because the element that the sequence tends to in the limit may not be in the set $\mathscr{X}$. An example of a sequence having a limit outside the defining set is given below. Certain metric spaces have the convenient property that every Cauchy sequence is convergent. Such spaces are called *complete* metric spaces.

Example 2.6. Let $C[T]$ denote the set of continuous, real time functions defined over an interval $T = \{t; a \leqslant t \leqslant b\}$ and let the associated metric be that given in (2.5b). We can show that this metric space is not complete by constructing a non-convergent Cauchy sequence in the manner shown in Figure 2.3.

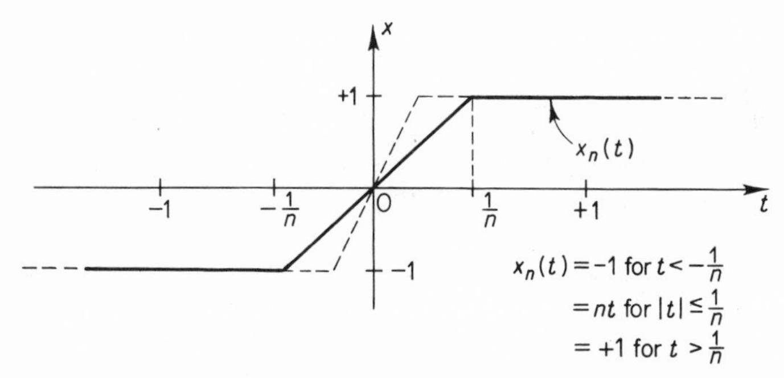

Figure 2.3. Cauchy sequence of continuous functions.

Letting $a = -1$, $b = +1$, we have

$$\begin{aligned} d_2(x_m, x_n) &= \left[\int_{-1}^{1} |x_m(t) - x_n(t)|^2 \, dt\right]^{\frac{1}{2}} \\ &= \sqrt{\frac{2}{3n}}\left(1 - \frac{n}{m}\right) \qquad \text{for } m > n \end{aligned} \tag{2.7}$$

Hence $\{x_1(t), x_2(t), \ldots\}$ is a Cauchy sequence (of functions) but it tends in the limit to the discontinuous function $\operatorname{sgn} t = t/|t|$. By means of this counterexample, we have established that the metric space $(C[T], d_2)$ is not complete.

Using the metric of (2.5c) with $C[T]$, we see that the sequence shown in Figure 2.3 is not a Cauchy sequence, since

$$\begin{aligned} d_3(x_m, x_n) &= \sup\{|x_m(t) - x_n(t)|; -1 \leqslant t \leqslant +1\} \\ &= 1 - \frac{n}{m} \qquad \text{for } m > n \end{aligned} \tag{2.8}$$

Hence this sequence will not serve as a counterexample to prove that $(C[T], d_3)$ is not complete. In fact, we can show that $(C[T], d_3)$ is complete by the following argument. Let $\{x_n; n = 1, 2, 3, \ldots\}$ be any Cauchy sequence, then for any $\varepsilon > 0$ we have

$$d_3(x_m, x_n) = \sup\{|x_m(t) - x_n(t)|; t \in T\} < \varepsilon$$

for m and n sufficiently large. But this says that $|x_m(t) - x_n(t)| < \varepsilon$ for any $t \in T$; hence $\{x_n(t)\}$ is a Cauchy sequence in R for each t and converges to a limit which we call $x(t)$. We can say that $|x_n(t) - x(t)| < \varepsilon/3$, for sufficiently large n. We now have to show that x is a continuous function of t; i.e., for any $\varepsilon > 0$ and any $t_0 \in T$, we can find $\delta > 0$ such that

$$|x(t) - x(t_0)| < \varepsilon \qquad \text{if } |t - t_0| < \delta \tag{2.9}$$

Since x_n is a continuous function, we can find a δ which makes $|x_n(t) - x_n(t_0)| < \varepsilon/3$; then

$$\begin{aligned} |x(t) - x(t_0)| &= |[x(t) - x_n(t)] + [x_n(t) - x_n(t_0)] + [x_n(t_0) - x(t_0)]| \\ &\leqslant |x(t) - x_n(t)| + |x_n(t) - x_n(t_0)| + |x_n(t_0) - x(t_0)| < \varepsilon \end{aligned}$$

Thus $x(t)$ is continuous at any $t_0 \in T$ and $(C[T], d_3)$ is a complete metric space. ∎

One of the important consequences of establishing metric spaces is that the ordinary concept of continuity discussed above can be generalized to apply to arbitrary mappings from one metric space into another. Let $f\colon (\mathscr{X}, d_1) \to (\mathscr{Y}, d_2)$, then we say that f is *continuous at* x_0 if for any $\varepsilon > 0$ there exists a $\delta > 0$ such that

$$d_1(x, x_0) < \delta \Rightarrow d_2(y, y_0) < \varepsilon; \qquad x \in \mathscr{X} \text{ and } y \in \mathscr{Y} \tag{2.10}$$

where $y = f(x)$ and $y_0 = f(x_0)$. If f is continuous at each point in its domain then f is called a *continuous* mapping.

Example 2.7. To illustrate this more general concept of continuity, we consider a mapping from the space of real time functions into R, i.e., a functional. For the space of time functions we use the metric d_2 in (2.5) and the usual metric for R, (2.2). The mapping is defined by

$$f_\varphi(x) = \int_{-\infty}^{\infty} x(t)\varphi(t)\,dt \tag{2.11}$$

We have, for any x_0,

$$\begin{aligned} d(f_\varphi(x), f_\varphi(x_0)) &= |f_\varphi(x) - f_\varphi(x_0)| \\ &= \left| \int_{-\infty}^{\infty} \{x(t) - x_0(t)\}\varphi(t)\,dt \right| \end{aligned} \tag{2.12}$$

Applying a particular form of the *Schwarz inequality* (Section 2.5) to (2.12) gives the inequality

$$\left| \int_{-\infty}^{\infty} \{x(t) - x_0(t)\}\varphi(t)\,dt \right| \leqslant \left[\int_{-\infty}^{\infty} \{x(t) - x_0(t)\}^2\,dt \right]^{\frac{1}{2}} \left[\int_{-\infty}^{\infty} \varphi^2(t)\,dt \right]^{\frac{1}{2}} \tag{2.13}$$

Thus, if φ is square-integrable,

$$\left[\int_{-\infty}^{\infty} \varphi^2(t)\, dt\right]^{\frac{1}{2}} < K,$$

where K is a real positive number, then

$$|f_\varphi(x) - f_\varphi(x_0)| < K\, d_2(x, x_0) < \varepsilon \qquad \text{if } d_2(x, x_0) < \delta = \frac{\varepsilon}{K} \tag{2.14}$$

and f_φ is a continuous functional. ∎

Exercise 2.4. To illustrate the fact that continuity is strongly dependent on the metric spaces involved, show that *any* mapping from a space using the metric in Example 2.5 into any other metric space is continuous.

Two additional properties of metric spaces, which are helpful in analysis, are the properties of *separability* and *compactness.* Roughly speaking, these properties provide useful insight to the extent or complexity of a metric space which contains an infinite number of elements. A metric space $(\mathscr{X}, d)$ is separable if, for any $\varepsilon > 0$, we can find a *countable* sequence of elements in $\mathscr{X}$, $\{x_1, x_2, \ldots\}$, such that $d(x, x_i) < \varepsilon$ for some i and any $x \in \mathscr{X}$. The metric space is compact if we can find a *finite* sequence $\{x_1, x_2, \ldots, x_{n(\varepsilon)}\}$ such that $d(x, x_i) < \varepsilon$ for some i; $1 \leqslant i \leqslant n(\varepsilon)$ and any $x \in \mathscr{X}$. We can think of a compact space as one capable of being "covered" by a finite set of "spheres" of radius ε. A separable space may be "bigger" than a compact space, but at least it can be "covered" by a countable set of spheres.

2.3 *Linear Spaces*

Our next step in the development of signal spaces is to add a simple algebraic structure to the sets of signals under consideration. This structure is provided by a *linear space*, which is defined in the following manner.

A linear space is a set of elements (called *vectors* and indicated by boldface symbols) with the following properties:

A. For each pair of vectors $\mathbf{x}$ and $\mathbf{y}$ in the set, there is a corresponding vector in the set $\mathbf{x} + \mathbf{y}$ called the *sum* of $\mathbf{x}$ and $\mathbf{y}$, such that

   (a) Addition is commutative: $\mathbf{x} + \mathbf{y} = \mathbf{y} + \mathbf{x}$.

   (b) Addition is associative: $\mathbf{x} + (\mathbf{y} + \mathbf{z}) = (\mathbf{x} + \mathbf{y}) + \mathbf{z}$.

   (c) The set contains a unique vector $\mathbf{0}$ (called the *origin*) such that

   $$\mathbf{x} + \mathbf{0} = \mathbf{x} \text{ for each } \mathbf{x}. \tag{2.15A}$$

   (d) For each $\mathbf{x}$, there is a unique vector $(-\mathbf{x})$ such that $\mathbf{x} + (-\mathbf{x}) = \mathbf{0}$.

B. There is a set of elements (called *scalars*) which form a *field* and an operation (called *scalar multiplication*) such that for every scalar α and every vector $\mathbf{x}$ there is a vector $\alpha\mathbf{x}$, and

(a) Multiplication by scalars is associative: $\alpha(\beta\mathbf{x}) = \alpha\beta\mathbf{x}$. (2.15B)

(b) $1\mathbf{x} = \mathbf{x}$ and $0\mathbf{x} = \mathbf{0}$ for each $\mathbf{x}$.

(c) $\alpha(\mathbf{x} + \mathbf{y}) = \alpha\mathbf{x} + \alpha\mathbf{y}$.

(d) $(\alpha + \beta)\mathbf{x} = \alpha\mathbf{x} + \beta\mathbf{x}$.

} distributive laws

Readers familiar with modern algebra will recognize that (2.15A) constitutes the postulates for a commutative group under the operation denoted by (+). The second part of the definition introduces another operation and specifies its interaction with the first operation. A field is any set of elements which essentially exhibits a double commutative group structure, with addition and multiplication operations, except that the additive identity 0 does not have a multiplicative inverse.[3] The field of scalars thus contains an additive identity element 0 and a multiplicative identity element 1. We see that a linear space involves two different additive identities, one for vectors and one for scalars, both called zero. Symbolically, these zeros will be distinct since the vector zero is in boldface, whereas the scalar zero is not.

It would simplify matters, and be adequate for our purposes, to identity the scalars with either the set of real numbers R or the set of complex numbers C, both of which are fields. The more general definition of a linear space is presented, however, because there are important signal theory applications (especially in coding theory) which involve finite fields of scalars. For example, the binary set $\{0, 1\}$ along with the conventional modulo-2 arithmetic operations forms a finite field, and linear spaces using these scalars have been extensively studied from the standpoint of communication system applications.[2]

If the real numbers are used as the system of scalars, then the resulting linear space is called a *real linear space*. Using the complex numbers for scalars, we have a *complex linear space*. The vector obtained by taking the sum of n particular vectors, each multiplied by a scalar coefficient

$$\mathbf{x} = \sum_{i=1}^{n} \alpha_i \mathbf{x}_i \tag{2.16}$$

is called a *linear combination*. It is easy to see that the set of all linear combinations of $\{\mathbf{x}_1, \mathbf{x}_2, \dots, \mathbf{x}_n\}$ forms a linear space. Furthermore, if we take a subset of

$$\{\mathbf{x}_1, \mathbf{x}_2, \dots, \mathbf{x}_n\}, \qquad \text{e.g., } \{\mathbf{x}_1, \mathbf{x}_2, \dots, \mathbf{x}_m\}; \qquad m < n$$

then the set of linear combinations forms a linear space which is a subset of the linear space formed from linear combinations of $\{\mathbf{x}_1, \mathbf{x}_2, \ldots, \mathbf{x}_n\}$. Such a subset is called a *linear subspace.* A set of vectors $\{\mathbf{x}_i; i = 1, 2, \ldots, n\}$ is said to be *linearly independent* if the relation

$$\sum_{i=1}^{n} \alpha_i \mathbf{x}_i = \mathbf{0} \tag{2.17}$$

can only be satisfied if each of the scalars α_i is zero. In other words, a vector in a linearly independent set cannot be expressed as a linear combination of the other vectors in the set. Let M be the space of linear combinations of the n linearly independent vectors $\{\mathbf{x}_i; i = 1, 2, \ldots, n\}$. Each vector in M is a unique linear combination of the $\{\mathbf{x}_i\}$ (a unique set of scalar coefficients). M is said to be an *n-dimensional* linear space. The set $\{\mathbf{x}_i\}$ is called a *basis* for M, and M is said to be *spanned* by this basis. Any set of n linearly independent vectors in M will serve as a basis for M; hence a linear space does not have a unique basis.

Two examples of particular types of linear spaces will suffice to introduce the spaces we shall use in signal analysis problems.

Example 2.8. The set of n-tuples of scalars R^n or C^n forms an n-dimensional linear space. Let $\mathbf{x} = \{\alpha_1, \alpha_2, \ldots, \alpha_n\}$ and $\mathbf{y} = \{\beta_1, \beta_2, \ldots, \beta_n\}$. Vector addition is defined by

$$\mathbf{x} + \mathbf{y} = \{\alpha_1 + \beta_1, \alpha_2 + \beta_2, \ldots, \alpha_n + \beta_n\} \tag{2.18}$$

and scalar multiplication by

$$\alpha\mathbf{x} = \{\alpha\alpha_1, \alpha\alpha_2, \ldots, \alpha\alpha_n\} \tag{2.19}$$

It is clear that each vector can be expressed as the linear combination

$$\mathbf{x} = \sum_{i=1}^{n} \alpha_i \mathbf{e}_i \tag{2.20}$$

where the n linearly independent vectors $\{\mathbf{e}_i\}$ are given by

$$\begin{aligned} \mathbf{e}_1 &= \{1, 0, 0, \ldots, 0\} \\ \mathbf{e}_2 &= \{0, 1, 0, \ldots, 0\} \\ &\cdot \quad \cdot \\ &\cdot \quad \cdot \\ &\cdot \quad \cdot \\ \mathbf{e}_n &= \{0, 0, 0, \ldots, 1\} \end{aligned} \tag{2.21}$$

▌

Representation of Finite-dimensional Vectors

Now let M be an arbitrary n-dimensional linear space spanned by the basis $\{\mathbf{u}_i; i = 1, 2, \ldots, n\}$. Any $\mathbf{x} \in M$ can be expressed uniquely as

$$\mathbf{x} = \sum_{i=1}^{n} \alpha_i \mathbf{u}_i \tag{2.22}$$

The ordered sequence of scalar coefficients $\{\alpha_i\}$ can be interpreted as an n-tuple. Thus there is a one-to-one correspondence between vectors in the arbitrary space M and the space of n-tuples, and R^n or C^n can be used as a model for any real or complex n-dimensional space. We say that the n-tuple $\boldsymbol{\alpha} = \{\alpha_i\}$ is a *representation* (in R^n or C^n) for $\mathbf{x}$ relative to the basis $\{\mathbf{u}_i\}$. It is important to remember that a representation has no meaning by itself but must always be referred to a particular basis. A different n-tuple could represent the same $\mathbf{x}$ relative to a different basis.

Example 2.9. The set of real or complex time functions defined on an interval $T = \{t; a \leqslant t \leqslant b\}$ is a linear space where vector addition and scalar multiplication are defined pointwise; i.e.,

$$\left.\begin{aligned} \mathbf{z} &= \mathbf{x} + \mathbf{y} \Rightarrow z(t) = x(t) + y(t) \\ \mathbf{z} &= \alpha\mathbf{x} \Rightarrow z(t) = \alpha x(t) \end{aligned}\right\} \quad \text{for all } t \in T \tag{2.23}$$

This space is referred to as a *function space*. In most cases of interest, these spaces are infinite dimensional. It is usually a simple matter to establish this fact by constructing infinite sequences of functions in the space, any finite number of which are linearly independent. The problem of mapping signals from their natural settings in function spaces to the more manageable finite-dimensional spaces is the subject of Chapter 3. ∎

Exercise 2.5. Show that the subset of a function space defined by $\{\mathbf{x}; x(0) = 0\}$ is itself a linear space.

Exercise 2.6. Show that $C[T]$ used in Example 2.6 is a linear space.

2.4 Normed Linear Spaces

We now combine the geometrical concepts associated with metric spaces with the algebraic concepts associated with linear spaces. This is accomplished by assigning a real number reflecting the "size" of any element in a linear space. This number is called the *norm* of a vector (denoted by $\|\mathbf{x}\|$) and can be defined in terms of any mapping from the linear space into the real line which satisfies the following properties:

$$\begin{aligned} &\text{a.} \quad \|\mathbf{x}\| \geqslant 0 \quad \text{and} \quad \|\mathbf{x}\| = 0 \quad \text{if, and only if, } \mathbf{x} = \mathbf{0} \\ &\text{b.} \quad \|\mathbf{x} + \mathbf{y}\| \leqslant \|\mathbf{x}\| + \|\mathbf{y}\| \\ &\text{c.} \quad \|\alpha\mathbf{x}\| = |\alpha|\,\|\mathbf{x}\| \end{aligned} \tag{2.24}$$

From these properties, it is easy to show that

$$d(\mathbf{x}, \mathbf{y}) = \|\mathbf{x} - \mathbf{y}\| \tag{2.25}$$

is a metric (2.1) and this metric is implied when we refer to a normed linear space as a metric space. Note that the norm of a vector is its distance from the origin. A normed linear space which is also complete as a metric space is called a *Banach space.*

In each of the examples in Section 2.1, except Example 2.5, the metrics can be interpreted as being induced by a norm. Thus, for example, we can define a norm on R^n or C^n by

$$\|\mathbf{x}\| = \left[\sum_{i=1}^{n} |\alpha_i|^2\right]^{\frac{1}{2}} \tag{2.26}$$

and for real or complex time functions defined on T, we can define

$$\|\mathbf{x}\| = \left[\int_T |x(t)|^2\, dt\right]^{\frac{1}{2}} \tag{2.27}$$

This is the norm we choose for signal representation, partly because of the convenience of physically interpreting the square of the norm as signal energy (1.5) and partly because this norm arises naturally in the more highly structured linear spaces introduced in the following section. The set of all functions for which the norm in (2.27) is bounded is called the L^2 *space,* denoted by $L^2(T)$.[1] The origin in this space is the function which is zero almost everywhere over the interval T.

Exercise 2.7. Show that the mapping $f\colon \mathscr{X} \to R$ defined by $f(\mathbf{x}) = \|\mathbf{x}\|$ is continuous.

2.5 Inner Product Spaces

The final step in the development of signal spaces is to supply additional geometrical structure in the form of an *inner product* relationship between pairs of vectors. We shall henceforth deal with complex linear spaces, since the real spaces can always be treated as a special case. The inner product is a mapping of ordered pairs of vectors in the linear space into the complex

[1] At this point we introduce a more convenient notation for defining time intervals. It is conventional to write

$$[a, b] = \{t;\, a \leqslant t \leqslant b\}, \qquad [a, b) = \{t;\, a \leqslant t < b\}$$
$$(a, b] = \{t;\, a < t \leqslant b\}, \qquad (a, b) = \{t;\, a < t < b\}$$

Accordingly, various types of L^2 function spaces will be denoted by $L^2(-\infty, \infty)$, $L^2[0, \infty)$, $L^2[-1, +1]$, etc.

plane. This mapping, with images denoted by $(\mathbf{x}, \mathbf{y})$ in C, satisfies the following properties.[2]

$$\begin{aligned}
&\text{a.}\quad (\mathbf{x}, \mathbf{y}) = (\mathbf{y}, \mathbf{x})^* \\
&\text{b.}\quad (\alpha\mathbf{x} + \beta\mathbf{y}, \mathbf{z}) = \alpha(\mathbf{x}, \mathbf{z}) + \beta(\mathbf{y}, \mathbf{z}) \\
&\text{c.}\quad (\mathbf{x}, \mathbf{x}) \geqslant 0 \quad \text{and} \quad (\mathbf{x}, \mathbf{x}) = 0 \qquad \text{if, and only if, } \mathbf{x} = \mathbf{0}
\end{aligned} \tag{2.28}$$

From (2.28a and b) we see that $(\alpha\mathbf{x}, \mathbf{y}) = \alpha(\mathbf{x}, \mathbf{y})$, $(\mathbf{x}, \alpha\mathbf{y}) = \alpha^*(\mathbf{x}, \mathbf{y})$, and that $(\mathbf{x}, \mathbf{x})$ is real. The inner product is sometimes referred to as the *scalar product* or *dot product*. An important consequence of the definition of the inner product is that

$$\|\mathbf{x}\| = (\mathbf{x}, \mathbf{x})^{\frac{1}{2}} \tag{2.29}$$

is a valid norm for the linear space. Properties (a) and (c) in (2.24) defining a norm are easily seen to be satisfied. Property (b), the triangle inequality, requires more effort. We first establish a very useful relationship, known as the *Schwarz inequality*,

$$|(\mathbf{x}, \mathbf{y})|^2 \leqslant (\mathbf{x}, \mathbf{x})(\mathbf{y}, \mathbf{y}) \tag{2.30}$$

To show this, we use property (c) in (2.28) with the vector $\mathbf{x} + \alpha\mathbf{y}$, where α is any scalar.

$$\begin{aligned}
0 &\leqslant (\mathbf{x} + \alpha\mathbf{y}, \mathbf{x} + \alpha\mathbf{y}) \\
&= (\mathbf{x}, \mathbf{x}) + \alpha(\mathbf{y}, \mathbf{x}) + \alpha^*(\mathbf{x}, \mathbf{y}) + |\alpha|^2 (\mathbf{y}, \mathbf{y})
\end{aligned} \tag{2.31}$$

In particular, (2.31) holds for $\alpha = -(\mathbf{x}, \mathbf{y})/(\mathbf{y}, \mathbf{y})$, giving

$$(\mathbf{x}, \mathbf{x}) - \frac{|(\mathbf{x}, \mathbf{y})|^2}{(\mathbf{y}, \mathbf{y})} \geqslant 0 \tag{2.32}$$

from which (2.30) follows. Note that the equality in (2.30) holds if $\mathbf{x} = \alpha\mathbf{y}$, for any α. Now for the triangle inequality, using $\|\mathbf{x}\|^2 = (\mathbf{x}, \mathbf{x})$,

$$\begin{aligned}
\|\mathbf{x} + \mathbf{y}\|^2 &= (\mathbf{x} + \mathbf{y}, \mathbf{x} + \mathbf{y}) \\
&= (\mathbf{x}, \mathbf{x}) + (\mathbf{x}, \mathbf{y}) + (\mathbf{y}, \mathbf{x}) + (\mathbf{y}, \mathbf{y}) \\
&\leqslant \|\mathbf{x}\|^2 + \|\mathbf{y}\|^2 + 2\,\|\mathbf{x}\|\,\|\mathbf{y}\| \\
&= (\|\mathbf{x}\| + \|\mathbf{y}\|)^2
\end{aligned}$$

Hence,

$$\|\mathbf{x} + \mathbf{y}\| \leqslant \|\mathbf{x}\| + \|\mathbf{y}\| \tag{2.33}$$

and (2.29) gives a valid norm. The inner product thus induces a norm which in turn induces a metric, by (2.25), so that an inner product space is a metric space with a particular metric implied. An inner product space which is also complete, as a metric space, is called a *Hilbert space*.

It is sometimes conceptually helpful to think of the inner product as providing an angular relationship between vectors. Since the Schwarz

[2] The asterisk denotes the complex conjugate.

inequality can be rewritten as

$$|(\mathbf{x}, \mathbf{y})| \leqslant \|\mathbf{x}\| \, \|\mathbf{y}\| \tag{2.34}$$

we can define a real angle θ between $\mathbf{x}$ and $\mathbf{y}$ as

$$\cos \theta = \frac{\operatorname{Re}(\mathbf{x}, \mathbf{y})}{\|\mathbf{x}\| \, \|\mathbf{y}\|} \tag{2.35}$$

For analysis, however, we really only use the concept of orthogonality between pairs of vectors. We say that $\mathbf{x}$ and $\mathbf{y}$ are *orthogonal* if, and only if, $(\mathbf{x}, \mathbf{y}) = 0$. The difficulty with (2.35) applied to complex spaces is apparent since we could have $\theta = \pm\pi/2$ with $(\mathbf{x}, \mathbf{y}) \neq 0$. On the other hand, if we replaced $\operatorname{Re}(\mathbf{x}, \mathbf{y})$ with $|(\mathbf{x}, \mathbf{y})|$ in (2.35), we would not generate second- and third-quadrant angles.

The spaces we shall be primarily concerned with have inner products given by

$$(\mathbf{x}, \mathbf{y}) = \sum_{i=1}^{n} \alpha_i \beta_i^*; \qquad \mathbf{x}, \mathbf{y} \in C^n \tag{2.36}$$

and

$$(\mathbf{x}, \mathbf{y}) = \int_T x(t) y^*(t) \, dt; \qquad \mathbf{x}, \mathbf{y} \in L^2(T) \tag{2.37}$$

Exercise 2.8. Let $\mathbf{x}$ and $\mathbf{y}$ be unit norm vectors in a real inner product space. Show that $\mathbf{x} + \mathbf{y}$ and $\mathbf{x} - \mathbf{y}$ are orthogonal. Is the orthogonality maintained for a complex space? What is the angle between $\mathbf{x} + \mathbf{y}$ and $\mathbf{x} - \mathbf{y}$, as defined by (2.35), for the complex case?

Exercise 2.9. For an inner product space with $\|\mathbf{x}\|^2 = (\mathbf{x}, \mathbf{x})$, verify the *parallelogram equality*

$$\|\mathbf{x} + \mathbf{y}\|^2 + \|\mathbf{x} - \mathbf{y}\|^2 = 2\,\|\mathbf{x}\|^2 + 2\,\|\mathbf{y}\|^2$$

Exercise 2.10. Show that an inner product, in a complex space, satisfies the *polarization identity*

$$4(\mathbf{x}, \mathbf{y}) = \|\mathbf{x} + \mathbf{y}\|^2 - \|\mathbf{x} - \mathbf{y}\|^2 + j\,\|\mathbf{x} + j\mathbf{y}\|^2 - j\,\|\mathbf{x} - j\mathbf{y}\|^2$$

Exercise 2.11. Let $\mathscr{X}$ be a Banach space whose norm satisfies the parallelogram equality, then define an inner product according to the polarization identity. Show that this is a valid inner product and, hence, $\mathscr{X}$ is actually a Hilbert space.

Exercise 2.12. Give an example of a normed linear space where the norm does not satisfy the parallelogram equality.

Example 2.10. Time–ambiguity Functions: An important application of inner products appears in connection with the characterization of the time fluctuation properties of finite-energy signals. If a time function is characterized by very rapid fluctuations, then we would expect that relatively small time translations of the signal would produce a substantial shift of the signal point in the relevant signal space, say $L^2(-\infty, \infty)$. On the other hand, a slowly fluctuating signal is essentially unchanged by small time translations and the signal point moves only a small distance. To express this separation in signal space, we let $\mathbf{x}_\tau$ be a time-shifted (by τ seconds) version of $\mathbf{x}$; i.e., $x_\tau(t) = x(t + \tau)$. Now we have

$$\begin{aligned} d^2(\mathbf{x}, \mathbf{x}_\tau) &= \|\mathbf{x} - \mathbf{x}_\tau\|^2 = (\mathbf{x} - \mathbf{x}_\tau, \mathbf{x} - \mathbf{x}_\tau) \\ &= (\mathbf{x}, \mathbf{x}) + (\mathbf{x}_\tau, \mathbf{x}_\tau) - (\mathbf{x}_\tau, \mathbf{x}) - (\mathbf{x}, \mathbf{x}_\tau) \\ &= \|\mathbf{x}\|^2 + \|\mathbf{x}_\tau\|^2 - 2\,\mathrm{Re}\,(\mathbf{x}, \mathbf{x}_\tau) \end{aligned} \tag{2.38}$$

The energy content of the signal is not affected by the time shift; $\|\mathbf{x}\|^2 = \|\mathbf{x}_\tau\|^2$; hence

$$\begin{aligned} d^2(\mathbf{x}, \mathbf{x}_\tau) &= 2[\|\mathbf{x}\|^2 - \mathrm{Re}\,(\mathbf{x}, \mathbf{x}_\tau)] \\ &= 2[r_x(0) - r_x(\tau)] \end{aligned} \tag{2.39}$$

where we have defined

$$\begin{aligned} r_x(\tau) &= \mathrm{Re}\,(\mathbf{x}, \mathbf{x}_\tau) \\ &= \mathrm{Re} \int_{-\infty}^{\infty} x(t)x^*(t + \tau)\, dt \end{aligned} \tag{2.40}$$

Hence, for each signal $\mathbf{x}$, there is a real function of the time shift τ which characterizes the separation in signal space due to the time shift. For a rapidly fluctuating signal, we would expect $r_x(\tau)$ to drop off rapidly with τ. For a narrow signal pulse, it is clear that $r_x(\tau)$ is narrow but the signal might also have a long duration and still produce a narrow $r_x(\tau)$ as shown in Figure 2.4. In many applications, e.g., radar systems, it is desirable to use

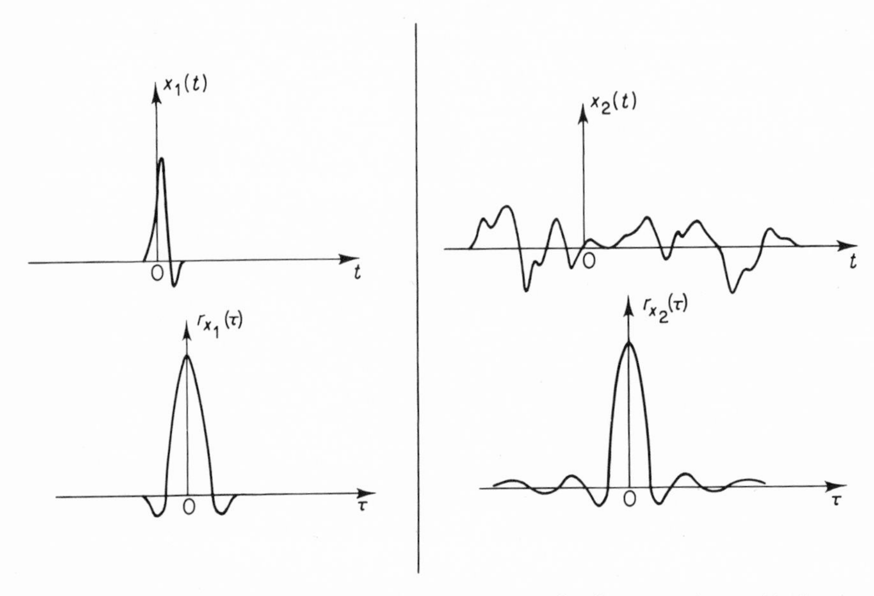

Figure 2.4. Two signals which have approximately the same time-ambiguity characteristics.

signals which produce a narrow $r_x(\tau)$ so that an accurate time location of the signal can be obtained by measurements on the signal.[5] Accordingly, we shall refer to $r_x(\tau)$ as the *time-ambiguity function* for the signal $x(t)$. Small ambiguity means that the distance (2.39) is large. The function $r_x(\tau)$ is also sometimes called the *autocorrelation function* for $x(t)$ by analogy with a similar characterization for random processes. The autocorrelation function is defined in Chapter 7, and in Chapter 8 several relationships between autocorrelation functions and time-ambiguity functions are presented. For real signals, it is a simple matter to take the Fourier transform of (2.40),

$$R_x(f) = \int_{-\infty}^{\infty} r_x(\tau)e^{-j2\pi f\tau}\,d\tau = |X(f)|^2 \tag{2.41}$$

and we see that signals exhibiting the same time ambiguity have Fourier transforms that can differ by an arbitrary phase factor (see Exercise 1.5). ▮

Exercise 2.13. Show that the time-ambiguity function for an arbitrary finite-energy signal is continuous (as a function of τ). For real finite-energy signals, show that $r_x(-\tau) = r_x(\tau)$ and $r_x(\tau) \leqslant r_x(0)$.

Exercise 2.14. Evaluate and sketch graphs of the time-ambiguity functions for

a. $x(t) = 1; \quad 0 \leqslant t \leqslant T$
$\quad = 0; \quad$ otherwise

b. $x(t) = \dfrac{\sin 2\pi Wt}{2\pi Wt}; \quad -\infty < t < \infty$

c. $x(t) = e^{-at}; \quad t \geqslant 0$
$\quad = 0; \quad t < 0$

Hint: (2.41) may be helpful.

Obtaining the Representation for a Point in an Inner Product Space

The most significant feature of an inner product space as a signal space is that a direct relationship between a signal and its representation is provided. Suppose M is an arbitrary n-dimensional space spanned by the basis $\{\mathbf{u}_i;\ i = 1, 2, \ldots, n\}$, then $\mathbf{x} \in M$ is given by

$$\mathbf{x} = \sum_{i=1}^{n} \alpha_i \mathbf{u}_i \tag{2.42}$$

Taking inner products on both sides of (2.42),

$$(\mathbf{x}, \mathbf{u}_j) = \sum_{i=1}^{n} (\mathbf{u}_i, \mathbf{u}_j)\alpha_i; \qquad j = 1, 2, \ldots, n \tag{2.43}$$

This is a set of n simultaneous, linear, scalar equations which can be solved for the n-tuple $\boldsymbol{\alpha} = \{\alpha_1, \alpha_2, \ldots, \alpha_n\}$ providing the representation (in C^n) for $\mathbf{x}$ relative to the basis $\{\mathbf{u}_i\}$. An alternative approach, which is notationally more convenient, is to establish another basis $\{\mathbf{v}_i\}$ for M related to $\{\mathbf{u}_i\}$ in a pairwise orthogonal manner according to

$$(\mathbf{u}_i, \mathbf{v}_j) = \delta_{ij}; \qquad i, j = 1, 2, \ldots, n \tag{2.44}$$

where δ_{ij} is the *Kronecker delta* defined by $\delta_{ij} = 1$ for $i = j$, $\delta_{ij} = 0$ for $i \neq j$. Now, from (2.42),

$$(\mathbf{x}, \mathbf{v}_j) = \sum_{i=1}^{n} \alpha_i(\mathbf{u}_i, \mathbf{v}_j) \Rightarrow \alpha_j = (\mathbf{x}, \mathbf{v}_j); \qquad j = 1, 2, \ldots, n \tag{2.45}$$

Bases which satisfy (2.44) are called *reciprocal bases*, and we have

$$\mathbf{x} = \sum_{i=1}^{n} (\mathbf{x}, \mathbf{v}_i)\mathbf{u}_i = \sum_{i=1}^{n} (\mathbf{x}, \mathbf{u}_i)\mathbf{v}_i \tag{2.46}$$

for any $\mathbf{x} \in M$ and any pair of reciprocal bases for M. Yet another approach, which has many attractive features, is to use an *orthonormal set* as a basis for M. The set $\{\mathbf{u}_i; i = 1, 2, \ldots, n\}$ is said to be an orthonormal set if it is self-reciprocal, i.e., if the vectors are mutually orthogonal and have unit norm.

$$(\mathbf{u}_i, \mathbf{u}_j) = \delta_{ij} \tag{2.47}$$

Then, for any $\mathbf{x} \in M$,

$$\mathbf{x} = \sum_{i=1}^{n} (\mathbf{x}, \mathbf{u}_i)\mathbf{u}_i \tag{2.48}$$

With an orthonormal basis for M, we have not only a one-to-one correspondence between vectors in M and their representation in C^n but an equality of inner products in both spaces. For

$$\mathbf{x} = \sum_{i=1}^{n} \alpha_i \mathbf{u}_i \quad \text{and} \quad \mathbf{y} = \sum_{i=1}^{n} \beta_i \mathbf{u}_i$$

we have

$$\begin{aligned}(\mathbf{x}, \mathbf{y}) &= \left(\sum_{i=1}^{n} \alpha_i \mathbf{u}_i, \sum_{j=1}^{n} \beta_j \mathbf{u}_j\right) \\ &= \sum_{i=1}^{n}\sum_{j=1}^{n} \alpha_i \beta_j^*(\mathbf{u}_i, \mathbf{u}_j) = \sum_{i=1}^{n} \alpha_i \beta_i^* = (\boldsymbol{\alpha}, \boldsymbol{\beta})\end{aligned} \tag{2.49}$$

Because of the convenience of an orthonormal basis, procedures for constructing such a basis are of interest. One such procedure, which is computationally advantageous because of its iterative nature, is the *Gram-Schmidt orthonormalization procedure*. Given a set of n linearly independent vectors in M, $\{\mathbf{v}_i; i = 1, 2, \ldots, n\}$, the orthonormal set $\{\mathbf{u}_i\}$ is generated by normalizing the $\{\mathbf{w}_i\}$ obtained by

$$\begin{aligned}
\mathbf{w}_1 &= \mathbf{v}_1 \\
\mathbf{w}_2 &= \mathbf{v}_2 - (\mathbf{v}_2, \mathbf{u}_1)\mathbf{u}_1 \\
\mathbf{w}_3 &= \mathbf{v}_3 - (\mathbf{v}_3, \mathbf{u}_2)\mathbf{u}_2 - (\mathbf{v}_3, \mathbf{u}_1)\mathbf{u}_1 \\
&\cdot \\
&\cdot \\
&\cdot \\
\mathbf{w}_i &= \mathbf{v}_i - \sum_{k=1}^{i-1} (\mathbf{v}_i, \mathbf{u}_k)\mathbf{u}_k \\
&\cdot \\
&\cdot \\
&\cdot
\end{aligned} \tag{2.50a}$$

where

$$\mathbf{u}_i = \frac{\mathbf{w}_i}{\|\mathbf{w}_i\|}; \qquad i = 1, 2, \ldots, n \tag{2.50b}$$

It is important to notice that distinct orthonormal sets will be generated by reorderings of the set $\{\mathbf{v}_i\}$ in the Gram-Schmidt procedure.

Exercise 2.15. Show that the Gram-Schmidt procedure (2.50) produces an orthonormal set. *Hint:* Prove by induction, assuming that at the ith stage of the process, the $\{\mathbf{u}_1, \mathbf{u}_2, \ldots, \mathbf{u}_i\}$ are an orthonormal set; then show that this implies $\mathbf{w}_{i+1}$ is orthogonal to each of these vectors. Also, derive the relationship

$$\|\mathbf{w}_i\|^2 = \|\mathbf{v}_i\|^2 - \sum_{k=1}^{i-1} |(\mathbf{v}_i, \mathbf{u}_k)|^2$$

Exercise 2.16. *Change of basis:* Let $\mathbf{x} \in M$ be represented by $\boldsymbol{\alpha} \in C^n$ relative to the basis $\{\mathbf{u}_i; i = 1, 2, \ldots, n\}$. Let $\{\mathbf{w}_i; i = 1, 2, \ldots, n\}$ be any other basis for M, and let $\boldsymbol{\beta} \in C^n$ be the representation for $\mathbf{x}$ relative to this new basis; i.e.,

$$\mathbf{x} = \sum_{i=1}^{n} \alpha_i \mathbf{u}_i = \sum_{i=1}^{n} \beta_i \mathbf{w}_i$$

Show that $\boldsymbol{\beta}$ is related to $\boldsymbol{\alpha}$ by a matrix multiplication

$$\boldsymbol{\beta} = \boldsymbol{\Gamma}\boldsymbol{\alpha} \Rightarrow \beta_i = \sum_{j=1}^{n} \gamma_{ij}\alpha_j; \qquad i = 1, 2, \ldots, n$$

where the elements of the $n \times n$ matrix $\boldsymbol{\Gamma}$ are $\gamma_{ij} = (\mathbf{u}_j, \mathbf{z}_i)$ and $\{\mathbf{z}_i; i = 1, 2, \ldots, n\}$ is the reciprocal basis for $\{\mathbf{w}_i; i = 1, 2, \ldots, n\}$.

Exercise 2.17. Give the representation $\boldsymbol{\beta} \in C^n$ for

$$\mathbf{x} = \sum_{i=1}^{n} \alpha_i \mathbf{u}_i$$

relative to the basis $\{\mathbf{w}_i; i = 1, 2, \ldots, n\}$, where

$$\mathbf{w}_1 = \mathbf{u}_1$$
$$\mathbf{w}_2 = \mathbf{u}_2 + \mathbf{u}_1$$
$$\vdots$$
$$\mathbf{w}_i = \sum_{j=1}^{i} \mathbf{u}_j$$

Hint: $\mathbf{u}_i = \mathbf{w}_i - \mathbf{w}_{i-1}$ for $i \geqslant 2$.

2.6 Linear Functionals

Having introduced various types of linear spaces to be used as signal spaces, we turn our attention to properties of mappings from these spaces into numerical values. These mappings are of practical interest since their physical counterparts are measurements on signals. The class of linear measurements is especially important from the standpoint of identification and representation of signals. The primary motivation for developing linear signal spaces and endowing them with a particular geometric structure is that there results a remarkable correspondence between the set of linear measurements on signals and the set of signals itself. Various aspects of this correspondence will be examined in this section.

A mapping from a complex linear space $\mathscr{X}$ into the complex scalars $f: \mathscr{X} \to C$ which has the property that

$$f(\alpha\mathbf{x} + \beta\mathbf{y}) = \alpha f(\mathbf{x}) + \beta f(\mathbf{y}) \tag{2.51}$$

for any $\alpha, \beta \in C$ and any $\mathbf{x}, \mathbf{y} \in \mathscr{X}$ is called a *linear functional.* If $\mathscr{X}$ is an inner product space, then from property (b) of (2.28), we see that $\mathbf{f}_\varphi$ defined by[3]

$$f_\varphi(\mathbf{x}) = (\mathbf{x}, \boldsymbol{\varphi}) \tag{2.52}$$

is a linear functional. Furthermore, if $\|\boldsymbol{\varphi}\|$ is bounded, $\|\boldsymbol{\varphi}\| < K$, then $\mathbf{f}_\varphi$ is a continuous linear functional. This follows directly from the Schwarz inequality, since

$$|f_\varphi(\mathbf{x}) - f_\varphi(\mathbf{x}_0)| = |(\mathbf{x} - \mathbf{x}_0, \boldsymbol{\varphi})| \leqslant \|\mathbf{x} - \mathbf{x}_0\| \, \|\boldsymbol{\varphi}\| < K \, \|\mathbf{x} - \mathbf{x}_0\| \tag{2.53}$$

for any $\mathbf{x}_0 \in \mathscr{X}$. This is the same result as obtained in Example 2.7 except in more general terms.

A more significant result is that if $\mathscr{X}$ is complete, i.e., a Hilbert space, then *any* continuous linear functional can be expressed as an inner product

[3] The reason for indicating $\mathbf{f}_\varphi$ as a vector quantity will be apparent shortly.

as in (2.52). For each continuous linear functional there is a unique vector $\boldsymbol{\varphi} \in \mathscr{X}$. The proof of this statement is presented in most texts on functional analysis.[3],[4] A fundamental part of the proof involves the fact that the set of vectors which a linear functional maps into zero is a linear subspace of $\mathscr{X}$. Let M_f denote this subspace for the functional $\mathbf{f}$; i.e., $M_f = \{\mathbf{x}; f(\mathbf{x}) = 0\}$. Now choose a non-zero vector $\mathbf{x}_0$ orthogonal to M_f.

$$(\mathbf{y}, \mathbf{x}_0) = 0 \qquad \text{for any } \mathbf{y} \in M_f \tag{2.54}$$

If there is no such $\mathbf{x}_0$, we conclude that $\mathbf{f} = \mathbf{0}$. Otherwise, $f(\mathbf{x}_0) \neq 0$ and for any $\mathbf{x} \in \mathscr{X}$ the vector

$$\mathbf{y} = f(\mathbf{x})\mathbf{x}_0 - f(\mathbf{x}_0)\mathbf{x} \tag{2.55}$$

is in M_f since $f(\mathbf{y}) = f(\mathbf{x})f(\mathbf{x}_0) - f(\mathbf{x}_0)f(\mathbf{x}) = 0$. Combining (2.55) and (2.54), we find

$$\begin{aligned} f(\mathbf{x}) &= \frac{f(\mathbf{x}_0)}{(\mathbf{x}_0, \mathbf{x}_0)}(\mathbf{x}, \mathbf{x}_0) \\ &= (\mathbf{x}, \boldsymbol{\varphi}) \qquad \text{with } \boldsymbol{\varphi} = \frac{[f(\mathbf{x}_0)]^*}{(\mathbf{x}_0, \mathbf{x}_0)}\mathbf{x}_0 \end{aligned} \tag{2.56}$$

An example of this construction for a linear functional on R^2 is illustrated in Figure 2.5.

The correspondence between linear functionals and vectors can be developed further by noting that the set of linear functionals on a linear space $\mathscr{X}$ in itself forms a linear space. Vector addition and scalar multiplication for functionals is on a pointwise basis; i.e.,

$$\mathbf{f} = \alpha\mathbf{f}_1 + \beta\mathbf{f}_2 \Rightarrow f(\mathbf{x}) = \alpha f_1(\mathbf{x}) + \beta f_2(\mathbf{x}) \qquad \text{for all } \mathbf{x} \in \mathscr{X} \tag{2.57}$$

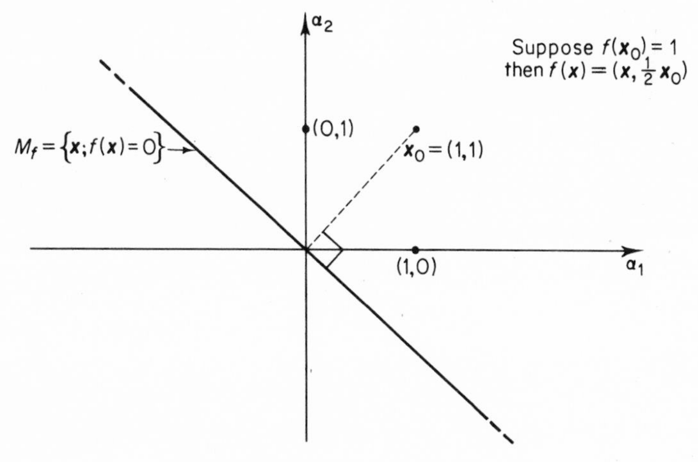

Figure 2.5. A linear functional **f** on R^2.

This space can be normed by the following definition:

$$\|\mathbf{f}\| = \sup \{|f(\mathbf{x})|; \|\mathbf{x}\| \leqslant 1, \mathbf{x} \in \mathscr{X}\} \tag{2.58}$$

or, equivalently, by

$$\|\mathbf{f}\| = \inf \{K; |f(\mathbf{x})| \leqslant K \|\mathbf{x}\|, \mathbf{x} \in \mathscr{X}\} \tag{2.59}$$

If the norm of a functional is bounded, we say that the functional is *bounded.* (Note that this does not imply that $|f(\mathbf{x})|$ is bounded.) A bounded linear functional is continuous since we have $|f(\mathbf{x}) - f(\mathbf{x}_0)| = |f(\mathbf{x} - \mathbf{x}_0)| \leqslant \|\mathbf{f}\| \, \|\mathbf{x} - \mathbf{x}_0\|$ for any $\mathbf{x}_0 \in \mathscr{X}$. Continuity at $\mathbf{x}_0$ implies continuity since $\|\mathbf{f}\|$ is independent of $\mathbf{x}_0$. Conversely, a continuous linear functional is bounded since continuity at the origin gives

$$\|\mathbf{x}\| < \delta \Rightarrow |f(\mathbf{x})| < \varepsilon \Rightarrow |f(\mathbf{y})| < \frac{2\varepsilon}{\delta}$$

with $\|\mathbf{y}\| < 2$. *Hence boundedness and continuity of a linear functional are equivalent.*

For a linear functional expressed as an inner product, we see that

$$\|\mathbf{f}_\varphi\| = \|\boldsymbol{\varphi}\| \tag{2.60}$$

using the Schwarz inequality and noting that the equality holds for $\mathbf{x}$ proportional to $\boldsymbol{\varphi}$.

If we restrict our consideration to the space of all continuous linear functionals defined on a Hilbert space $\mathscr{X}$ (this is a subspace of the space of all linear functionals on $\mathscr{X}$), then we find that this space is a Hilbert space which is very simply related to $\mathscr{X}$. This space is called the *conjugate space* $\mathscr{X}^*$. We have observed that there is a one-to-one correspondence between the elements $\mathbf{f}_\varphi \in \mathscr{X}^*$ and $\boldsymbol{\varphi} \in \mathscr{X}$. Furthermore, the corresponding scalars are simply complex conjugates since $\mathbf{f}_{\alpha\varphi} = \alpha^* \mathbf{f}_\varphi$. This suggests

$$(\mathbf{f}_\varphi, \mathbf{f}_\psi) = (\boldsymbol{\varphi}, \boldsymbol{\psi})^* = (\boldsymbol{\psi}, \boldsymbol{\varphi}) \tag{2.61}$$

which is easily shown to be a valid definition for an inner product in $\mathscr{X}^*$. The norm induced by this inner product $\|\mathbf{f}_\varphi\| = (\mathbf{f}_\varphi, \mathbf{f}_\varphi)^{\frac{1}{2}} = \|\boldsymbol{\varphi}\|$ is consistent with that of (2.60). Finally, if $\{\mathbf{u}_i\}$ is a basis for $\mathscr{X}$, then $\{\mathbf{f}_{v_i}\}$ is a basis for $\mathscr{X}^*$, where $\{\mathbf{v}_i\}$ is the reciprocal basis $[(\mathbf{u}_i, \mathbf{v}_j) = \delta_{ij}]$. Hence an arbitrary continuous linear functional $\mathbf{f}$ can be expressed as the linear combination

$$\mathbf{f} = \sum_i f(\mathbf{u}_i) \mathbf{f}_{v_i} \tag{2.62}$$

With some signal spaces, say $L^2(T)$, we shall have occasion to employ linear functionals which are not continuous (bounded). The primary example of this is the functional on $L^2(T)$ corresponding to the sampling operation on a time function. It is clear that $f(\mathbf{x}) = x(t_0)$ describes a linear functional, but it

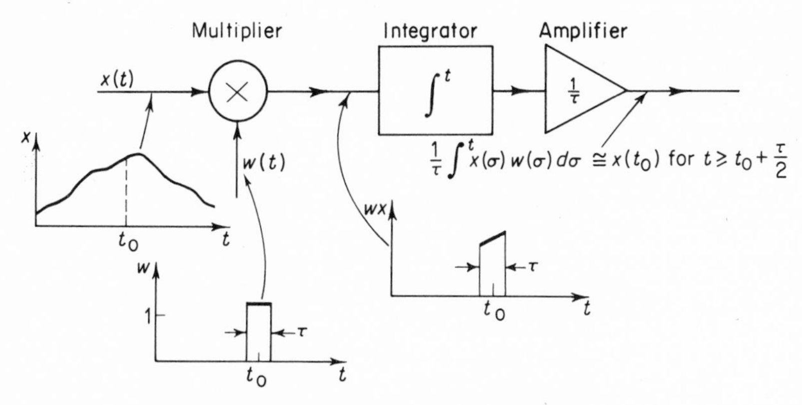

Figure 2.6. Implementation of a sampler.

is also clear that there exist square-integrable time functions which are not bounded for all $t \in T$. Consider, for example, $x(t) = |t|^{-\frac{1}{3}}$, which is in $L^2(-1, 1)$ but is not bounded at $t = 0$. We can still retain the inner product representation for the functional, in this case by defining the δ-function (not in L^2) such that

$$f(\mathbf{x}) = x(t_0) = \int_T x(t)\,\delta(t - t_0)\,dt; \qquad t_0 \in T \tag{2.63}$$

As a practical matter, the physical implementation of the sampling operation will be continuous because of the impossibility of realizing an infinitely narrow sampling gate. A typical implementation of a sampler is shown in Figure 2.6 where the signal is multiplied by a rectangular gate function which is narrow compared to the fluctuation periods of the signal to be sampled. The gate is usually realized as a switch which is closed during $t_0 - \tau/2 \leqslant t \leqslant t_0 + \tau/2$ and open otherwise. The integral of the nearly rectangular signal out of the gate is approximately proportional to $x(t_0)$.

In a similar manner, an arbitrary linear functional for real signals can be implemented by a multiplier and integrator as shown in Figure 2.7. It is assumed, of course, that either the signal or the gate function eventually becomes small so that after some finite time the integrator output represents the value of the functional.

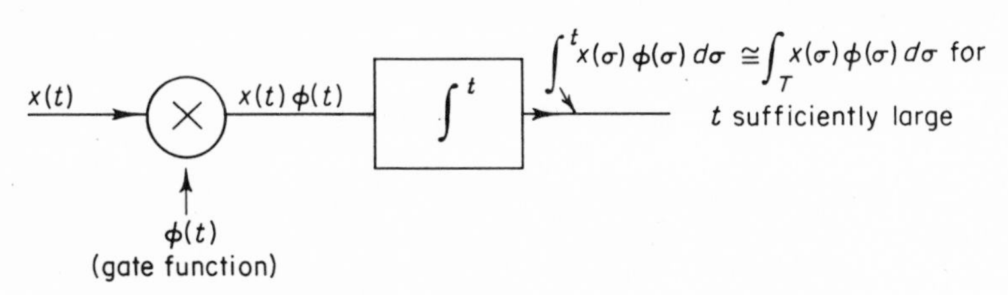

Figure 2.7. Implementation of arbitrary linear functional, $f_\varphi(\mathbf{x}) = (\mathbf{x}, \boldsymbol{\varphi})$.

There is an alternative implementation of a linear functional which has the advantage of replacing the multiplication operation with a sampling operation as shown in Figure 2.8. Since the output of the linear, time-invariant network at $t = t_0$ is given by the convolution integral

$$y(t_0) = \int x(\sigma)h(t_0 - \sigma)\, d\sigma \tag{2.64}$$

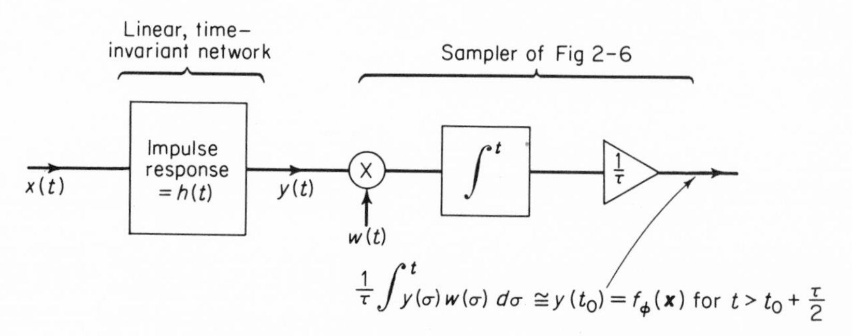

Figure 2.8. Alternative implementation of $f_\varphi(\mathbf{x}) = (\mathbf{x}, \boldsymbol{\varphi})$.

the desired inner product is obtained if the impulse response of the network is given by

$$h(t) = \varphi(t_0 - t) \tag{2.65}$$

which is a time-reversed and delayed version of the gate function in Figure 2.7. Since $h(t)$ for a physical network is non-zero only for positive t, an additional delay may have to be incorporated in the sampling time.

Exercise 2.18. We have shown that the conjugate space $\mathscr{X}^*$ is an inner product space. If it is to be a Hilbert space, it must be complete. Consider an arbitrary Cauchy sequence in $\mathscr{X}^*$ and show that it converges to a point in $\mathscr{X}^*$.
Hints: (1) For $\mathbf{f}_n, \mathbf{f}_m \in \mathscr{X}^*$, $\|\mathbf{f}_n - \mathbf{f}_m\| < \varepsilon \Rightarrow |f_n(\mathbf{x}) - f_m(\mathbf{x})| < \varepsilon \|\mathbf{x}\|$ for each $\mathbf{x} \in \mathscr{X}$. (2) Suppose that the sequence of numbers $\{f_n(\mathbf{x})\}$ converges to a number, call it $f(\mathbf{x})$, for each $\mathbf{x}$. Show that this implies that $\mathbf{f}$, as a mapping, is linear and continuous. (3) $|f(\mathbf{x}) - f(\mathbf{x}_0)| \leqslant |f(\mathbf{x}) - f_n(\mathbf{x})| + |f_n(\mathbf{x}) - f_n(\mathbf{x}_0)| + |f_n(\mathbf{x}_0) - f(\mathbf{x}_0)|$.

REFERENCES

1. R. W. Hamming, "Error Detecting and Error Correcting Codes," *Bell Sys. Tech. Jour.*, Vol. 29, pp. 147–60 (April, 1950).

2. W. W. Peterson, *Error Correcting Codes*, John Wiley & Sons, 1961.
3. G. F. Simmons, *Introduction to Topology and Modern Analysis*, McGraw-Hill, 1963.
4. N. I. Akhiezer and I. M. Glazman, *Theory of Linear Operators in Hilbert Space*, Vol. 1, Frederick Ungar, 1961.
5. P. M. Woodward, *Probability and Information Theory, with Applications to Radar*, Pergamon Press, 1953.

DISCRETE SIGNAL REPRESENTATIONS

3

3.1 Subspaces of $L^2(T)$

Applying the concepts introduced in the preceding chapters, we now consider the problem of associating a *numerical representation* with an arbitrary signal of finite energy, i.e., a (perhaps complex) time function $\mathbf{x} \in L^2(T)$. This amounts to finding a suitable mapping from $L^2(T)$ into C^n where n is usually chosen on the basis of a compromise between accuracy and economy of representation. From the relative dimensionality of $L^2(T)$ and C^n, it is clear that the mapping must exhibit a many-to-one nature and some sort of approximation is implied since each signal in $L^2(T)$ cannot have a distinct representation in C^n. It is natural to think of this mapping in terms of an equivalence relation whereby $L^2(T)$ is partitioned in such a way that the equivalence sets have a one-to-one correspondence with points in C^n.

The standard approach to this problem is to first select a particular n-dimensional subspace of $L^2(T)$. Suppose $\{\boldsymbol{\varphi}_i; i = 1, 2, \ldots, n\}$ is a linearly independent set of functions in $L^2(T)$; i.e.,

$$\sum_{i=1}^{n} \alpha_i \varphi_i(t) = 0 \qquad \text{(almost everywhere) for } t \in T \tag{3.1}$$

if, and only if, $\alpha_i = 0$ for each i. Let M_n denote the linear subspace spanned by these functions. Now if the signal to be represented, $\mathbf{x}$, happens to lie in M_n, then it has a unique expression as a linear combination of the $\{\boldsymbol{\varphi}_i\}$,

$$x(t) = \sum_{i=1}^{n} \alpha_i \varphi_i(t); \qquad \mathbf{x} \in M_n, \qquad t \in T \tag{3.2}$$

and the n-tuple $\boldsymbol{\alpha} = \{\alpha_1, \alpha_2, \ldots, \alpha_n\}$ provides the desired representation in C^n. Since $L^2(T)$ is an inner product space, with inner products given by

$$(\mathbf{x}, \mathbf{y}) = \int_T x(t)y^*(t)\,dt \tag{3.3}$$

the relationship between $\mathbf{x}$ and $\boldsymbol{\alpha}$ can be established according to (2.43). Using matrix notation,

$$\begin{bmatrix} (\boldsymbol{\varphi}_1, \boldsymbol{\varphi}_1) & (\boldsymbol{\varphi}_2, \boldsymbol{\varphi}_1) & \cdots & (\boldsymbol{\varphi}_n, \boldsymbol{\varphi}_1) \\ (\boldsymbol{\varphi}_1, \boldsymbol{\varphi}_2) & (\boldsymbol{\varphi}_2, \boldsymbol{\varphi}_2) & \cdots & (\boldsymbol{\varphi}_n, \boldsymbol{\varphi}_2) \\ \cdot & & & \\ \cdot & & & \\ \cdot & & & \\ (\boldsymbol{\varphi}_1, \boldsymbol{\varphi}_n) & (\boldsymbol{\varphi}_2, \boldsymbol{\varphi}_n) & \cdots & (\boldsymbol{\varphi}_n, \boldsymbol{\varphi}_n) \end{bmatrix} \begin{bmatrix} \alpha_1 \\ \alpha_2 \\ \cdot \\ \cdot \\ \cdot \\ \alpha_n \end{bmatrix} = \begin{bmatrix} (\mathbf{x}, \boldsymbol{\varphi}_1) \\ (\mathbf{x}, \boldsymbol{\varphi}_2) \\ \cdot \\ \cdot \\ \cdot \\ (\mathbf{x}, \boldsymbol{\varphi}_n) \end{bmatrix}$$

or

$$\mathbf{G}\boldsymbol{\alpha} = \boldsymbol{\beta} \Rightarrow \boldsymbol{\alpha} = \mathbf{G}^{-1}\boldsymbol{\beta} \tag{3.4}$$

with

$$\boldsymbol{\beta} = \{(\mathbf{x}, \boldsymbol{\varphi}_i);\quad i = 1, 2, \ldots, n\}$$

Employing an alternate notation, the reciprocal basis $\{\boldsymbol{\theta}_i; i = 1, 2, \ldots, n\}$ for M_n can be expressed as linear combinations of the $\{\boldsymbol{\varphi}_i\}$,

$$\theta_j(t) = \sum_{k=1}^{n} \gamma_{jk}\varphi_k(t) \tag{3.5}$$

where

$$(\boldsymbol{\varphi}_i, \boldsymbol{\theta}_j) = \sum_{k=1}^{n} \gamma_{jk}^*(\boldsymbol{\varphi}_i, \boldsymbol{\varphi}_k) = \delta_{ij}$$

or, in matrix notation,

$$\boldsymbol{\Gamma}^*\mathbf{G} = \mathbf{I} \Rightarrow \boldsymbol{\Gamma} = [\mathbf{G}^{-1}]^* \tag{3.6}$$

Now using the reciprocal basis, (3.4) is equivalent to

$$\alpha_i = (\mathbf{x}, \boldsymbol{\theta}_i);\qquad i = 1, 2, \ldots, n \tag{3.7}$$

although a matrix inversion is required in either case.

Exercise 3.1. Show that the representation for $\mathbf{x} \in M_n$ given by (3.2) and (3.4) is unique by virtue of the linear independence of the basis functions.

Signals Outside M_n (Projection Theorem)

Now the question remains as to how we go about representing a signal which is not contained in M_n. Since $L^2(T)$ is a metric space, it seems quite reasonable to associate with an arbitrary $\mathbf{x}$ the vector in M_n, call it $\hat{\mathbf{x}}$, which is closest to $\mathbf{x}$. In this way, each vector in M_n generates an equivalence set given by

$$S_{\hat{x}} = \{\mathbf{x} \in L^2(T);\ \|\mathbf{x} - \hat{\mathbf{x}}\| \leqslant \|\mathbf{x} - \tilde{\mathbf{x}}\| \text{ for any } \tilde{\mathbf{x}} \in M_n\} \tag{3.8}$$

and every vector in $S_{\hat{x}}$ is to be represented by the n-tuple for $\hat{\mathbf{x}}$. It turns out, rather conveniently, that this representation is also given by (3.7), even for an arbitrary $\mathbf{x} \in L^2(T)$. This result comes from the *projection theorem*: For any $\mathbf{x} \in L^2(T)$, there is a unique $\hat{\mathbf{x}}$ in M_n, given by

$$\hat{\mathbf{x}} = \sum_{i=1}^{n} (\mathbf{x}, \boldsymbol{\theta}_i)\boldsymbol{\varphi}_i \tag{3.9}$$

such that $\mathbf{x} - \hat{\mathbf{x}}$ is orthogonal to every vector in M_n and furthermore $\|\mathbf{x} - \hat{\mathbf{x}}\| < \|\mathbf{x} - \tilde{\mathbf{x}}\|$, where $\tilde{\mathbf{x}}$ is any other vector in M_n. To show this, we first note that $\mathbf{x} - \hat{\mathbf{x}}$ is orthogonal to each of the basis vectors $\{\boldsymbol{\theta}_i\}$ of M_n. From (3.9) and (3.5),

$$\begin{aligned}(\mathbf{x} - \hat{\mathbf{x}}, \boldsymbol{\theta}_i) &= (\mathbf{x}, \boldsymbol{\theta}_i) - \sum_{j=1}^{n}(\mathbf{x}, \boldsymbol{\theta}_j)(\boldsymbol{\varphi}_j, \boldsymbol{\theta}_i)\\ &= (\mathbf{x}, \boldsymbol{\theta}_i) - (\mathbf{x}, \boldsymbol{\theta}_i) = 0; \quad i = 1, 2, \ldots, n\end{aligned} \tag{3.10}$$

from which it follows that $\mathbf{x} - \hat{\mathbf{x}}$ is orthogonal to any vector in M_n. To show that $\|\mathbf{x} - \hat{\mathbf{x}}\|$ is a minimum, we consider an arbitrary $\tilde{\mathbf{x}} \in M_n$, and

$$\begin{aligned}\|\mathbf{x} - \tilde{\mathbf{x}}\|^2 &= \|(\mathbf{x} - \hat{\mathbf{x}}) - (\tilde{\mathbf{x}} - \hat{\mathbf{x}})\|^2\\ &= (\mathbf{x} - \hat{\mathbf{x}}, \mathbf{x} - \hat{\mathbf{x}}) - (\mathbf{x} - \hat{\mathbf{x}}, \tilde{\mathbf{x}} - \hat{\mathbf{x}})\\ &\quad - (\tilde{\mathbf{x}} - \hat{\mathbf{x}}, \mathbf{x} - \hat{\mathbf{x}}) + (\tilde{\mathbf{x}} - \hat{\mathbf{x}}, \tilde{\mathbf{x}} - \hat{\mathbf{x}})\end{aligned}$$

Since $\tilde{\mathbf{x}} - \hat{\mathbf{x}} \in M_n$, the middle terms vanish and

$$\|\mathbf{x} - \tilde{\mathbf{x}}\|^2 = \|\mathbf{x} - \hat{\mathbf{x}}\|^2 + \|\tilde{\mathbf{x}} - \hat{\mathbf{x}}\|^2 \tag{3.11}$$

which is clearly minimized by making $\tilde{\mathbf{x}} = \hat{\mathbf{x}}$. We call $\hat{\mathbf{x}}$ the *orthogonal projection* of $\mathbf{x}$ on M_n and $\boldsymbol{\eta} = \mathbf{x} - \hat{\mathbf{x}}$ the *error* resulting from approximating $\mathbf{x}$ by $\hat{\mathbf{x}}$. The closeness of approximation is numerically characterized by the norm of $\boldsymbol{\eta}$ and, letting $\tilde{\mathbf{x}} = \mathbf{0}$ in (3.11), we have

$$\|\boldsymbol{\eta}\|^2 = \|\mathbf{x}\|^2 - \|\hat{\mathbf{x}}\|^2 \tag{3.12}$$

It is now clear that the equivalence relation corresponding to the partition of $L^2(T)$ indicated in (3.8) can be expressed as

$$\mathbf{x} \sim \mathbf{y} \Rightarrow (\mathbf{x}, \boldsymbol{\varphi}_i) = (\mathbf{y}, \boldsymbol{\varphi}_i); \qquad i = 1, 2, \ldots, n \tag{3.13}$$

corresponding to the introductory Example 1.7 illustrating equivalence relations and partitions. Pursuing further the concepts introduced in this example we note that (3.13) could be expressed as

$$\mathbf{x} \sim \mathbf{y} \Rightarrow \mathbf{x} - \mathbf{y} \in M \tag{3.14}$$

where

$$M = \{\mathbf{x}; (\mathbf{x}, \tilde{\mathbf{x}}) = 0 \text{ for all } \tilde{\mathbf{x}} \in M_n\}$$

It is easy to show that M in (3.14) is a linear subspace. M is called the *orthogonal complement* of M_n because any vector in $L^2(T)$ can be uniquely expressed as the sum of a vector in M_n and a vector in M. Furthermore, the two vectors in the sum are orthogonal; i.e., for any $\mathbf{x}$,

$$\mathbf{x} = \hat{\mathbf{x}} + \mathbf{z}; \qquad \hat{\mathbf{x}} \in M_n, \mathbf{z} \in M, (\hat{\mathbf{x}}, \mathbf{z}) = 0 \tag{3.15}$$

Finally, it is customary to refer to $L^2(T)$ as the *direct sum* $L^2(T) = M_n \oplus M$ of the subspaces M_n and M. Notice that M and M_n contain only the zero vector in common. These concepts can be summarized diagrammatically as in Figure 3.1.

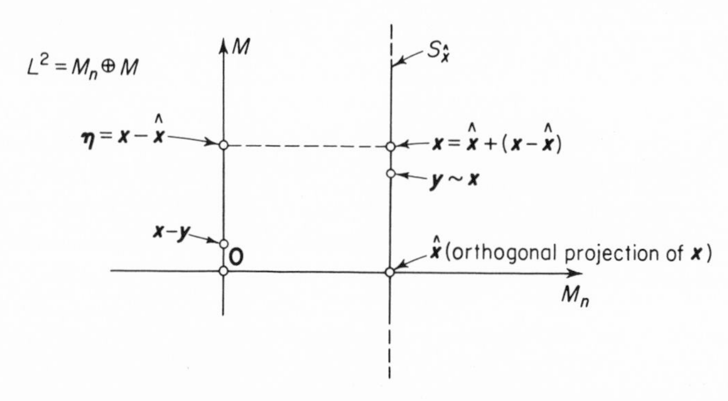

Figure 3.1. Interpretation of the orthogonal projection of $\mathbf{x} \in L^2(T)$ on a finite-dimensional subspace.

Exercise 3.2. *Direct sums and projections:* We can say that a linear space L is the *sum* of two subspaces $L = M + N$ if L is the space spanned by the largest set of linearly independent vectors taken from both M and N. Then any $\mathbf{z} \in L$ can be expressed as $\mathbf{z} = \mathbf{x} + \mathbf{y}$ with $\mathbf{x} \in M$ and $\mathbf{y} \in N$. Show that this decomposition $\mathbf{z} = \mathbf{x} + \mathbf{y}$ is unique if, and only if, $M \cap N = \{\mathbf{0}\}$, i.e., if M and N contain only the zero vector in common. In this case, L is said to be the *direct sum* of M and N; $L = M \oplus N$. Note that M and N are not necessarily orthogonal. If $L = M \oplus N$ and $\mathbf{z} = \mathbf{x} + \mathbf{y}$ is any vector in L, then $\mathbf{x} \in M$ is called the *projection of* $\mathbf{z}$ *onto* M *along* N (see Figure 3.2). Suppose $\{\boldsymbol{\varphi}_i; i = 1, 2, \ldots\}$ is a basis for L and that $\{\boldsymbol{\theta}_i; i = 1, 2, \ldots\}$ is the reciprocal basis. Let M be spanned by $\{\boldsymbol{\varphi}_i; i = 1, 2, \ldots, n\}$ and N be spanned by $\{\boldsymbol{\varphi}_i; i = n+1, n+2, \ldots\}$. Show that the projection of $\mathbf{z}$ onto M along N is given by

$$\mathbf{x} = \sum_{i=1}^{n} (\mathbf{z}, \boldsymbol{\theta}_i)\boldsymbol{\varphi}_i$$

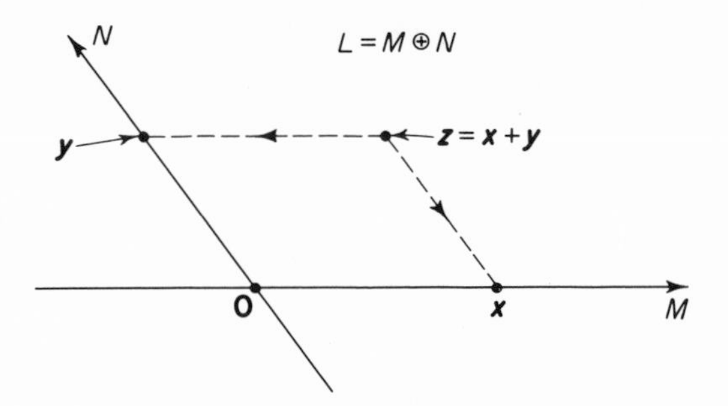

Figure 3.2. Non-orthogonal projections.

but that $\{\boldsymbol{\theta}_i; i = 1, 2, \ldots, n\}$ is not necessarily the reciprocal basis for M. The concept of non-orthogonal projections will be useful in certain signal detection problems discussed in Chapter 10.

Example 3.1. Let M_3 be a subspace of $L^2(T)$; $T = [0, \infty)$, spanned by the real exponential functions $\{e^{-t}, e^{-2t}, e^{-3t}\}$. Suppose we want to find the best approximation in M_3 to the rectangular pulse $x(t) = 1$ for $0 \leqslant t \leqslant \frac{1}{2}$; $x(t) = 0$ for $t > \frac{1}{2}$. We have

$$(\boldsymbol{\varphi}_k, \boldsymbol{\varphi}_m) = \int_0^\infty e^{-(k+m)t}\, dt$$
$$= \frac{1}{k + m}; \qquad k, m = 1, 2, 3$$

Hence, the matrix $\mathbf{G}$ in (3.4) becomes

$$\mathbf{G} = \begin{bmatrix} \frac{1}{2} & \frac{1}{3} & \frac{1}{4} \\ \frac{1}{3} & \frac{1}{4} & \frac{1}{5} \\ \frac{1}{4} & \frac{1}{5} & \frac{1}{6} \end{bmatrix} \Rightarrow \mathbf{G}^{-1} = \begin{bmatrix} 72 & -240 & 180 \\ -240 & 900 & -720 \\ 180 & -720 & 600 \end{bmatrix}$$

The reciprocal basis, from (3.5), is

$$\theta_1(t) = 72e^{-t} - 240e^{-2t} + 180e^{-3t}$$
$$\theta_2(t) = -240e^{-t} + 900e^{-2t} - 720e^{-3t}$$
$$\theta_3(t) = 180e^{-t} - 720e^{-2t} + 600e^{-3t}$$

and

$$\alpha_i = (\mathbf{x}, \boldsymbol{\theta}_i) = \int_0^{\frac{1}{2}} \theta_i(t)\, dt \Rightarrow \begin{cases} \alpha_1 = -0.912 \\ \alpha_2 = 3.57 \\ \alpha_3 = -1.36 \end{cases}$$

Hence $\hat{\mathbf{x}}(t) = -0.912e^{-t} + 3.57e^{-2t} - 1.36e^{-3t}$ and $\boldsymbol{\alpha} = \{-0.912, 3.57, -1.36\}$ is the representation for $x(t)$ in R^3. The error in this approximation amounts to $\|\boldsymbol{\eta}\|^2 = \|\mathbf{x}\|^2 - \|\hat{\mathbf{x}}\|^2 \Rightarrow \|\boldsymbol{\eta}\| = 0.289$. ∎

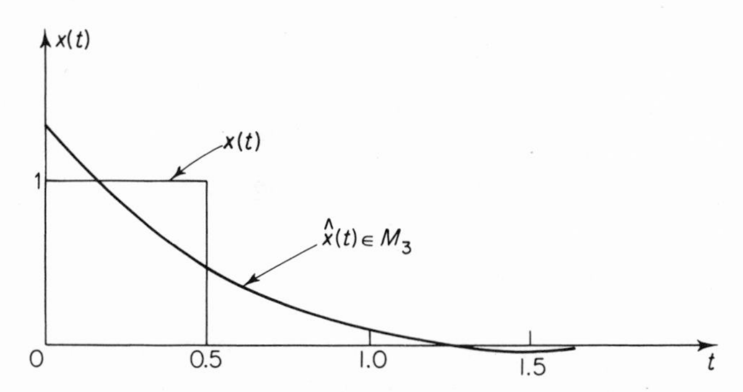

Figure 3.3. Approximation of rectangular pulse, $\mathbf{x}$, by $\hat{\mathbf{x}} \in M_3$.

Example 3.2. It is not always necessary to perform a matrix inversion, which becomes quite difficult for large n, in order to generate a reciprocal basis. Extending the previous example, we show a method for finding the reciprocal basis for M_n spanned by $\{e^{-kt}; k = 1, 2, \ldots, n\}$ with $T = [0, \infty)$ and arbitrary n. We want $\{\boldsymbol{\theta}_m\}$ such that

$$(\boldsymbol{\varphi}_k, \boldsymbol{\theta}_m) = \int_0^\infty e^{-kt}\theta_m(t)\,dt = \Theta_m(s)\Big|_{s=k} = \delta_{km} \tag{3.16}$$

where $\Theta_m(s)$ is the Laplace transform[1] of $\theta_m(t)$. Since $\theta_m(t)$ is, in general, a linear combination of the $\{\boldsymbol{\varphi}_k; k = 1, 2, \ldots, n\}$, then we know that $\Theta_m(s)$ is a rational function of s with poles at $s = -1, -2, -3, \ldots, -n$, for each m. But from (3.16) we see that $\Theta_m(s)$ must have zeros at $s = 1, 2, 3, \ldots, n$; except at $s = m$. Thus $\Theta_m(s)$ can be expressed as

$$\Theta_m(s) = \frac{K_m \prod_{k=1}^{n} (s - k)}{(s - m) \prod_{k=1}^{n} (s + k)} \tag{3.17}$$

Also, from (3.16), we have $\Theta_m(m) = 1$ and the constant K_m can be evaluated. If we define

$$C(j, n) \triangleq \prod_{\substack{k=1 \\ \neq j}}^{n} (j - k) \Rightarrow K_m = \frac{(-1)^n C(-m, n)}{C(m, n)} \tag{3.18}$$

then taking the inverse transform of $\Theta_m(s)$ by the standard method of residues,

$$\theta_m(t) = \sum_{l=1}^{n} \gamma_{ml}\varphi_l(t) \tag{3.19}$$

[1] For time functions defined on $T = [0, \infty)$, it is often convenient to use the Laplace transform defined by $X(s) = \int_0^\infty x(t)e^{-st}\,dt$, where s is a complex variable. This allows for a characterization of signals in terms of pole-zero configurations (in the s-plane) and for evaluating signal properties by means of the calculus of residues. It is assumed that the reader is familiar with these concepts. Chapter 9 in [1] may be consulted for reference.

where

$$\gamma_{ml} = (s + l)\Theta_m(s)\Big|_{s=-l} = \frac{C(-m, n)C(-l, n)}{(l + m)C(m, n)C(l, n)}$$

▮

Exercise 3.3. Check the $\theta_i(t)$ in Example 3.1 using (3.19).

3.2 Complete Orthonormal Sets

In the foregoing, we have used the projection theorem to find the best representation for an arbitrary signal relative to a given finite-dimensional subspace M_n, but we have avoided the question as to how one goes about selecting a suitable subspace. For a given subspace, it is clear that there is a large subset of $L^2(T)$ which is not adequately represented, e.g., points in $L^2(T)$ which are orthogonal to M_n. The problem of finding an optimum subspace is meaningful only relative to some suitably restricted subset (usually compact) of $L^2(T)$. A problem of this type is examined in Chapter 6. In the present section, we consider some preliminary aspects of convergence, as the dimension of the subspace is made arbitrarily large, and present some examples of methods which have been frequently employed for generating sets of basis functions without particular regard for the optimality of the subspaces spanned by these sets.

In most texts on functional analysis it is shown[2],[3] that the $L^2(T)$ space is *complete* and *separable*. From these properties it can be shown that approximation by means of orthogonal projection,

$$\mathbf{x} \sim \mathbf{x}_n = \sum_{i=1}^{n} (\mathbf{x}, \boldsymbol{\varphi}_i)\boldsymbol{\varphi}_i \tag{3.20}$$

can be made arbitrarily close, by choosing n sufficiently large, for any $\mathbf{x} \in L^2(T)$. In (3.20) the $\boldsymbol{\varphi}_i$ are selected from a countably infinite, orthonormal set of functions which satisfies certain conditions.

We first note, from (3.12) with $\hat{\mathbf{x}} = \mathbf{x}_n$, that

$$\|\mathbf{x} - \mathbf{x}_n\|^2 = \|\mathbf{x}\|^2 - \sum_{i=1}^{n} |(\mathbf{x}, \boldsymbol{\varphi}_i)|^2$$

and, hence

$$\sum_{i=1}^{n} |(\mathbf{x}, \boldsymbol{\varphi}_i)|^2 \leqslant \|\mathbf{x}\|^2 \text{ for any } n \tag{3.21}$$

The relation (3.21) is known as *Bessel's inequality* and it establishes the fact that the expansion coefficients $(\mathbf{x}, \boldsymbol{\varphi}_i)$ are square-summable for any $\mathbf{x} \in L^2(T)$. It also shows that $\{\mathbf{x}_n\}$ in (3.20) is a Cauchy sequence since, for any $\varepsilon > 0$ with n_0 sufficiently large,

$$\|\mathbf{x}_n - \mathbf{x}_m\|^2 = \sum_{i=m+1}^{n} |(\mathbf{x}, \boldsymbol{\varphi}_i)|^2 < \varepsilon^2; \qquad n > m > n_0 \tag{3.22}$$

and since $L^2(T)$ is complete, $\{\mathbf{x}_n\}$ converges to some point in $L^2(T)$. The sequence converges to $\mathbf{x}$ if $\{\boldsymbol{\varphi}_i; i = 1, 2, \ldots\}$ is a *complete orthonormal set.* An orthonormal set is said to be complete if there are no additional non-zero orthogonal vectors which can be added to the set. It is the separability of $L^2(T)$ that ensures that complete orthonormal sets will be countable. It is apparent that a complete orthonormal set is for $L^2(T)$ the counterpart of a basis for a finite-dimensional space. An arbitrary infinite orthonormal set is not necessarily complete; for example,

$$\left\{\sqrt{2W}\,\frac{\sin 2\pi W[t - (i/2W)]}{2\pi W[t - (i/2W)]}; \qquad i = 0, \pm 1, \pm 2, \ldots\right\}$$

used in (1.34) for time-series representation is orthonormal but not complete in $L^2(-\infty, \infty)$ since functions with bandwidth greater than W are not in the subspace spanned by this set. We remark here that the concept of completeness of an orthonormal set is not to be confused with completeness for a metric space.

For a complete orthonormal set, the Bessel inequality (3.21) becomes

$$\sum_{i=1}^{\infty} |(\mathbf{x}, \boldsymbol{\varphi}_i)|^2 = \|\mathbf{x}\|^2 \qquad \text{for any } \mathbf{x} \in L^2(T) \tag{3.23}$$

which is also known as the *completeness relation.* From these facts, we are able to say that, for any $\varepsilon > 0$, there is an n_0 such that

$$\left\|\mathbf{x} - \sum_{i=1}^{n} (\mathbf{x}, \boldsymbol{\varphi}_i)\boldsymbol{\varphi}_i\right\| < \varepsilon; \qquad n > n_0 \tag{3.24}$$

for any $\mathbf{x} \in L^2(T)$; i.e., if we use subspaces M_n, spanned by the first n elements of a complete orthonormal set, for the representation of an arbitrary signal, then the norm of the error can be made arbitrarily small by choosing n sufficiently large. The drawback is that n_0 depends on $\mathbf{x}$ so that we cannot uniformly bound the error. Nevertheless, as a practical matter, this approach has certain merits, which accounts for its extensive usage. First of all, there are many complete orthonormal sets for which formulas for the nth element of the set have been derived and are presented in standard texts. Secondly, there are the advantages of characterizing M_n in terms of an orthonormal basis, foremost of which is that inner products in M_n and C^n correspond [see (2.49)]. Thirdly, also a result of the self-reciprocal nature of the orthonormal basis, the projection of $\mathbf{x}$ on M_{n+1} does not need to be wholly recomputed if the projection on M_n is known; only $(\mathbf{x}, \boldsymbol{\varphi}_{n+1})$ need be evaluated. This provides a considerable saving of effort if, on the basis of error evaluation, it is decided that the dimension of the representation must be increased. In fact, with orthonormal bases, we usually tend to think of orthogonal projections as a set of individual projections, each onto the one-dimensional space spanned by the ith basis vector and giving directly the ith element in the representation.

In the examples to follow, most of the sets are generated from sets of relatively elementary functions by means of the Gram-Schmidt procedure (2.50a and b) which, because of the iterative nature of the process, is not limited in application to finite sets. Following the discussion of the preceding paragraph, it is easy to give a geometrical interpretation of the Gram-Schmidt procedure; at the ith step of the process, a subspace M_{i-1} spanned by an orthonormal basis has been created. Select a new function, not in M_{i-1}, and subtract off its projection on M_{i-1}. The result is a function orthogonal to M_{i-1}, and after normalization it becomes the ith basis function for M_i.

Exercise 3.4. For the exponential basis functions, ordered as in Example 3.2, use the Gram-Schmidt procedure to construct the first few elements of an orthonormal set. Results can be checked in reference [6], page 463.

Exercise 3.5. Let $\{\boldsymbol{\varphi}_i; i = 1, 2, \ldots\}$ be a complete orthonormal set in $L^2(T)$. For any $\mathbf{x}$ and $\mathbf{y}$ in $L^2(T)$, verify *Parseval's relation*

$$(\mathbf{x}, \mathbf{y}) = \sum_{i=1}^{\infty} (\mathbf{x}, \boldsymbol{\varphi}_i)(\boldsymbol{\varphi}_i, \mathbf{y})$$

This is a somewhat more general version of (3.23).

In constructing basis functions for signal representation, the concept of a *weighted norm* is often used. It may turn out, in a particular situation, that in evaluating approximation error it is desirable to focus attention on a particular region of the time domain relative to others. For example, the integral

$$\left[\int_T w(t)\,|x(t) - x_n(t)|^2\,dt\right]^{\frac{1}{2}} \tag{3.25}$$

where $w(t)$ is a non-negative function defined on T, may actually be a more significant measure of approximation error than $\|\mathbf{x} - \mathbf{x}_n\|$. This suggests the possibility of redefining a norm for $L^2(T)$. In fact, if $w(t)$ is real and positive except possibly at a countable number of points in T, then

$$(\mathbf{x}, \mathbf{y})_w = \int_T w(t)x(t)y^*(t)\,dt \tag{3.26}$$

can be seen to satisfy the requirements (2.28) for an inner product. A set of functions $\{\boldsymbol{\varphi}_i\}$ is said to be *orthonormal with respect to the weight function*, $w(t)$ if $(\boldsymbol{\varphi}_i, \boldsymbol{\varphi}_j)_w = \delta_{ij}$. This generality in defining inner products has not been mentioned previously because it amounts simply to a modification of basis functions according to

$$\psi_i(t) = \sqrt{w(t)}\,\varphi_i(t); \qquad i = 1, 2, \ldots \tag{3.27}$$

where the $\{\boldsymbol{\psi}_i\}$ are orthonormal in the usual sense if the $\{\boldsymbol{\varphi}_i\}$ are orthonormal with respect to $w(t)$.

3.3 Examples of Complete Orthonormal Sets

In this section we list some of the better known orthonormal sets for a prescribed interval and a prescribed weight function $w(t)$. The completeness and orthogonality of these sets are demonstrated in other references.[3],[4],[5]

Complex Sinusoids

For $T = [-1, +1]$ and $w(t) = 1$, the complex sinusoidal functions $\{e^{j\pi nt};\ n = 0, \pm 1, \pm 2, \ldots\}$ are orthogonal; hence only normalization is required to provide a complete orthonormal set. The expansion

$$x(t) \sim \frac{1}{\sqrt{2}} \sum_{k=-n}^{n} \alpha_k e^{j(\pi k t)} \tag{3.28}$$

with

$$\alpha_k = \frac{1}{\sqrt{2}} \int_{-1}^{1} x(t) e^{-j(\pi k t)}\, dt$$

is the familiar *Fourier-series* representation for functions duration-limited to $[-1, 1]$, or periodic with period 2. We note that this representation is valid for any finite interval since an arbitrary finite interval can be mapped into $[-1, 1]$ by the appropriate translation and scaling of the time variable.

Legendre Polynomials

For $T = [-1, +1]$ and $w(t) = 1$, we can generate another orthonormal set by applying the Gram-Schmidt procedure to the sequence $\{1, t, t^2, t^3, \ldots\}$. This results in the normalized polynomials

$$\varphi_0(t) = \frac{1}{\sqrt{2}}; \qquad \varphi_1(t) = \sqrt{\tfrac{3}{2}}\,t; \qquad \varphi_2(t) = \sqrt{\tfrac{5}{2}}\left(\tfrac{3}{2}t^2 - \tfrac{1}{2}\right)$$

$$\varphi_3(t) = \sqrt{\tfrac{7}{2}}\left(\tfrac{5}{2}t^3 - \tfrac{3}{2}t\right); \qquad \varphi_4(t) = \sqrt{\tfrac{9}{2}}\left(\tfrac{35}{8}t^4 - \tfrac{15}{4}t^2 + \tfrac{3}{8}\right)$$

$$\vdots$$

$$\varphi_n(t) = \sqrt{\frac{2n+1}{2}}\, P_n(t) \tag{3.29}$$

where the $\{P_n(t)\}$ are the *Legendre polynomials.* These polynomials are also characterized by

$$P_n(t) = \frac{1}{2^n n!} \frac{d^n}{dt^n} (t^2 - 1)^n \tag{3.30}$$

and by the recursion formula

$$nP_n(t) = (2n - 1)tP_{n-1}(t) - (n - 1)P_{n-2}(t) \tag{3.31}$$

The n zeros of $P_n(t)$ are all real and lie inside the interval $[-1, 1]$. The mapping of $[-1, 1]$ into any finite interval, mentioned previously, is appropriate here also.

Chebyshev Polynomials

For $T = [-1, +1]$ and $w(t) = [1 - t^2]^{-\frac{1}{2}}$, the polynomials

$$\varphi_n(t) = 2^n(2\pi)^{-\frac{1}{2}}T_n(t); \qquad n = 0, 1, 2, \ldots \tag{3.32}$$

form an orthonormal set. The T_n are the *Chebyshev polynomials* given by

$$T_0(t) = 1; \qquad T_n(t) = \frac{1}{2^{n-1}} \cos(n \cos^{-1} t); \qquad n \geqslant 1 \tag{3.33}$$

which exhibit the interesting property of having the smallest maximum absolute value over $[-1, 1]$, i.e., the least deviation from zero, of any nth degree polynomial with leading coefficient of 1. For $n \geqslant 3$, the T_n satisfy the recursion formula

$$T_n(t) = tT_{n-1}(t) - \tfrac{1}{4}T_{n-2}(t) \tag{3.34}$$

Laguerre Functions

For $T = [0, \infty)$ and $w(t) = e^{-t}$; the polynomials

$$\varphi_n(t) = \frac{1}{n!} L_n(t); \qquad n = 0, 1, 2, \ldots \tag{3.35}$$

form an orthonormal set. The L_n are the *Laguerre polynomials* given by

$$L_n(t) = e^t \frac{d^n}{dt^n}(t^n e^{-t}) \tag{3.36}$$

with n real zeros in $[0, \infty)$. The L_n satisfy the recursion formula

$$L_n(t) = (2n - 1 - t)L_{n-1}(t) - (n - 1)^2 L_{n-2}(t) \tag{3.37}$$

The *Laguerre functions*

$$\psi_n(t) = \frac{e^{-\frac{t}{2}}}{n!} L_n(t)$$

orthonormal over $[0, \infty)$ with unit weight, can also be obtained by applying the Gram-Schmidt procedure to $\{t^n e^{-\frac{t}{2}}; n = 0, 1, 2, \ldots\}$. The Laguerre functions have a special significance in practical applications in that they can be generated as impulse responses of relatively simple finite-order physical networks.[6] We can show that the functions

$$f_n(t) = \psi_n(2pt) = \frac{e^{-pt}}{n!} L_n(2pt); \qquad t \geqslant 0 \tag{3.38}$$

where p is a real, positive parameter allowing the flexibility of time-domain scaling, have Laplace transforms given by

$$F_n(s) = \frac{(s - p)^n}{(s + p)^{n+1}} \tag{3.39}$$

From this expression it follows that the nth Laguerre function is the impulse response of the cascade combination of a single-pole network and n all-pass

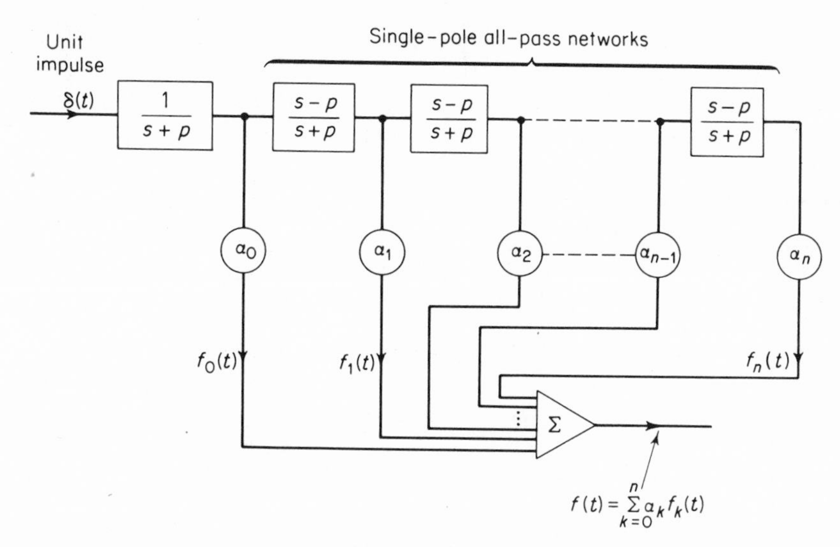

Figure 3.4. Generation of Laguerre functions and linear combinations.

networks, each with a pole at $s = -p$, as shown in Figure 3.4. By tapping off signals at the network junctions and summing with appropriate weighting factors, the overall network, known as a *transversal filter*, has an impulse response given as an arbitrary linear combination of Laguerre functions. Reference [6] also shows network realizations for other orthonormal sets of impulse responses.

Legendre Functions

Using the mapping $\tau = 1 - 2e^{-2pt}$, the interval $[-1, 1]$ for τ goes into $[0, \infty)$ for t, and a new set of orthonormal functions on $[0, \infty)$ can be generated from the Legendre polynomials. (p is an arbitrary, real, positive parameter.) The *Legendre functions*

$$\pi_n(t) = [2p(2n+1)]^{\frac{1}{2}} e^{-pt} P_n(1 - 2e^{-2pt}) \tag{3.40}$$

are an orthonormal set, with unit weight, over $[0, \infty)$. Laplace transforms of these functions have poles at $s = -p, -3p, -5p, \ldots$.

Chebyshev Functions

Using $\tau = 1 - 2e^{-2pt}$, the Chebyshev polynomials give the *Chebyshev functions*

$$\rho_n(t) = 2^n \left(\frac{p}{\pi}\right)^{\frac{1}{2}} T_n(1 - 2e^{-2pt}) \tag{3.41}$$

which are orthonormal, with weight function $w(t) = (e^{2pt} - 1)^{-\frac{1}{2}}$, over $[0, \infty)$. Their Laplace transforms have poles at $s = 0, -2p, -4p, -6p, \ldots$.

A more extensive treatment of the problem of finding sets orthonormal over $[0, \infty)$ which are characterized in terms of the pole and zero positions of their Laplace transforms is presented in references [7] and [8].

Hermite Functions

For $T = (-\infty, \infty)$ and $w(t) = e^{-t^2}$, the polynomials

$$\varphi_n(t) = (2^n n! \sqrt{\pi})^{-\frac{1}{2}} H_n(t); \qquad n = 0, 1, 2, \ldots \tag{3.42}$$

form an orthonormal set. The H_n are the *Hermite polynomials* given by

$$H_n(t) = (-1)^n e^{t^2} \frac{d^n}{dt^n}(e^{-t^2}) \tag{3.43}$$

with the recursion formula

$$H_n(t) = 2tH_{n-1}(t) - 2(n-1)H_{n-2}(t) \tag{3.44}$$

Also the *Hermite functions*

$$\psi_n(t) = (2^n n! \sqrt{\pi})^{-\frac{1}{2}} e^{-\frac{t^2}{2}} H_n(t) \tag{3.45}$$

orthonormal with unit weight over $(-\infty, \infty)$, can be generated by applying the Gram-Schmidt procedure to the set

$$\{t^n e^{-\frac{t^2}{2}}; n = 0, 1, 2, \ldots\}.$$

Walsh Functions

For $T = [0, 1]$ and $w(t) = 1$, we can find a complete orthonormal set of "square-wave" functions. For defining the functions, a double indexing scheme is helpful. The functions $\varphi_n^{(k)}(t)$ are called *Walsh functions*[9]–[11]

defined by

$$\varphi_0(t) = 1; \quad 0 \leqslant t \leqslant 1; \qquad \varphi_1(t) = \begin{cases} 1; & 0 \leqslant t < \frac{1}{2} \\ -1; & \frac{1}{2} < t \leqslant 1 \end{cases}$$

$$\varphi_2^{(1)}(t) = \begin{cases} 1; & 0 \leqslant t < \frac{1}{4},\ \frac{3}{4} < t \leqslant 1 \\ -1; & \frac{1}{2} < t < \frac{3}{4} \end{cases}$$

$$\varphi_2^{(2)}(t) = \begin{cases} 1; & 0 \leqslant t < \frac{1}{4},\ \frac{1}{2} < t < \frac{3}{4} \\ -1; & \frac{1}{4} < t < \frac{1}{2},\ \frac{3}{4} < t \leqslant 1 \end{cases}$$

$$\cdots \qquad (3.46)$$

$$\left.\begin{aligned} \varphi_{m+1}^{(2k-1)}(t) &= \varphi_m^{(k)}(2t); \quad 0 \leqslant t < \tfrac{1}{2} \\ &= (-1)^{k+1}\varphi_m^{(k)}(2t-1); \quad \tfrac{1}{2} < t \leqslant 1 \\ \varphi_{m+1}^{(2k)}(t) &= \varphi_m^{(k)}(2t); \quad 0 \leqslant t < \tfrac{1}{2} \\ &= (-1)^{k}\varphi_m^{(k)}(2t-1); \quad \tfrac{1}{2} < t \leqslant 1 \end{aligned}\right\} \quad \begin{aligned} &\text{for } m = 1, 2, 3, \ldots \\ &\text{and } k = 1, 2, 3, \ldots, 2^{m-1} \end{aligned}$$

This set has a considerable practical importance due to the piecewise constant nature of the functions and the fact that the functions take on only two values (+1 and −1). Generation of such a signal is readily accomplished with digital logic circuitry. Also, multiplication of signals with these functions, e.g., in forming inner products, is especially simple since all that is needed is a polarity-reversing switch actuated at the proper instants. Note that the functions $\varphi_m^{(2^{m-1})}(t)$ correspond to the conventional form of a square wave.

We can convert to a single-variable indexing scheme which is consistent with the foregoing orthonormal sets by arranging the functions in (3.46) so that the nth function has n zero crossings in $0 < t < 1$ (n sign changes). This is accomplished by defining

$$\omega_0(t) = \varphi_0(t), \qquad \omega_1(t) = \varphi_1(t), \qquad \omega_n(t) = \varphi_m^{(k)}(t) \qquad (3.47)$$

where $n = 2^{m-1} + k - 1$ with $k \leqslant 2^{m-1}$. Graphs for the first 16 Walsh functions are shown in Figure 3.5.

3.4 Implementation of the Signal Resolution Operation

To summarize the process of providing an approximate numerical representation for an arbitrary signal of finite energy, we select a set of basis functions $\{\boldsymbol{\varphi}_i\}$ (with reciprocal basis $\{\boldsymbol{\theta}_i\}$) and establish the equivalence

$$\mathbf{x} \sim \sum_{i=1}^{n} \alpha_i \boldsymbol{\varphi}_i \qquad (3.48a)$$

with

$$\alpha_i = f_{\theta_i}(\mathbf{x}) = (\mathbf{x}, \boldsymbol{\theta}_i); \qquad i = 1, 2, \ldots, n \qquad (3.48b)$$

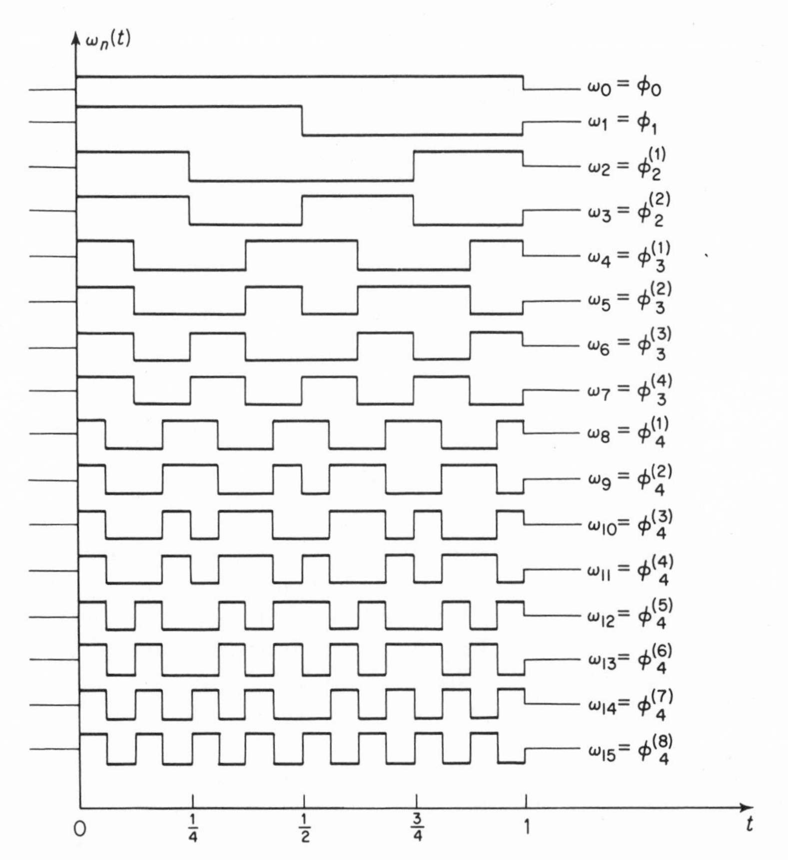

Figure 3.5. Walsh functions indexed according to number of sign changes in $0 < t < 1$.

By analogy with conventional usage of Fourier transforms, we can consider (3.48) as a *transform pair*. In (3.48a) the signal is represented as a particular linear combination of the $\{\boldsymbol{\varphi}_i\}$; whereas in (3.48b) we have the formula, in terms of a set of linear functionals, for resolving any signal into the components of its representation. This resolution operation may be considered the inverse of representation, and it is this operation which we want to implement in terms of physical devices. Using the implementation for a linear functional shown in Figure 2.7, we obtain a structure which may be thought of as a generalized version of the familiar laboratory instrument known as a *waveform analyzer*. Alternatively, the structure in Figure 3.6 could be realized by replacing the multiplier-integrator combinations with filter-sampler combinations as indicated in Figure 2.8.

In many cases the locally generated "reference" waveforms cannot exactly reproduce the desired set $\{\boldsymbol{\theta}_i\}$. For example, in certain optical signal

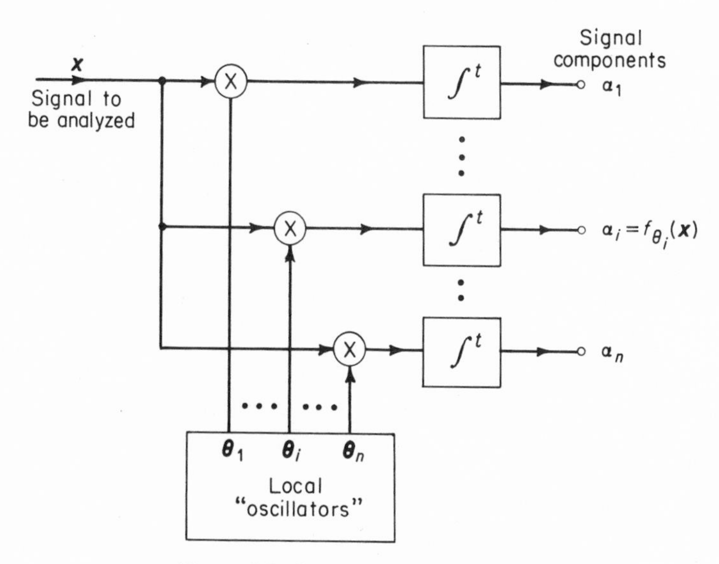

Figure 3.6. Generalized waveform analyzer.

processing systems, the multiplication operation is implemented by illuminating a variable density transparency. In this situation it is desirable that the multiplier functions be non-negative, e.g., a set of real exponentials. As another example, it is sometimes desirable to use "on-off" multiplier functions so that the multiplier can be implemented as a switch. It may turn out that the designer simply has no choice and must work with a given set of waveforms. This leads us to consider a problem of the following type: Given set of devices corresponding to the linear functionals $\{\mathbf{f}_{u_1}, \mathbf{f}_{u_2}, \ldots, \mathbf{f}_{u_m}\}$, how can we best use these devices to implement an arbitrary linear functional $\mathbf{f}_\theta$? To answer this question, we make use of the correspondence between signals and linear functionals discussed in Section 2.6. The problem becomes one of minimum distance approximation in the conjugate space by taking the orthogonal projection of $\mathbf{f}_\theta \in [L^2(T)]^*$ on the subspace of $[L^2(T)]^*$ spanned by $\{\mathbf{f}_{u_1}, \mathbf{f}_{u_2}, \ldots, \mathbf{f}_{u_m}\}$. The approximation of $\mathbf{f}_\theta$ by a linear combination of the $\{\mathbf{f}_{u_i}\}$ is implemented as a parallel bank of the devices discussed in Section 2.6, with an adjustable gain element a_i in each path as shown in Figure 3.7. The desired gain settings minimize $\left\| \mathbf{f}_\theta - \sum_{k=1}^{m} a_k \mathbf{f}_{u_k} \right\|$ and are expressed in terms of inner products in the conjugate space as defined in (2.61). Taking the orthogonal projection, we have

$$\mathbf{f}_\theta \sim \sum_{k=1}^{m} a_k \mathbf{f}_{u_k} \tag{3.49a}$$

where

$$a_k = (\mathbf{f}_\theta, \mathbf{f}_{v_k}) = (\mathbf{v}_k, \boldsymbol{\theta}) \tag{3.49b}$$

and $\{\mathbf{v}_k; k = 1, 2, \ldots, m\}$ is the reciprocal basis for the subspace, N_m spanned by $\{\mathbf{u}_k; k = 1, 2, \ldots, m\}$. It is easy to see that this approximation for $\mathbf{f}_\theta$ corresponds exactly to minimum distance approximation of $\boldsymbol{\theta}$ by projection on N_m. Noting equivalence of norms, we have

$$\left\| \boldsymbol{\theta} - \sum_{k=1}^{m} a_k \mathbf{u}_k \right\| = \varepsilon \Rightarrow \left\| \mathbf{f}_\theta - \sum_{k=1}^{m} a_k \mathbf{f}_{u_k} \right\| = \varepsilon$$

$$\Rightarrow \left| f_\theta(\mathbf{x}) - \sum_{k=1}^{m} a_k f_{u_k}(\mathbf{x}) \right| \leqslant \varepsilon \tag{3.50}$$

for any signal $\mathbf{x}$ of unit energy in $L^2(T)$. We can also say that 100ε gives the maximum error in the mapping as a percentage relative to $\|\mathbf{x}\|$.

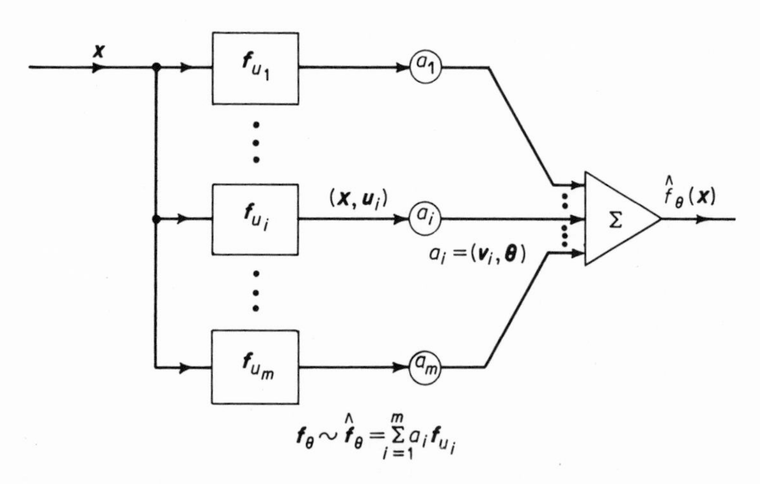

Figure 3.7. Approximation in conjugate space; implementation of $f_\theta(\mathbf{x}) = (\mathbf{x}, \boldsymbol{\theta})$ in terms of a given set of functionals.

With this result, the implementation of the signal resolution operation in terms of given devices is straightforward. We want to represent $\mathbf{x}$ by a point in M_n spanned by $\{\boldsymbol{\varphi}_i; i = 1, 2, \ldots, n\}$ but we can only form the inner products $(\mathbf{x}, \mathbf{u}_i)$, where the set $\{\mathbf{u}_i; i = 1, 2, \ldots, m\}$ is a basis for N_m (presumably, $n \leqslant m$). The arrangement which results in the least upper bound on error in evaluating the individual components, from (3.49), is

$$\tilde{\alpha}_i = \sum_{k=1}^{m} a_{ik}(\mathbf{x}, \mathbf{u}_k); \qquad i = 1, 2, \ldots, n \tag{3.51}$$

where

$$a_{ik} = (\mathbf{v}_k, \boldsymbol{\theta}_i)$$

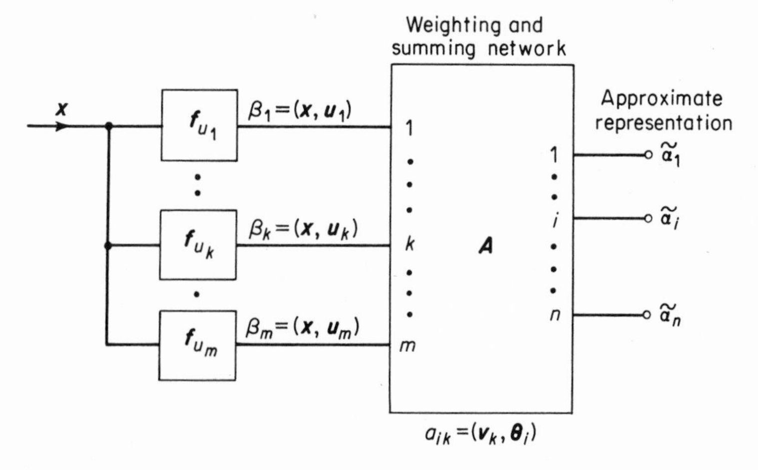

Figure 3.8. Modification of given wave analyzer for desired signal resolution.

The implementation of the wave analyzer uses the given linear functionals along with a weighting-summing network described by the matrix **A**, where the elements of **A** are the $\{a_{ik}\}$ in (3.51). This is shown schematically in Figure 3.8.

With some stretch of imagination, this approximation to the desired signal resolution can be expressed diagrammatically as shown in Figure 3.9. The signal representation is found by first obtaining the orthogonal projection of $\mathbf{x}$ on N_m, this represented by the m-tuple $\boldsymbol{\beta} = \{\beta_1, \beta_2, \ldots, \beta_m\}$. Next, this point in N_m is orthogonally projected on M_n, providing an approximate representation $\tilde{\mathbf{x}}$ in M_n for the desired representation which is the orthogonal projection $\hat{\mathbf{x}}$ of $\mathbf{x}$ directly on M_n. The approximate representation $\tilde{\boldsymbol{\alpha}} = \{\tilde{\alpha}_1, \tilde{\alpha}_2, \ldots, \tilde{\alpha}_n\}$ is obtained by matrix multiplication on the representation

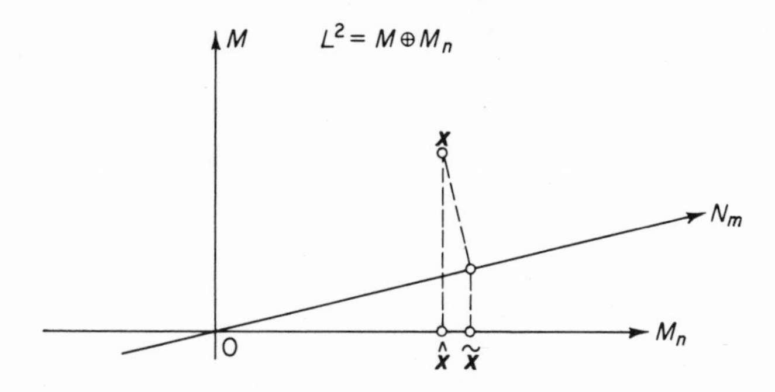

Figure 3.9. Interpretation of approximate signal resolution in terms of a given wave analyzer.

$\boldsymbol{\beta} = \{(\mathbf{x}, \mathbf{u}_i); i = 1, 2, \ldots, m\}$, for $\mathbf{x}$ relative to the basis $\{\mathbf{v}_i; i = 1, 2, \ldots, m\}$ for N_m.

$$\begin{bmatrix} \tilde{\alpha}_1 \\ \tilde{\alpha}_2 \\ \cdot \\ \cdot \\ \cdot \\ \tilde{\alpha}_n \end{bmatrix} = \begin{bmatrix} a_{11} & a_{12} & \cdots & a_{1m} \\ a_{21} & a_{22} & \cdots & a_{2m} \\ \cdot & & & \\ \cdot & & & \\ \cdot & & & \\ a_{n1} & a_{n2} & \cdots & a_{nm} \end{bmatrix} \begin{bmatrix} \beta_1 \\ \beta_2 \\ \cdot \\ \cdot \\ \cdot \\ \beta_m \end{bmatrix} \tag{3.52}$$

Of course, if M_n is a subspace of N_m, then the two projections are identical and the waveform analyzer corresponds exactly to the one that is desired. If M_n and N_m are identical, then $\mathbf{A}$ in (3.52) is non-singular and the relation simply reflects the effects of a *change of basis* for M_n (see Exercise 2.16).

Example 3.3. We shall illustrate the foregoing ideas with an example that is frequently encountered in signal analysis. Suppose that $\mathbf{x}$ can be considered duration-limited to the interval $[0, T)$ and that there is available a sampling switch which samples the signal at m uniformly spaced instants over this interval as illustrated in Figure 3.10. Suppose also that it is desired to measure the coefficients of a Fourier-series representation for the signal with period T. Hence

$$x(t) \sim \sum_{i=-\frac{n}{2}}^{\frac{n}{2}} \alpha_i e^{j\frac{2\pi i t}{T}} \qquad \text{with } n \leqslant m - 1 \tag{3.53}$$

is the desired representation. The basis functions are

$$\varphi_i(t) = e^{j\frac{2\pi i t}{T}}; \qquad \theta_i(t) = \frac{1}{T} e^{j\frac{2\pi i t}{T}} \tag{3.54}$$

Assuming that the switch rotor does not contact more than one pole at any instant and that the time between contacts is essentially zero, the reference waveforms may be taken as the orthogonal set of rectangular functions

$$u_k(t) = u_1\left(t - (k-1)\frac{T}{m}\right); \qquad k = 1, 2, \ldots, m$$

where

$$\begin{aligned} u_1(t) &= 1 \qquad \text{for } 0 \leqslant t < \frac{T}{m} \\ &= 0 \qquad \text{otherwise} \end{aligned} \tag{3.55}$$

It follows that the set reciprocal to $\{\mathbf{u}_k\}$ is given by

$$v_k(t) = \frac{m}{T} u_k(t); \qquad k = 1, 2, \ldots, m \tag{3.56}$$

Conversion of the sample values $\beta_k = (\mathbf{x}, \mathbf{u}_k)$ into Fourier coefficients approximating $\alpha_i = (\mathbf{x}, \boldsymbol{\theta}_i)$ is accomplished by the matrix $\mathbf{A}$ given in (3.51).

$$\tilde{\alpha}_i = \sum_{k=1}^{m} a_{ik}\beta_k; \qquad i = 0, \pm 1, \pm 2, \ldots, \pm \frac{n}{2} \tag{3.57}$$

where

$$a_{ik} = (\mathbf{v}_k, \boldsymbol{\theta}_i) = \frac{m}{T^2}\int_0^T u_k(t)e^{-j\frac{2\pi i t}{T}}\,dt$$

$$= \frac{m}{T^2}\int_{\frac{(k-1)T}{m}}^{\frac{kT}{m}} e^{-j\frac{2\pi i t}{T}}\,dt = \frac{1}{T}\frac{\sin(\pi i/m)}{\pi i/m}e^{-j\frac{2\pi i(k-1/2)}{m}} \tag{3.58}$$

Since complex weighting factors cannot be realized physically, we instead determine separately the real and imaginary parts of $\tilde{\alpha}_i = p_i + jq_i$. The equivalent representation (for $x(t)$ real) is given by the trigonometric series

$$x(t) \sim p_0 + \sum_{i=1}^{n/2} 2p_i \cos\frac{2\pi i t}{T} - \sum_{i=1}^{n/2} 2q_i \sin\frac{2\pi i t}{T} \tag{3.59}$$

The advantage of starting with the complex form of the Fourier series is that the derivation of (3.58) is much easier. The measurement of the coefficients in (3.59) is implemented by weighting-summing networks corresponding to $(n/2) \times m$ matrices $\mathbf{B}$ and $\mathbf{C}$ in Figure 3.10. These matrices were obtained

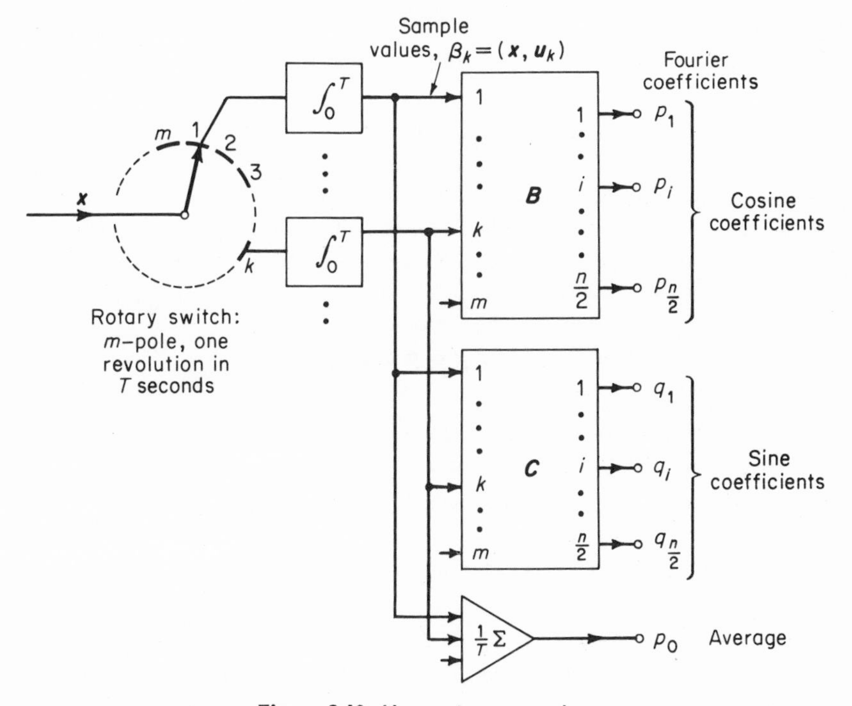

Figure 3.10. Harmonic wave analyzer.

from the real and imaginary parts of **A**.

$$p_i = \sum_{k=1}^{m} b_{ik}\beta_k; \qquad q_i = \sum_{k=1}^{m} c_{ik}\beta_k$$

$$p_0 = \sum_{k=1}^{m} \alpha_{0k}\beta_k = \frac{1}{T}\sum_{k=1}^{m}\beta_k$$

where

$$b_{ik} = \frac{1}{T}\frac{\sin(\pi i/m)}{\pi i/m}\cos\frac{2\pi i(k-\frac{1}{2})}{m}$$

$$c_{ik} = \frac{-1}{T}\frac{\sin(\pi i/m)}{\pi i/m}\sin\frac{2\pi i(k-\frac{1}{2})}{m} \tag{3.60}$$

$$i = 1, 2, \ldots, \frac{n}{2}; \qquad k = 1, 2, \ldots, m$$

I

Exercise 3.6. In implementing the harmonic wave analyzer described in Example 3.3, we would naturally be interested in the accuracy with which the individual Fourier coefficients can be determined assuming ideal behavior of the switch, integrators, and weighting-summing networks. Considering all signals in $L^2(T)$ of unit energy, show that the error in the ith Fourier coefficient is bounded by

$$|\alpha_i - \tilde{\alpha}_i|^2 \leqslant \frac{1}{T}\left[1 - \left(\frac{\sin(\pi i/m)}{\pi i/m}\right)^2\right]$$

Hint:

$$|\alpha_i - \tilde{\alpha}_i| = \left|(\mathbf{x}, \boldsymbol{\theta}_i) - \sum_{k=1}^{m}(\mathbf{v}_k, \boldsymbol{\theta}_i)(\mathbf{x}, \mathbf{u}_k)\right|$$

$$\leqslant \|\mathbf{x}\| \cdot \left\|\boldsymbol{\theta}_i - \sum_{k=1}^{m}(\mathbf{v}_k, \boldsymbol{\theta}_i)\mathbf{u}_k\right\|$$

Suppose we require that $|\alpha_i - \tilde{\alpha}_i| \leqslant 0.01/\sqrt{T}$, for $i = 0, \pm 1, \pm 2, \ldots, \pm n/2$, and $\|\mathbf{x}\| \leqslant 1$; how large must m be? How do the results change if, keeping the same number of contacts and the same revolution rate, the dwell time on each contact is cut in half?

REFERENCES

1. A. Papoulis, *The Fourier Integral and Its Applications*, McGraw-Hill, 1962.
2. F. Riesz and B. Sz.-Nagy, *Functional Analysis*, Frederick Ungar, 1955.
3. N. I. Akhiezer and I. M. Glazman, *Theory of Linear Operators in Hilbert Space*, Vol. 1, Frederick Ungar, 1961.
4. R. Courant and D. Hilbert, *Methods of Mathematical Physics*, Vol. 1, Interscience, 1953.
5. N. I. Akhieser, *Theory of Approximation*, Frederick Ungar, 1956.
6. Y. W. Lee, *Statistical Theory of Communication*, John Wiley & Sons, 1960.

7. W. H. Kautz, "Transient Synthesis in the Time Domain," *Trans. IRE*, Vol. CT-1, No. 3, pp. 29–39, September, 1954.
8. T. Y. Young and W. H. Huggins, "Complementary Signals and Orthogonalized Exponentials," *Trans. IRE*, Vol. CT-9, No. 4, pp. 362–370, December, 1962.
9. J. L. Walsh, "A Closed Set of Normal Orthogonal Functions," *American Jour. Math.*, 45, pp. 5–24 (1923).
10. J. L. Hammond and R. S. Johnson, "Review of Orthogonal Square-Wave Functions and their Application to Linear Networks," *Jour. Franklin Institute*, 273, pp. 211–25 (March, 1962).
11. H. F. Harmuth, "A Generalized Concept of Frequency and Some Applications," *Trans. IEEE*, Vol. IT-14, No. 3, pp. 375–382, May, 1968.

INTEGRAL TRANSFORMS FOR SIGNAL REPRESENTATION

4

4.1 Continuous Representations

The signal representations considered in the previous chapters were called *discrete* because of the feature of *countability* of the representation. The primary motivation for this type of representation lies in the relatively direct means for obtaining an approximate *finite* representation required in numerical analysis and physical implementation of signal measurements. On the other hand, for more theoretical considerations, such as the representation of linear operators discussed in Chapter 5, it is highly desirable to employ exact signal representations which can be easily manipulated and do not require constant attention to approximation error. For broad classes of signals, such as $L^2(-\infty, \infty)$, the exact discrete representation is apt to be quite cumbersome and rather remote from simple physical interpretation because of the more or less contrived nature of the complete orthonormal sets employed. The intuitively appealing time-series representation using the basis $\{\varphi_i(t) = \varphi(t - iT); i = 0, \pm 1, \pm 2, \ldots\}$ is not complete in $L^2(-\infty, \infty)$, no matter what the interpolating function $\varphi(t)$ may be.

For practical signal classes, however, the approximation error can be made as small as desired by choosing T sufficiently small. We are naturally led to examining the possibilities of signal representation by making the parameter iT continuously distributed over the time axis rather than at isolated, uniformly spaced points. The generalization of this idea of *continuous* representation leads to a variety of *integral transforms* for signal representation.[1] In this context, "continuous" refers not to the usual concept of continuity of a mapping but rather to the denseness of the set of parameters which characterizes the basis for signal representation.

The purpose of this chapter is to examine some of the general aspects of integral transform representation by analogy or extention of the ideas of discrete representation. Certain mathematical difficulties are encountered in attempting to make this analogy complete. Despite these difficulties, the physical insight gained from this approach is of considerable value. For mathematical rigor we must rely on the more specialized treatments for particular types of integral transforms.[2] The Fourier and Hilbert transforms, extensively used in signal analysis problems, are conveniently interpreted from this viewpoint. In fact, most introductory treatments of the Fourier integral use the Fourier series as a starting point. The idea that a large class of integral transforms could be similarly developed by extention from discrete bases offers many intriguing possibilities which, for the most part, remain to be exploited in practical applications.

4.2 Basis Kernels and Reciprocal Basis Kernels

The continuous analog of representation in a finite-dimensional space as presented in Chapter 3 is obtained by replacing the discrete variable i, which indexes the basis functions in (3.2) by the variable $s \in S$. The set S will usually be taken as some specified interval on the real line. The basis is now expressed as a function $\varphi(t, s)$ of two variables, and (3.2) becomes

$$x(t) = \int_S u(s)\varphi(t, s)\, ds; \qquad t \in T \tag{4.1a}$$

where $u(s)$ is the continuous representation for $x(t)$ analogous to the α_i in (3.2). The function $u(s)$ is a "density" function characterizing the decomposition of $x(t)$, relative to $\varphi(t, s)$, for different localized regions in the s-domain. Borrowing from standard terminology for integral equations, $\varphi(t, s)$ is referred to as the *basis kernel* for the integral transform representation of the signal. Pursuing the analogy further, we attempt to express $u(s)$ in terms of a linear functional on $\mathbf{x}$ for each value of s. Thus (3.7) becomes

$$u(s) = \int_T x(t)\theta(s, t)\, dt; \qquad s \in S \tag{4.1b}$$

where the function $\theta(s, t)$ is called the *reciprocal basis kernel.* If such a reciprocal basis exists, then (4.1a and b), taken together, are referred to as a *transform pair.*

We shall not require that $\varphi(t, s)$ and $\theta(s, t)$ be square-integrable functions of t. In fact, this is precisely where the analogy breaks down. It will become apparent that we shall have to broaden our conception of a function space if it is to include both the basis and the reciprocal basis. This is true even if the class of signals under consideration is restricted to $L^2(T)$. In extending the space, however, we shall try to retain the characterization of linear functionals in terms of inner products as they are defined on $L^2(T)$.

Substitution of (4.1b) into (4.1a) and interchanging the order of integrations leads to a condition that must be satisfied by reciprocal kernels.

$$x(t) = \int_S \int_T x(\tau)\theta(s, \tau)\varphi(t, s)\, d\tau\, ds$$

$$= \int_T I(t, \tau)x(\tau)\, d\tau = f_I(\mathbf{x}) \tag{4.2a}$$

where

$$I(t, \tau) \triangleq \int_S \varphi(t, s)\theta(s, \tau)\, ds \tag{4.2b}$$

Considering (4.2a) as a linear functional on $\mathbf{x}$ for fixed values of the parameter t, we see that it is the sampling functional described in connection with (2.63). There we noted that the functional was not bounded (or continuous); hence we do not expect to find an $L^2(T)$ function (of τ) which satisfies (4.2a) for all $\mathbf{x}$. Worse yet, we cannot find any function, in the ordinary sense, which satisfies (4.2a) for arbitrary $\mathbf{x}$. We overcome this difficulty by insisting on a symbolic representation of the functional $\mathbf{f}_I$ in terms of an inner product of functions. This is the basis for the modern theories of *distributions* and *generalized functions*[3]–[6] which provide the rigorous framework for the broader function spaces which we require. Thus we have, symbolically,

$$f_I(\mathbf{x}) = x(t) = \int_T \delta(t - \tau)x(\tau)\, d\tau; \qquad t \in T \tag{4.3}$$

where $\delta(t)$ is a generalized function called the *Dirac δ-function*. Despite the lack of rigor, the physical interpretation of the δ-function as an idealized form of impulse function needs no apology and has long been used by engineers and physicists in practical applications. The algebraic operations of differentiation, integration, and convolution can be defined for generalized functions in a manner consistent with ordinary functions. We can define a "time derivative" for a linear functional $f_\varphi(\mathbf{x}) = (\mathbf{x}, \boldsymbol{\varphi})$ on real time functions in the following manner.

First, let $x_\varepsilon(t) = x(t - \varepsilon)$; $\varepsilon > 0$, be a time-shifted version of the argument function $\mathbf{x}$. Now we define the time derivative of $\mathbf{f}_\varphi$ and, hence, the time derivative of the generalized function $\boldsymbol{\varphi}$ as

$$f_{\dot{\varphi}}(\mathbf{x}) \triangleq \lim_{\varepsilon \to 0} \frac{1}{\varepsilon} [f_\varphi(\mathbf{x}_\varepsilon) - f_\varphi(\mathbf{x})]$$

$$= -f_\varphi\left[\lim_{\varepsilon \to 0} \frac{1}{\varepsilon} (\mathbf{x} - \mathbf{x}_\varepsilon)\right] = -f_\varphi(\dot{\mathbf{x}})$$

Hence we have

$$f_{\dot{\varphi}}(\mathbf{x}) \triangleq -f_\varphi(\dot{\mathbf{x}})$$

or, symbolically,

$$(\mathbf{x}, \dot{\boldsymbol{\varphi}}) = -(\dot{\mathbf{x}}, \boldsymbol{\varphi}) \tag{4.4}$$

We use relation (4.4) as the definition of the derivative of a generalized function. Using integration by parts, we see that the definition is consistent with that for ordinary functions, provided either $\boldsymbol{\varphi}$ or $\mathbf{x}$ vanishes at the end points of the integration interval.

Similarly, for higher-order derivatives we have

$$(\mathbf{x}, \boldsymbol{\varphi}^{(n)}) = (-1)^n(\mathbf{x}^{(n)}, \boldsymbol{\varphi}) \tag{4.5}$$

so that, in particular,

$$\int_T \delta^{(n)}(t - t_0)x(t)\, dt = (-1)^n x^{(n)}(t_0); \qquad t_0 \in T \tag{4.6}$$

This property of the δ-function and its derivatives will be frequently used in the remaining chapters.

Exercise 4.1. For any real constant a, show that

$$\delta(at - t_0) = \frac{1}{|a|}\, \delta\left(t - \frac{t_0}{a}\right)$$

Now returning to (4.2), for a given basis kernel we require that the reciprocal basis satisfy

$$\int_S \varphi(t, s)\theta(s, \tau)\, ds = \delta(t - \tau) \tag{4.7a}$$

By a similar argument, substitution of (4.1a) into (4.1b) leads to the additional requirement that

$$\int_T \theta(s, t)\varphi(t, \sigma)\, dt = \delta(s - \sigma) \tag{4.7b}$$

be satisfied by $\varphi(t, s)$ and $\theta(s, t)$ in order that they be reciprocal basis kernels. This relation (4.7) is the continuous analog of (3.5). The analogy between discrete and continuous representations is summarized in Table 4.1.

Table 4.1 Analogous Quantities in Discrete and Continuous Signal Representation

Discrete	*Continuous*
$x(t) = \sum_i \alpha_i \varphi_i(t); \quad t \in T$	$x(t) = \int_S u(s)\varphi(t, s)\, ds; \quad t \in T$
$\alpha_i = (\mathbf{x}, \boldsymbol{\theta}_i); \quad i = 1, 2, \ldots$	$u(s) = \int_T x(t)\theta(s, t)\, dt; \quad s \in S$
$(\boldsymbol{\varphi}_i, \boldsymbol{\theta}_j) = \delta_{ij}$	$\begin{cases} \int_S \varphi(t, s)\theta(s, \tau)\, ds = \delta(t - \tau) \\ \int_T \theta(s, t)\varphi(t, \sigma)\, dt = \delta(s - \sigma) \end{cases}$

Examination of (4.7) shows why $\boldsymbol{\varphi}$ and $\boldsymbol{\theta}$ cannot both be $L^2(T)$ functions of t (or $L^2(S)$ functions of s). Physically, this can be interpreted as follows. Suppose that $\boldsymbol{\theta}$ is in some sense "well-behaved," e.g., continuous and integrable, then $u(s)$ in (4.1b) is a "smoothed" version of $x(t)$. This smoothing operation must be undone in the inverse transform (4.1a) by means of a highly irregular $\boldsymbol{\varphi}$. It is clear that either $\boldsymbol{\varphi}$ or $\boldsymbol{\theta}$, or both, will exhibit some kind of singular behavior; e.g., they may contain δ-functions and their derivatives, or non-integrable functions as in the case of the Fourier transform.

It is often helpful to view the discrete representation as a special case of integral transform representation. If $\mathbf{x}$ lies in the subspace spanned by

$$\{\varphi(t, s_i);\, i = 1, 2, \ldots\}$$

i.e., if

$$x(t) = \sum_i \alpha_i \varphi(t, s_i) \tag{4.8a}$$

then the transform is

$$\begin{aligned} u(s) &= \int_T \sum_i \alpha_i \varphi(t, s_i)\theta(s, t)\, dt \\ &= \sum_i \alpha_i\, \delta(s - s_i) \end{aligned} \tag{4.8b}$$

In this case we have the expected result that the density function representing $\mathbf{x}$ relative to $\varphi(t, s)$ is concentrated entirely at the discrete points $s = s_i$. For some signals, the density function may consist of a mixture of δ-functions and ordinary functions. In these situations it is often convenient to decompose the signal into a "discrete part" and a "continuous part."

Some types of basis kernel have the property that transforms of $L^2(T)$ functions (of t) will always be $L^2(S)$ functions (of s). An even stronger correspondence between t-domain and s-domain functions is provided by a *self-reciprocal* kernel where

$$\varphi(t, s) = \theta^*(s, t) \tag{4.9}$$

corresponding to an orthonormal basis in the discrete case. Let $u(s)$ and $v(s)$ be transforms of $x(t)$ and $y(t)$, respectively; then

$$\begin{aligned}(\mathbf{u}, \mathbf{v}) &= \int_S u(s)v^*(s)\, ds \\ &= \int_S \int_T \int_T x(t)y^*(\tau)\theta(s, t)\theta^*(s, \tau)\, d\tau\, dt\, ds\end{aligned}$$

Now if $\boldsymbol{\varphi}$ is self-reciprocal (4.9), then

$$\begin{aligned}(\mathbf{u}, \mathbf{v}) &= \int_T \int_T x(t)y^*(\tau)\, \delta(t - \tau)\, dt\, d\tau \\ &= \int_T x(t)y^*(t)\, dt = (\mathbf{x}, \mathbf{y})\end{aligned} \tag{4.10}$$

and inner products have been preserved in going from t-domain to s-domain functions. This property is possessed by many of the commonly used transforms.

4.3 Fourier, Hilbert, and Other Integral Transforms

Fourier Transforms

For $T = (-\infty, \infty)$ and $S = (-\infty, \infty)$, the basis kernels

$$\begin{aligned}\varphi(t, s) &= e^{j2\pi st} \\ \theta(s, t) &= e^{-j2\pi st}\end{aligned} \tag{4.11}$$

produce the Fourier transform pair. From the Fourier integral theorem,[2] we conclude that, as generalized functions, the following convergence relation is valid.

$$\lim_{W \to \infty} \int_{-W}^{W} e^{j2\pi st}\, ds = \lim_{W \to \infty} \frac{\sin 2\pi W t}{\pi t} = \delta(t) \tag{4.12}$$

Using (4.12), it is clear that the requirements (4.7a and b) for reciprocal basis kernels are satisfied by (4.11). The parameter s characterizes the frequency of the individual basis functions and is usually written as f. Also, it is conventional to denote Fourier transforms by the capitalized version of the letter denoting the time function. Hence, using conventional notation, the

Fourier transform pair is

$$x(t) = \int_{-\infty}^{\infty} X(f) e^{j2\pi ft}\, df \tag{4.13a}$$

$$X(f) = \int_{-\infty}^{\infty} x(t) e^{-j2\pi ft}\, dt \tag{4.13b}$$

Comparing (4.11) with (4.9), we see that this basis kernel is self-reciprocal; hence, from (4.10),

$$(\mathbf{X}, \mathbf{Y}) = (\mathbf{x}, \mathbf{y}) \tag{4.14}$$

This relation, which will be frequently used in later chapters, is known as *Parseval's relation.* It is very often used for establishing time-frequency duality properties.

Exercise 4.2. Using Parseval's relation, verify the following statements [for t, $f \in (-\infty, \infty)$].

a. $x(t) = y(t)z(t) \Rightarrow X(f) = \int_{-\infty}^{\infty} Y(f - \nu) Z(\nu)\, d\nu$

b. $x(t) = \delta(t) \Rightarrow X(f) = 1$

c. $x(t) = y(t - t_0) \Rightarrow X(f) = Y(f) e^{-j2\pi t_0 f}$

d. $x(t)$ is real $\Rightarrow X(f) = X^*(-f)$

e. $(\mathbf{x}^{(n)}, \mathbf{x}) = (\mathbf{Y}, \mathbf{X})$ where $Y(f) = (j2\pi f)^n X(f)$

Write the time-frequency dual of the statements above.

Exercise 4.3. Show that

$$x(t) = \sum_{k=-\infty}^{\infty} g(t) h(t - k\tau) \Rightarrow X(f) = \frac{1}{\tau} \sum_{\ell=-\infty}^{\infty} H\left(\frac{\ell}{\tau}\right) G\left(f - \frac{\ell}{\tau}\right)$$

Hint: Make a Fourier-series expansion of the periodic factor

$$\sum_{k=-\infty}^{\infty} h(t - k\tau)$$

What is the time-frequency dual relationship? This relationship (and its dual) is extremely useful in signal analysis problems involving time sequences of signal pulses. Use the relationship to establish the following properties of infinite sequences of δ-functions.

a. $x(t) = \sum_{k=-\infty}^{\infty} \delta(t - k\tau) \Rightarrow X(f) = \frac{1}{\tau} \sum_{\ell=-\infty}^{\infty} \delta\left(f - \frac{\ell}{\tau}\right)$

b. $x(t) = \sum_{k=-\infty}^{\infty} (-1)^k \delta(t - k\tau) \Rightarrow X(f) = \frac{1}{\tau} \sum_{\ell=-\infty}^{\infty} \delta\left(f - \frac{1}{2\tau} - \frac{\ell}{\tau}\right)$

Exercise 4.4. Using the relationship in Exercise 4.3, verify the following versions of the *Poisson sum formula.*

a. $$\sum_{k=-\infty}^{\infty} h(k\tau) = \frac{1}{\tau} \sum_{\ell=-\infty}^{\infty} H\left(\frac{\ell}{\tau}\right)$$

b. $$\sum_{k=-\infty}^{\infty} h(t - k\tau) = \frac{1}{\tau} \sum_{\ell=-\infty}^{\infty} H\left(\frac{\ell}{\tau}\right) e^{j\frac{2\pi \ell t}{\tau}}$$

c. $$\sum_{k=-\infty}^{\infty} h(k\tau)e^{-j2\pi k\tau f} = \frac{1}{\tau} \sum_{\ell=-\infty}^{\infty} H\left(f - \frac{\ell}{\tau}\right)$$

Basis Kernels with Difference Argument

Following the suggestion in Section 4.1 for time-series representation, we can generate a basis kernel by taking a single, basic waveform and considering all possible time translates of it. Hence, with some abuse of notation, we can write

$$\varphi(t, s) = \varphi(t - s) \tag{4.15}$$

so that the kernel is actually a function of the single variable $t - s$. Using $T, S = (-\infty, \infty)$, $x(t)$ is expressed as the *convolution* of its density function with the basic waveform.

$$x(t) = \int_{-\infty}^{\infty} u(s)\varphi(t - s)\, ds \tag{4.16}$$

The reciprocal kernel, if one exists, can be found simply by taking the Fourier transform of (4.16),

$$X(f) = U(f)\Phi(f) \tag{4.17}$$

Hence,

$$U(f) = \Theta(f)X(f)$$

where

$$\Theta(f) = \frac{1}{\Phi(f)} \tag{4.18}$$

and the reciprocal kernel has also a difference argument,

$$u(s) = \int_{-\infty}^{\infty} x(t)\theta(s - t)\, dt \tag{4.19}$$

As an example, we could let

$$\varphi(t - s) = \delta(t - s) \tag{4.20}$$

Then

$$\Phi(f) = 1 = \Theta(f)$$

and

$$u(s) = \int_{-\infty}^{\infty} x(t)\, \delta(s - t)\, dt = x(s) \tag{4.21}$$

This gives the rather trivial result that a time function can be represented by itself. Nevertheless, considering a signal to be a dense sequence of δ-functions often has conceptual value. It is sometimes used, for example, to give a physical interpretation of the convolution integral for characterizing the response of a linear system.

A much more interesting example results from choosing the basis

$$\varphi(t - s) = \frac{-1}{\pi(t - s)} \tag{4.22}$$

Then, with

$$\operatorname{sgn} f \triangleq \frac{f}{|f|}$$

$$\begin{aligned} \Phi(f) = j \operatorname{sgn} f &\Rightarrow \Theta(f) = -j \operatorname{sgn} f \\ &\Rightarrow \theta(s - t) = \frac{1}{\pi(s - t)} \end{aligned} \tag{4.23}$$

The resulting transform pair is called the *Hilbert transform pair*. It has become customary[1] to denote the Hilbert transform of $\mathbf{x}$ by $\hat{\mathbf{x}}$. Using this notation, we have

$$x(t) = \frac{1}{\pi} \int_{-\infty}^{\infty} \frac{\hat{x}(s)}{s - t}\, ds \tag{4.24a}$$

$$\hat{x}(s) = \frac{1}{\pi} \int_{-\infty}^{\infty} \frac{x(t)}{s - t}\, dt \tag{4.24b}$$

Because of the singularities in the integrands in (4.24a and b), we have to specify the meaning of the integral. In both cases, the integral is the Cauchy principal value; i.e. in (4.24b),

$$\int_{-\infty}^{\infty} \Rightarrow \lim_{\varepsilon \to 0} \left(\int_{-\infty}^{s-\varepsilon} + \int_{s+\varepsilon}^{\infty} \right); \qquad \varepsilon > 0$$

Comparing (4.22) and (4.23) with (4.9), we see that the basis is self-reciprocal; hence

$$(\hat{\mathbf{x}}, \hat{\mathbf{y}}) = (\mathbf{x}, \mathbf{y}) \tag{4.25}$$

An interesting and useful feature of Hilbert transforms is the simple relationship that exists between the Fourier transforms of the Hilbert pair.

$$\hat{X}(f) = -j \operatorname{sgn} f X(f) \tag{4.26a}$$

$$X(f) = j \operatorname{sgn} f \hat{X}(f) \tag{4.26b}$$

[1] We shall rely on context to distinguish the Hilbert transform and an orthogonal projection. This notation is conventional for both quantities.

The way this relationship is most frequently used in signal analysis problems is to create a complex signal $z(t)$, called the *analytic signal* or *pre-envelope*, which has a "one-sided" Fourier transform. If we let

$$z(t) = x(t) + j\hat{x}(t) \tag{4.27}$$

then, from (4.26a),

$$\begin{aligned} Z(f) &= 2X(f) \quad \text{for } f > 0 \\ &= 0 \qquad\quad\;\; \text{for } f < 0 \end{aligned} \tag{4.28}$$

A practical application of this property will be discussed in Section 4.4.

Example 4.1. Let us suppose that we want to represent time functions relative to a continuously translated, double-sided exponential waveform; i.e.,

$$\varphi(t - s) = e^{-a|t-s|}; \qquad a > 0$$

Since

$$\Phi(f) = \frac{2a}{(2\pi f)^2 + a^2}$$

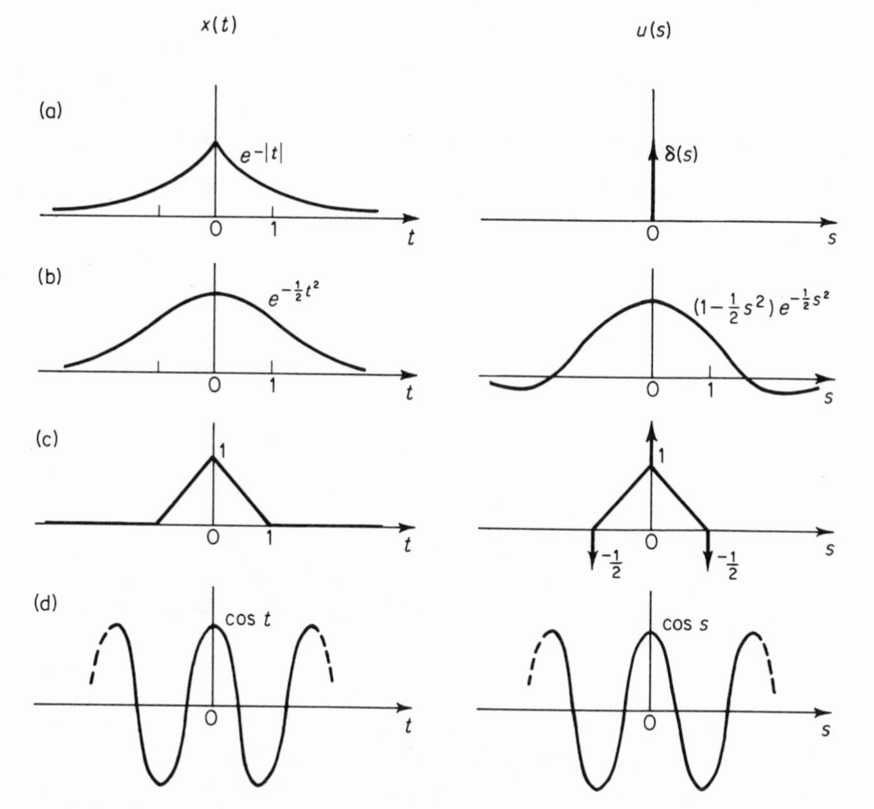

Figure 4.1. Graphs of transform pairs for the basis, $\varphi(t - s) = e^{-|t-s|}$.

we have

$$\Theta(f) = \frac{a}{2}\left[1 - \frac{1}{a^2}(j2\pi f)^2\right]$$

$$\Rightarrow \theta(t) = \frac{a}{2}\left[\delta(t) - \frac{1}{a^2}\ddot{\delta}(t)\right]$$

Hence, the transform pair resulting from this basis is

$$x(t) = \int_{-\infty}^{\infty} u(s)e^{-a|t-s|}\,ds$$

$$u(s) = \frac{a}{2}\left[x(s) - \frac{1}{a^2}\ddot{x}(s)\right]$$

Some specific transform pairs are sketched graphically for $a = 1$ in Figure 4.1. ∎

Exercise 4.5. Use (4.26) to show that

a. The Hilbert transform of a Hilbert transform is the negative of the original signal; i.e., $\hat{\hat{\mathbf{x}}} = -\mathbf{x}$.
b. A real signal and its Hilbert transform are orthogonal; i.e., $(\mathbf{x}, \hat{\mathbf{x}}) = 0$.

Exercise 4.6. Evaluate the Hilbert transforms of

a. $x(t) = \cos 2\pi f_0 t; \qquad -\infty < t < \infty$

b. $x(t) = \dfrac{1}{1 + t^2}; \qquad -\infty < t < \infty$

c. $x(t) = \dfrac{\sin 2\pi Wt}{2\pi Wt}; \qquad -\infty < t < \infty$

d. $x(t) = 1; \qquad |t| \leqslant T$
$ = 0; \qquad |t| > T$

Basis Kernels with Product Argument

Another way in which a single, basic waveform can be used to generate a basis is by continuous scaling of the width of the basic waveform. Hence

$$\varphi(t, s) = \varphi(st) \tag{4.29}$$

and the parameter s characterizes the reciprocal of the width of the basis function. The kernel for the Fourier transform is clearly of this type. In fact, kernels which are functions of the single variable st and which are also self-reciprocal are called *Fourier kernels*.[2] Another of the better known transforms of this type are the *Hankel transforms* with transform pairs given by

$$x(t) = \int_0^\infty u(s)(st)^{\frac{1}{2}} J_\nu(st)\, ds \tag{4.30a}$$

$$u(s) = \int_0^\infty x(t)(st)^{\frac{1}{2}} J_\nu(st)\, dt \tag{4.30b}$$

where J_ν is the νth-order Bessel function.

If we let $T, S = (0, \infty)$, the transform pairs, in general, will be of the form

$$x(t) = \int_0^\infty u(s)\varphi(st)\, ds; \qquad t \geqslant 0 \tag{4.31a}$$

$$u(s) = \int_0^\infty x(t)\theta(st)\, dt; \qquad s \geqslant 0 \tag{4.31b}$$

It is interesting to note that the reciprocal kernel, if one exists, can be found by a method similar to that for kernels with difference argument, (4.18). The analogous relationship for kernels with a product argument is expressed in terms of *Mellin transforms.*[2] The Mellin transform pair is given by

$$\mathscr{M}(\mathbf{x}; \nu) = \int_0^\infty x(t)t^{\nu-1}\, dt \tag{4.32a}$$

$$x(t) = \frac{1}{2\pi j}\int_{c-j\infty}^{c+j\infty} \mathscr{M}(\mathbf{x}; \nu)t^{-\nu}\, d\nu; \qquad t \geqslant 0 \tag{4.32b}$$

Now, taking the Mellin transform of (4.31a), we find

$$\begin{aligned}\mathscr{M}(\mathbf{x}; \nu) &= \int_0^\infty u(s)\int_0^\infty \varphi(st)t^{\nu-1}\, dt\, ds \\ &= \int_0^\infty u(s)s^{-\nu}\, ds \int_0^\infty \varphi(\sigma)\sigma^{\nu-1}\, d\sigma \\ &= \mathscr{M}(\mathbf{u}; 1-\nu)\mathscr{M}(\boldsymbol{\varphi}; \nu)\end{aligned} \tag{4.33}$$

Similarly, taking the Mellin transform of (4.31b), we find

$$\mathscr{M}(\mathbf{u}; \nu) = \mathscr{M}(\boldsymbol{\theta}; \nu)\mathscr{M}(\mathbf{x}; 1-\nu) \tag{4.34}$$

Combining (4.33) and (4.34), we see that

$$\mathscr{M}(\boldsymbol{\varphi}; \nu)\mathscr{M}(\boldsymbol{\theta}; 1-\nu) = 1 \tag{4.35}$$

and hence

$$\theta(t) = \frac{1}{2\pi j}\int_{c-j\infty}^{c+j\infty} \frac{t^{-\nu}\, d\nu}{\mathscr{M}(\boldsymbol{\varphi}; 1-\nu)} \tag{4.36}$$

Example 4.2. Suppose that we desire to represent signals $x(t)$ on the interval $(0, \infty)$ by a distribution of rectangular pulses of various widths. Then we let

$$\begin{aligned} \varphi(t) &= 1 \qquad \text{for } 0 < t \leqslant 1 \\ &= 0 \qquad t > 1 \end{aligned}$$

and we have

$$\mathscr{M}(\boldsymbol{\varphi}; \nu) = \int_0^1 t^{\nu-1}\, dt = \frac{1}{\nu}$$

Hence

$$\mathscr{M}(\boldsymbol{\theta}; \nu) = 1 - \nu \Rightarrow \theta(t) = \dot{\delta}(t-1)$$

since

$$\int_0^\infty \dot{\delta}(t-1)t^{\nu-1}\, dt = -\frac{d}{dt}[t^{\nu-1}]_{t=1} = 1 - \nu$$

Thus the density of rectangular pulses is given by

$$u(s) = \int_0^\infty x(t)\dot{\delta}(st-1)\, dt$$

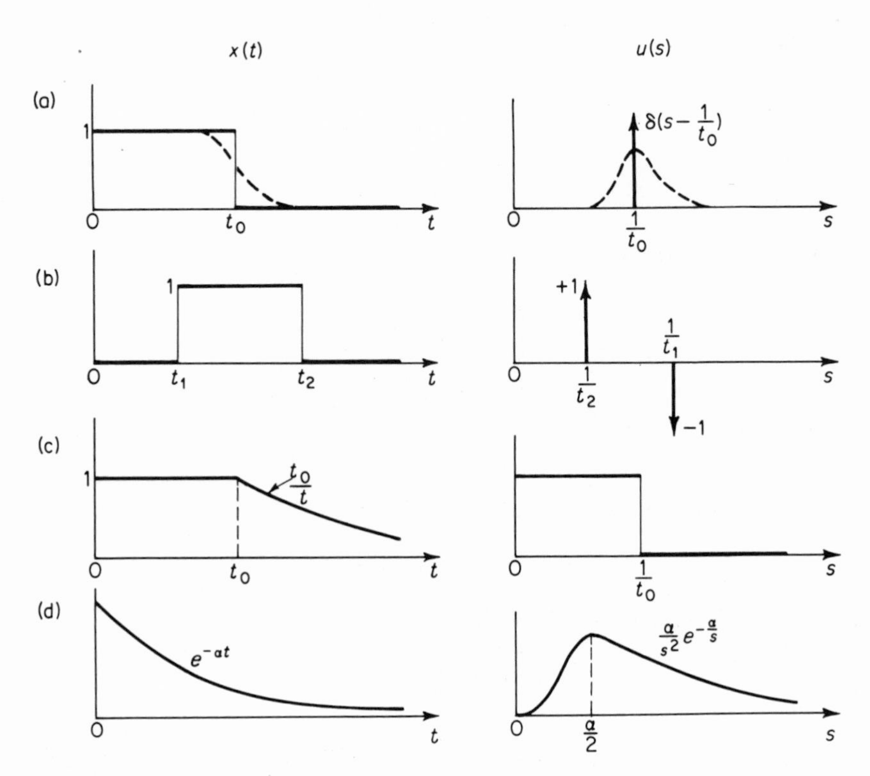

Figure 4.2. Graphs of transform pairs for the basis kernel, $\varphi(st) = 1$ for $0 < t \leqslant 1/s$; $= 0$ otherwise.

Let $st = \sigma$ in the integral above and we have

$$u(s) = \frac{1}{s}\int_0^\infty x\left(\frac{\sigma}{s}\right)\dot{\delta}(\sigma - 1)\,d\sigma$$

$$= -\frac{1}{s}\frac{d}{d\sigma}\left[x\left(\frac{\sigma}{s}\right)\right]_{\sigma=1} = -\frac{1}{s^2}\,\dot{x}\left(\frac{1}{s}\right)$$

Some of the resulting transform pairs for this example are shown in Figure 4.2. ∎

4.4 Representation of Bandpass Signals

The Complex Envelope Signal

In this section we take up some practical matters which are also well-suited for illustrating how Fourier transforms and Hilbert transforms are conventionally used in signal analysis problems. In many signal processing systems, some of the signals will be of the "bandpass" type. For want of a better term, signals which have Fourier transforms concentrated to some degree about a frequency removed from the origin (see Figure 4.3a) are called *bandpass signals*. On the other hand, signals with a substantial frequency composition over an interval including the origin are called *lowpass signals*. One reason that bandpass signals are so prevalent is that signal transmission media are always dispersive (non-uniform transmission characteristics over various regions of the frequency band). In order to utilize the medium with tolerably small signal distortion due to the dispersion, the transmission channel is normally split up into relatively narrow-band segments spanning frequency intervals of minimal dispersion. Both radio and guided-wave transmission systems employ this technique. This can be accomplished by using the basic lowpass signal to modulate the amplitude or phase of a sinusoidal carrier signal with a frequency suitably chosen relative to the transmission band. As far as signal representation goes, pure amplitude modulation of a sinusoidal carrier with constant frequency f_0 and phase θ presents no difficulties. If the bandpass signal is of the form $x(t) = u(t)\cos(2\pi f_0 t + \theta)$, where $u(t)$ is a lowpass signal, then representation of $\mathbf{x}$ corresponds directly to a representation of $\mathbf{u}$. If this signal is transmitted over a dispersive channel, however, then it cannot, in general, be expressed in the same simple form. We can say that the channel dispersion introduces some degree of amplitude-to-phase-modulation conversion.

The generalization of this technique for providing lowpass representations for arbitrary bandpass signals is the topic of this section. The method we present here is general and does not require the "narrow-band approximation" for validity. First, it may be helpful to say a few words about the motivation for going to lowpass representations. The signal representation techniques presented in earlier chapters are directly applicable to bandpass signals, but the representations are apt to be highly inefficient. For example, if we use a time-series representation (1.34) by sampling the bandpass signal at a rate exceeding twice the highest significant frequency component of the signal, then it is clear we are using many more samples in a given time interval than would be needed to represent some equivalent lowpass version of the signal. Another argument is that the better known basis functions for signal representation presented in Section 3.3 are all of a lowpass nature. The orthonormal sets described in that section had the property that the nth basis function has n zero crossings. Hence the signal space spanned by a truncated set would be best suited for what we are calling lowpass signals. Finally, we often find it desirable to simulate the operation of a bandpass system on an analog computer. The inherent lowpass nature and speed limitations of the analog computer make its operation in the lowpass mode a significant advantage. We shall show how a bandpass system can be simulated equivalently by two interconnected lowpass systems.

The starting point for developing these alternate signal representations is to consider the *analytic signal* or *pre-envelope*[7]–[9] corresponding to the real bandpass signal $x(t)$. The pre-envelope is a complex bandpass signal obtained by adding the Hilbert transform of $x(t)$ as the imaginary part.

$$\psi(t) = x(t) + j\hat{x}(t) \tag{4.37}$$

Recalling that $\|\hat{\mathbf{x}}\| = \|\mathbf{x}\|$ and $(\mathbf{x}, \hat{\mathbf{x}}) = 0$, we have the following useful relationship.

$$\|\boldsymbol{\psi}\| = \sqrt{2}\,\|\mathbf{x}\| \tag{4.38}$$

The most significant property of the pre-envelope is that its Fourier transform is "one-sided." From (4.28),

$$\begin{aligned} \Psi(f) &= 2X(f) && \text{for } f > 0 \\ &= 0 && \text{for } f < 0 \end{aligned} \tag{4.39}$$

Fortunately, the pre-envelope does not obscure the familiar concepts of envelope and phase shift for narrow-band signals. On the contrary, it provides a simple and intuitively satisfying definition for these quantities which can be extended to apply to arbitrary signals. The *envelope* $w(t)$ is taken to be the magnitude of the pre-envelope.

$$w(t) = |\psi(t)| = \sqrt{x^2(t) + \hat{x}^2(t)} \tag{4.40}$$

This quantity will closely approximate the output of a physical envelope detector for narrow-band signals. Obviously, $w(t)$ does not completely characterize the bandpass signal. The argument of $\boldsymbol{\psi}$, i.e., the imaginary part of the natural logarithm of $\boldsymbol{\psi}$, contains the remainder of the information about the signal. Actually, the time derivative of this quantity is often more useful from the standpoint of physical interpretation. We define the *instantaneous frequency* $f_i(t)$ for an arbitrary signal by

$$\begin{aligned} f_i(t) &= \frac{1}{2\pi} \operatorname{Im}\left[\frac{d}{dt} \ln \psi(t)\right] \\ &= \frac{1}{2\pi} \operatorname{Im}\left(\frac{\dot{\psi}(t)}{\psi(t)}\right) \\ &= \frac{1}{2\pi w^2(t)} [\dot{\hat{x}}(t)x(t) - \dot{x}(t)\hat{x}(t)] \end{aligned} \tag{4.41}$$

This function will correspond approximately to the output of a physical frequency discrimination circuit.

Now, to obtain equivalent lowpass representations, we simply translate the Fourier transform of ψ so that it is concentrated around the origin and appears as a lowpass function as illustrated in Figure 4.3(c).

We have, then, by definition,

$$\Gamma(f) = \Psi(f + f_0) \tag{4.42}$$

and hence

$$\gamma(t) = \psi(t)e^{-j2\pi f_0 t} \tag{4.43}$$

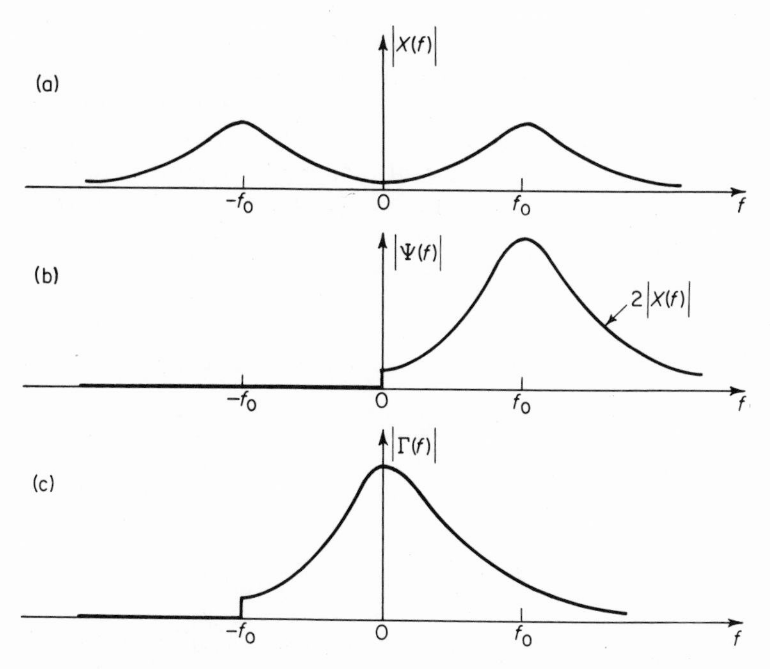

Figure 4.3. Fourier transform of a bandpass signal and its lowpass equivalent.

In this manner, the bandpass signal is related to the complex signal $\gamma(t)$ by

$$x(t) = \text{Re}\,[\gamma(t)e^{j2\pi f_0 t}] \tag{4.44}$$

It is clear that the complex representation is a straightforward extension of the familiar complex number, or phasor, representation for sinusoidal signals. The signal $\boldsymbol{\gamma}$ is called the *complex envelope* for the signal $\mathbf{x}$. In terms of the real and imaginary parts of $\boldsymbol{\gamma}$, the bandpass signal from (4.44) is given by

$$x(t) = u(t)\cos 2\pi f_0 t - v(t)\sin 2\pi f_0 t \tag{4.45}$$

where

$$\gamma(t) = u(t) + jv(t)$$

The real lowpass signals $\mathbf{u}$ and $\mathbf{v}$ are called the *in-phase* and *quadrature* components, respectively, of the bandpass signal. It is evident that the envelope is simply $|\gamma(t)|$, independently of the choice of f_0.

$$w(t) = |\psi(t)| = |\gamma(t)| = \sqrt{u^2(t) + v^2(t)} \tag{4.46}$$

The instantaneous frequency, in terms of the complex envelope, is

$$f_i(t) = f_0 + \frac{1}{2\pi}\,\text{Im}\left[\frac{\dot{\gamma}(t)}{\gamma(t)}\right] \tag{4.47}$$

In many situations the choice of f_0 is predetermined as, for example, in the case of a modulated sinusoidal carrier where it is natural and convenient to make f_0 the frequency of the unmodulated carrier. In other cases, the choice is more or less arbitrary and f_0 can be chosen to minimize the bandwidth of $\Gamma(f)$ with a view toward economy of representation. One way to do this is to choose f_0 as the "center of gravity" of the real, positive function $|\Psi(f)|^2$.[7] This is the value of f_0 that minimizes the quantity

$$\int_0^\infty (f - f_0)^2\,|\Psi(f)|^2\,df$$

Setting the derivative with respect to f_0 equal to zero and solving we obtain

$$f_0 = \frac{\int_0^\infty f\,|\Psi(f)|^2\,df}{\int_0^\infty |\Psi(f)|^2\,df} = \frac{1}{2\pi j}\,\frac{(j2\pi f\boldsymbol{\Psi}, \boldsymbol{\Psi})}{(\boldsymbol{\Psi}, \boldsymbol{\Psi})} \tag{4.48}$$

Using Parseval's relation (4.14) on (4.48),

$$f_0 = \frac{1}{2\pi j}\,\frac{(\dot{\boldsymbol{\psi}}, \boldsymbol{\psi})}{\|\boldsymbol{\psi}\|^2} = \frac{1}{2\pi}\,\frac{(\mathbf{x}, \dot{\hat{\mathbf{x}}})}{\|\mathbf{x}\|^2} \tag{4.49}$$

This definition of center frequency has also a sensible interpretation as a weighted time average of the instantaneous frequency. Using (4.41), (4.49), and (4.4), we can write

$$\int_{-\infty}^{\infty} w^2(t) f_i(t)\, dt = \frac{1}{2\pi} \int_{-\infty}^{\infty} [\dot{\hat{x}}(t) x(t) - \dot{x}(t) \hat{x}(t)]\, dt$$
$$= \frac{1}{\pi} (\mathbf{x}, \dot{\hat{\mathbf{x}}}) = 2\, \|\mathbf{x}\|^2 f_0 \tag{4.50}$$

We note that from (4.40)

$$w^2(t) = x^2(t) + \hat{x}^2(t)$$

and, hence,

$$\|\mathbf{w}\|^2 = \|\mathbf{x}\|^2 + \|\hat{\mathbf{x}}\|^2 = 2\, \|\mathbf{x}\|^2 \tag{4.51}$$

Combining (4.50) and (4.51), our interpretation of f_0 as "average" frequency takes on the intuitively satisfying (envelope weighted) form,

$$f_0 = \int_{-\infty}^{\infty} \frac{w^2(t)}{\|\mathbf{w}\|^2} f_i(t)\, dt \tag{4.52}$$

Exercise 4.7. For an arbitrary real signal $u_0(t)$, show that the complex envelope of $x(t) = u_0(t) \cos 2\pi f_0 t$ is given by

$$\gamma(t) = u_0(t) + \int_{f_0}^{\infty} [U_0(f) e^{j2\pi(f-2f_0)t} - U_0^*(f) e^{-j2\pi f t}]\, df$$

It follows that if $u_0(t)$ is band-limited to frequencies less than f_0; i.e., $U_0(f) = 0$ for $f > f_0$, then $\gamma(t) = u_0(t) + j0$. In this case, show that $\hat{x}(t) = u_0(t) \sin 2\pi f_0 t$ and that the complex envelope of $\hat{x}(t)$ is $0 - ju_0(t)$.

Exercise 4.8. Let $x_1(t)$ and $x_2(t)$ be real bandpass signals with complex envelopes $\gamma_1(t)$ and $\gamma_2(t)$. Show that

$$(\mathbf{x}_1, \mathbf{x}_2) = \tfrac{1}{2} \operatorname{Re} (\boldsymbol{\gamma}_1, \boldsymbol{\gamma}_2)$$
$$(\mathbf{x}_1, \hat{\mathbf{x}}_2) = -\tfrac{1}{2} \operatorname{Im} (\boldsymbol{\gamma}_1, \boldsymbol{\gamma}_2)$$

Show also that $\hat{x}_2(t)$ is a 90° phase-shifted version of $x_2(t)$; i.e., its complex envelope is $\gamma_2(t) e^{-j\frac{\pi}{2}} = -j\gamma_2(t)$.

Bandpass Filtering Operations

A great deal of the utility of the complex envelope representation would be lost if it were not possible to characterize the effects of bandpass filtering directly in terms of the complex envelope. It turns out, however, that bandpass filtering of a bandpass signal can be simply treated in terms of complex lowpass filtering of complex lowpass signals. Consider the bandpass filter in Figure 4.4 with transfer function $R(f)$.

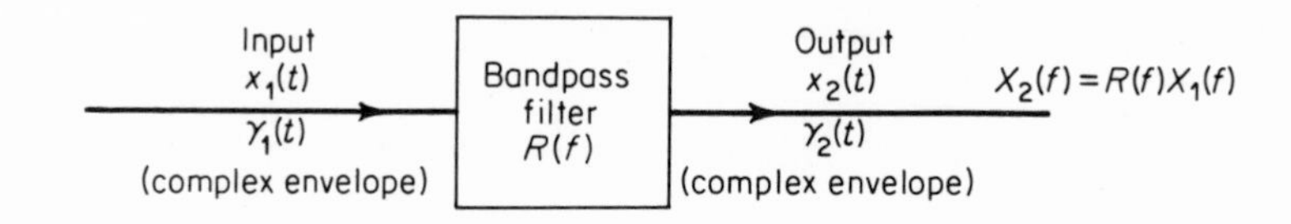

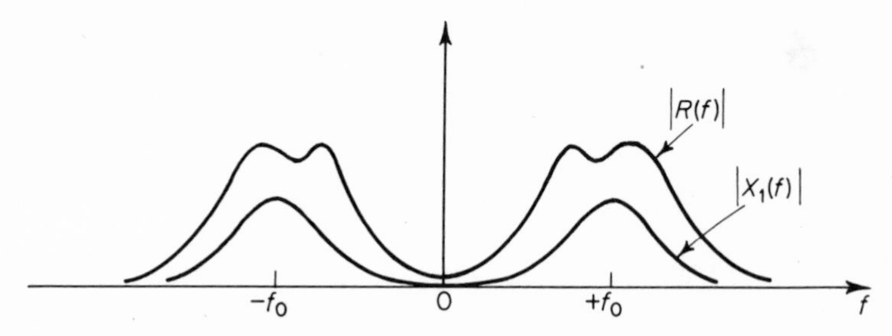

Figure 4.4. Bandpass filtering of bandpass signals.

The input signal is given by

$$x_1(t) = \text{Re}\,[\gamma_1(t)e^{j2\pi f_0 t}]$$
$$= \tfrac{1}{2}\gamma_1(t)e^{j2\pi f_0 t} + \tfrac{1}{2}\gamma_1^*(t)e^{-j2\pi f_0 t}$$

Hence,

$$X_1(f) = \tfrac{1}{2}\Gamma_1(f - f_0) + \tfrac{1}{2}\Gamma_1^*(-f - f_0) \tag{4.53}$$

The transfer function of the filter can be expressed in terms of $\Lambda(f)$ in a similar manner.

$$R(f) = \tfrac{1}{2}\Lambda(f - f_0) + \tfrac{1}{2}\Lambda^*(-f - f_0) \tag{4.54}$$

where

$$\Lambda(f - f_0) = (1 + \text{sgn}\, f)R(f)$$

This is equivalent to saying that the filter will be represented by the complex envelope $\lambda(t)$ of its impluse response $r(t)$. Now for the output signal,

$$\begin{aligned} X_2(f) &= R(f)X_1(f) \\ &= \tfrac{1}{4}\Lambda(f - f_0)\Gamma_1(f - f_0) + \tfrac{1}{4}\Lambda^*(-f - f_0)\Gamma_1^*(-f - f_0) \\ &= \tfrac{1}{2}\Gamma_2(f - f_0) + \tfrac{1}{2}\Gamma_2^*(-f - f_0) \end{aligned} \tag{4.55}$$

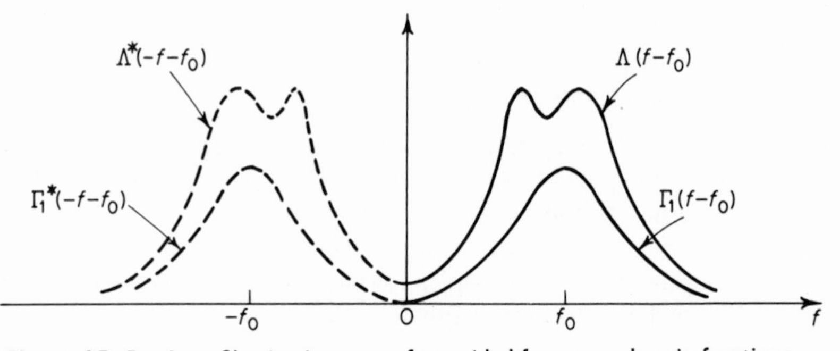

Figure 4.5. Bandpass filtering in terms of one-sided frequency domain functions.

where two of the four terms in the product of (4.53) and (4.54) have vanished due to the single-sided nature of $\Gamma_1(f)$ and $\Lambda(f)$ (see Figure 4.5).

From (4.55) it is evident that

$$\Gamma_2(f) = \tfrac{1}{2}\Lambda(f)\Gamma_1(f) \tag{4.56}$$

and thus that the complex envelope of the filtered signal is given simply by the convolution of the complex envelope of the input signal with the complex envelope of the filter impulse response.

$$\begin{aligned} \gamma_2(t) &= \tfrac{1}{2}\int_{-\infty}^{\infty} \gamma_1(\tau)\lambda(t-\tau)\, d\tau \\ \boldsymbol{\gamma}_2 &= \boldsymbol{\gamma}_1 \otimes \tfrac{1}{2}\boldsymbol{\lambda} \end{aligned} \tag{4.57}$$

It remains now to show how this complex lowpass filtering operation can be implemented in terms of real lowpass filters. We shall do this by considering the in-phase and quadrature components separately. First, what is meant by a real filter in this discussion is one whose response to a real signal is real. This means that the impulse response of the filter must be real which, in turn, implies that the transfer function, say $H(f)$, must exhibit a certain kind of symmetry about the origin; namely, $H^*(-f) = H(f)$. Now we decompose $\lambda(t)$ into its real and imaginary parts so that

$$\tfrac{1}{2}\lambda(t) = m(t) + jn(t) \tag{4.58}$$

where $m(t)$ and $n(t)$ are real and can be considered as impulse responses of real lowpass filters. The corresponding transfer functions are given by

$$\begin{aligned} M(f) &= \tfrac{1}{4}[\Lambda(f) + \Lambda^*(-f)] \\ N(f) &= \frac{1}{4j}[\Lambda(f) - \Lambda^*(-f)] \end{aligned} \tag{4.59}$$

i.e., by the symmetrical and antisymmetrical parts, respectively, in the sense noted above, of the complex filter transfer function $\Lambda(f)$. It follows that $M(f)$ and $N(f)$ are related directly to the parts of the bandpass function $R(f)$ which are symmetrical and antisymmetrical, respectively, about the "center" frequency f_0.

Now rewriting (4.57) in terms of real lowpass functions, we have

$$\begin{aligned} \boldsymbol{\gamma}_2 &= \mathbf{u}_2 + j\mathbf{v}_2 \\ &= (\mathbf{u}_1 + j\mathbf{v}_1) \otimes (\mathbf{m} + j\mathbf{n}) \\ &= [(\mathbf{u}_1 \otimes \mathbf{m}) - (\mathbf{v}_1 \otimes \mathbf{n})] + j[(\mathbf{v}_1 \otimes \mathbf{m}) + (\mathbf{u}_1 \otimes \mathbf{n})] \end{aligned} \tag{4.60}$$

or, in the frequency domain,

$$\begin{aligned} U_2(f) &= M(f)U_1(f) - N(f)V_1(f) \\ V_2(f) &= N(f)U_1(f) + M(f)V_1(f) \end{aligned} \tag{4.61}$$

To summarize this approach, a single-input, single-output bandpass filtering operation is replaced by a double-input, double-output lowpass filtering

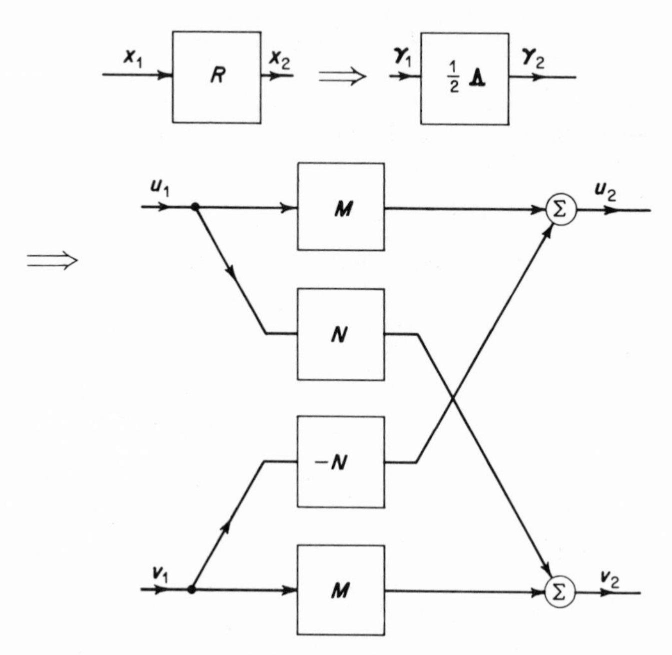

Figure 4.6. Equivalent lowpass filtering.

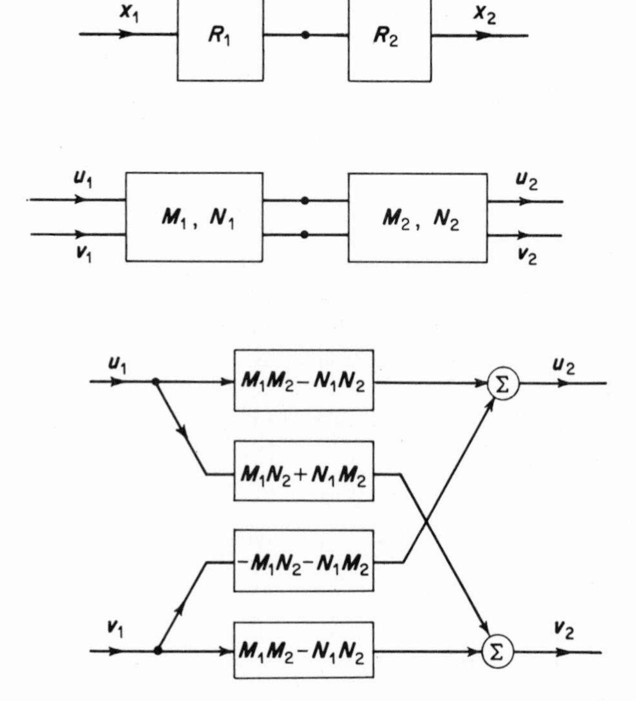

Figure 4.7. Lowpass equivalent of cascade combination of bandpass filters.

operation which characterizes separately the transfer of in-phase and quadrature components. The equivalent lowpass system is represented schematically by the lattice arrangement shown in Figure 4.6. A cascade combination of bandpass filters can be handled directly in terms of lowpass equivalents as shown in Figure 4.7.

We note that if a bandpass filter exhibits symmetry about f_0, then $N(f) = 0$ and there is no cross coupling between in-phase and quadrature components. This is the kind of transmission characteristic that is required in order to avoid amplitude-to-phase-modulation conversion on a bandpass signal.

The following examples illustrate the usefulness of the complex envelope representation for dealing with various modulation techniques. It is assumed that the reader has some acquaintance with the basic principle of double-sideband and single-sideband amplitude modulation (DSB-AM and SSB-AM).

Example 4.3. Conceptually, the simplest way to generate a SSB-AM signal is to first generate a DSB-AM signal by a multiplication operation and then use a bandpass filter with sharp cutoff at f_0 to eliminate one of the sidebands. This process is shown schematically in Figure 4.8.

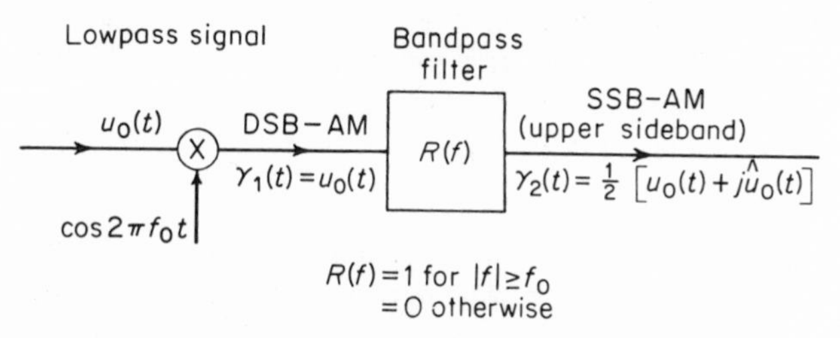

Figure 4.8. Generation of SSB-AM by bandpass filtering.

For the bandpass transfer function $R(f)$ shown in Figure 4.8, we have from (4.54)

$$\Lambda(f) = 1 + \operatorname{sgn} f$$

and, hence, from (4.59)

$$M(f) = \tfrac{1}{2}; \qquad N(f) = \frac{-j}{2} \operatorname{sgn} f$$

We shall assume that the modulating signal $u_0(t)$ has no frequency components above f_0 [$U(f) = 0$ for $|f| \geqslant f_0$]. Then $\gamma_1(t) = u_0(t) + j0$, and using (4.61) along with (4.26), we find

$$\gamma_2(t) = \tfrac{1}{2}u_0(t) + j\tfrac{1}{2}\hat{u}_0(t)$$

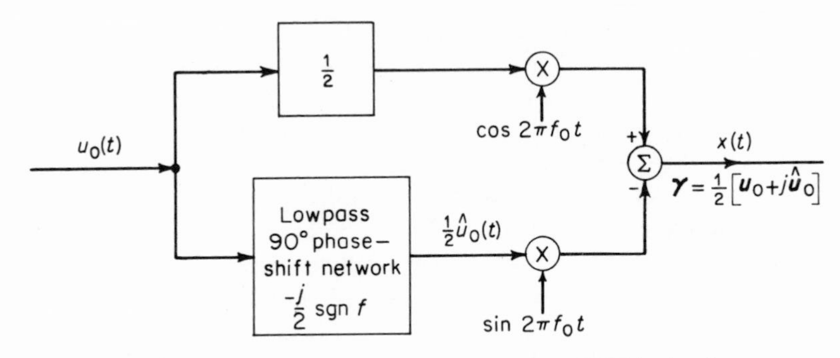

Figure 4.9. Hartley modulator for producing SSB-AM signal.

which implies that an upper sideband, SSB-AM signal retains the original modulating signal (at half amplitude) as the in-phase component and has the Hilbert transform of this signal added on in quadrature.

$$x_2(t) = \tfrac{1}{2}u_0(t)\cos 2\pi f_0 t - \tfrac{1}{2}\hat{u}_0(t)\sin 2\pi f_0 t$$

If the filter had instead rejected the upper sideband to produce a lower sideband, SSB-AM signal, the resulting signal would have the form

$$x_2(t) = \tfrac{1}{2}u_0(t)\cos 2\pi f_0 t + \tfrac{1}{2}\hat{u}_0(t)\sin 2\pi f_0 t$$

Example 4.4. The standard form for the SSB-AM signal presented in Example 4.3 suggests an alternative modulation scheme which replaces the bandpass filtering operation with a lowpass phase-shifting operation. This scheme, shown schematically in Figure 4.9, is called the *Hartley modulator.*

Example 4.5. A similar modulation scheme which uses two identical lowpass filters instead of the phase-shift network can be derived from the two-path configuration shown in Figure 4.10. Referring to the figure, we see

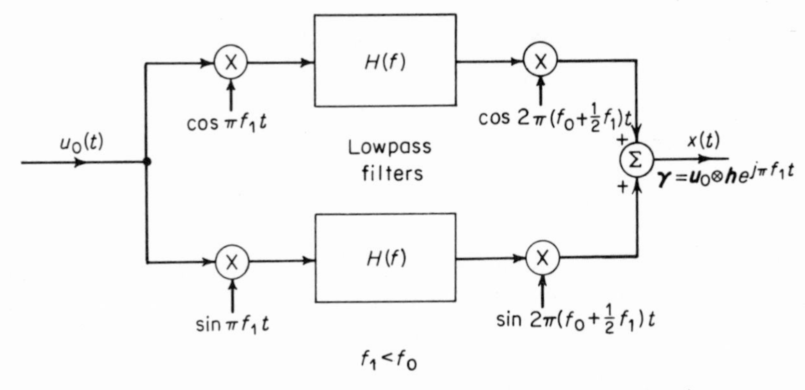

Figure 4.10. Two-path modulation scheme (Weaver modulator).

that the output signal is given by

$$x(t) = \cos 2\pi\left(f_0 + \frac{f_1}{2}\right)t \int_{-\infty}^{\infty} h(\tau)u_0(t-\tau)\cos \pi f_1(t-\tau)\, d\tau$$
$$+ \sin 2\pi\left(f_0 + \frac{f_1}{2}\right)t \int_{-\infty}^{\infty} h(\tau)u_0(t-\tau)\sin \pi f_1(t-\tau)\, d\tau$$

Using the familiar trigonometric identities for sines and cosines of sums and differences of angles, the expression above can be rewritten as

$$x(t) = \cos 2\pi f_0 t \int_{-\infty}^{\infty} h(\tau)u_0(t-\tau)\cos \pi f_1\tau\, d\tau$$
$$- \sin 2\pi f_0 t \int_{-\infty}^{\infty} h(\tau)u_0(t-\tau)\sin \pi f_1\tau\, d\tau$$

From this expression, it is clear that the complex envelope (relative to f_0) for $\mathbf{x}$ is simply

$$\boldsymbol{\gamma} = \mathbf{u}_0 \otimes \mathbf{g}$$

where

$$g(t) = h(t)e^{j\pi f_1 t}$$

This configuration becomes a SSB-AM modulator if we make $H(f)$ correspond to an ideal lowpass filter with sharp cutoff at a frequency $f_1/2$ ($H(f) = 1$, for $|f| \leqslant f_1/2$; $= 0$ otherwise). Now splitting $\mathbf{g}$ into its real and imaginary parts,

$$g(t) = g_1(t) + jg_2(t) = h(t)\cos \pi f_1 t + jh(t)\sin \pi f_1 t$$

$$G_1(f) = \frac{1}{2}\left[H\left(f - \frac{f_1}{2}\right) + H\left(f + \frac{f_1}{2}\right)\right]$$
$$= \tfrac{1}{2} \qquad \text{for } |f| \leqslant f_1$$
$$= 0 \qquad \text{otherwise}$$

$$G_2(f) = \frac{1}{2j}\left[H\left(f - \frac{f_1}{2}\right) - H\left(f + \frac{f_1}{2}\right)\right]$$
$$= \frac{-j}{2}\operatorname{sgn} f \qquad \text{for } |f| \leqslant f_1$$
$$= 0 \qquad \text{otherwise}$$

Now, assuming $u_0(t)$ is band-limited to frequencies less than f_1 ($U_0(f) = 0$ for $|f| > f_1$), we have

$$\gamma_2(t) = \tfrac{1}{2}[u_0(t) + j\hat{u}_0(t)]$$

resulting in an upper sideband, SSB-AM signal. This type of modulator has been called the *Weaver modulator*. ▌

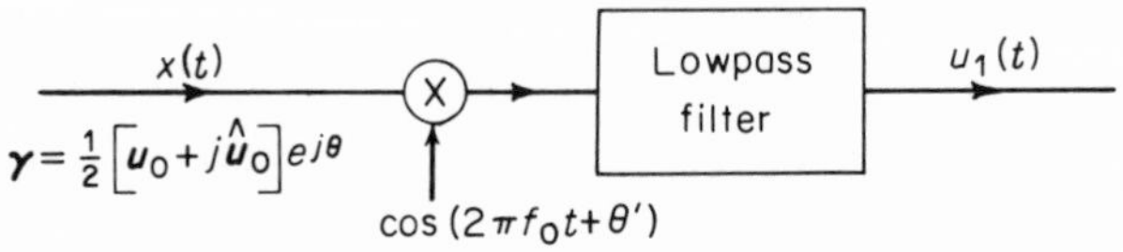

$x(t)$ — $\gamma = \frac{1}{2}\left[u_0 + j\hat{u}_0\right]e^{j\theta}$ — (X) — Lowpass filter — $u_1(t)$; reference $\cos(2\pi f_0 t + \theta')$

Figure 4.11. Coherent demodulator for AM signals.

It should be noted that the SSB-AM modulators in these examples are idealized and not physically realizable. This is due to the unrealizability of the transfer characteristics of the bandpass filter in the first example, the phase-shifter in the second example, and the ideal lowpass filters in the third example. These characteristics can be closely approximated by physical networks of sufficient complexity, however, and the three configurations shown have been used in practical applications.

To implement a demodulator for a SSB-AM signal, we require a circuit which can extract the in-phase component of the bandpass signal without interference from the quadrature component. The conventional scheme is a *coherent* or *homodyne* demodulator which employs a reference carrier signal locked in phase to the carrier used at the modulator. The actual received signal may have picked up a phase shift θ in transmission as indicated in Figure 4.11. The input to the lowpass filter is

$$\tfrac{1}{4}[u_0(t)\cos(\theta - \theta') - \hat{u}_0(t)\sin(\theta - \theta')] \\ + \tfrac{1}{4}[u_0(t)\cos(4\pi f_0 t + \theta + \theta') - \hat{u}_0(t)\sin(4\pi f_0 t + \theta + \theta')]$$

Assuming that $u_0(t)$ contains no frequencies above f_0 and that the lowpass filter has an ideal characteristic with sharp cutoff at f_0, then the second term above is rejected by the lowpass filter and the demodulated signal is

$$u_1(t) = \frac{u_0(t)}{4}\cos(\theta - \theta') - \frac{\hat{u}_0(t)}{4}\sin(\theta - \theta')$$

which, for $\theta' = \theta$, is the desired signal. Phase error in the reference carrier introduces an interference known as *quadrature distortion* as well as a reduction in amplitude of the desired component. Note that similar types of distortion can arise from the dispersive nature of the bandpass transmission channel.

Example 4.6. The ability of the coherent demodulator to reject the quadrature component suggests an alternative to SSB-AM for efficient bandwidth utilization (relative to DSB-AM). In this scheme, two independent lowpass signals, each band-limited to one-half the available bandpass width, are used to separately modulate the in-phase and quadrature part of the sinusoidal carrier.

$$x(t) = u_0(t)\cos 2\pi f_0 t - v_0(t)\sin 2\pi f_0 t$$

This signal can be interpreted as a superposition of two DSB-AM signals with the same carrier frequency and carrier phases differing by 90°. This modulation scheme is called *co-modulation* or *quadrature-carrier modulation.* The demodulator uses two coherent demodulators of the type shown in Figure 4.11 with a 90° phase shift between reference carriers. The complete system is shown in Figure 4.12.

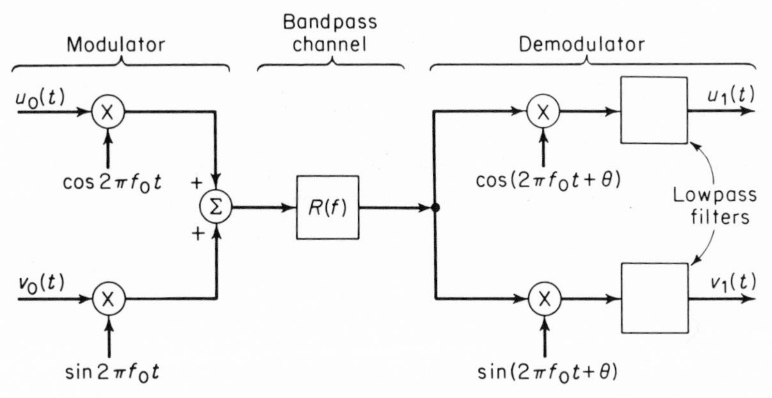

Figure 4.12. Co-modulation system.

The demodulated signals are given by

$$\mathbf{u}_1 = \tfrac{1}{2}[(\mathbf{m} \otimes \mathbf{u}_0) - (\mathbf{n} \otimes \mathbf{v}_0)]$$

$$\mathbf{v}_1 = \tfrac{1}{2}[(\mathbf{m} \otimes \mathbf{v}_0) + (\mathbf{n} \otimes \mathbf{u}_0)]$$

Hence, if the bandpass characteristic $R(f)$ is symmetrical about f_0 (except for a possible constant phase shift θ across the transmission band which can be compensated for by a shift in the reference carriers), then $\mathbf{n} = 0$ and there is no crosstalk between the individual signals. Otherwise there will be an interfering signal $-\frac{1}{2}(\mathbf{n} \otimes \mathbf{v}_0)$ on the first channel and an interference $\frac{1}{2}(\mathbf{n} \otimes \mathbf{u}_0)$ on the second channel. ∎

Exercise 4.9. Consider a single-resonator bandpass filter. Such a filter has a transfer function which exhibits an approximate symmetry about the center frequency, the approximation improving as the Q of the resonant circuit is increased.

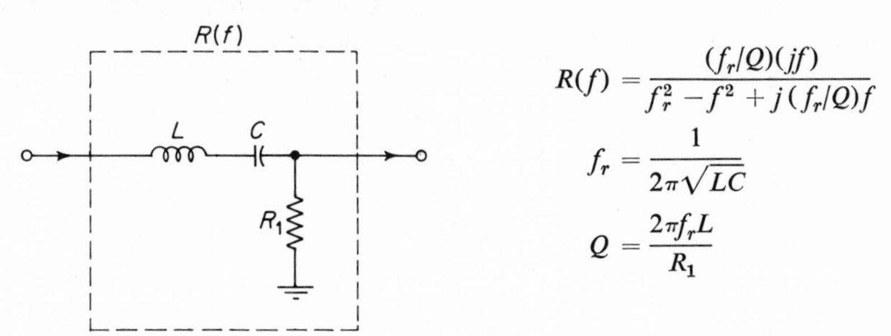

$$R(f) = \frac{(f_r/Q)(jf)}{f_r^2 - f^2 + j(f_r/Q)f}$$

$$f_r = \frac{1}{2\pi\sqrt{LC}}$$

$$Q = \frac{2\pi f_r L}{R_1}$$

Choose $f_0 = [\sqrt{1 - (1/4Q^2)}]f_r$ as the "center" frequency and evaluate the equivalent low pass transfer functions $M(f)$ and $N(f)$, (4.59), in the region $|f| < f_0$. Sketch graphically these lowpass functions for various values of Q. *Hint:* We can write

$$R(f) = \frac{f_r}{2Q}\left[\frac{1 + jK}{j(f - f_0) + (f_r/2Q)} + \frac{1 - jK}{j(f + f_0) + (f_r/2Q)}\right]$$

where

$$K \triangleq \frac{1}{\sqrt{4Q^2 - 1}}$$

The symmetric and antisymmetric parts (about f_0) of the first term can be obtained by inspection.

Repeat the investigation for a distributed-parameter resonator, i.e., with the series combination of L and C replaced by a lossless transmission line with open-circuit termination, characteristic impedance R_0, and length equal to a quarter wavelength at $f = f_0$.

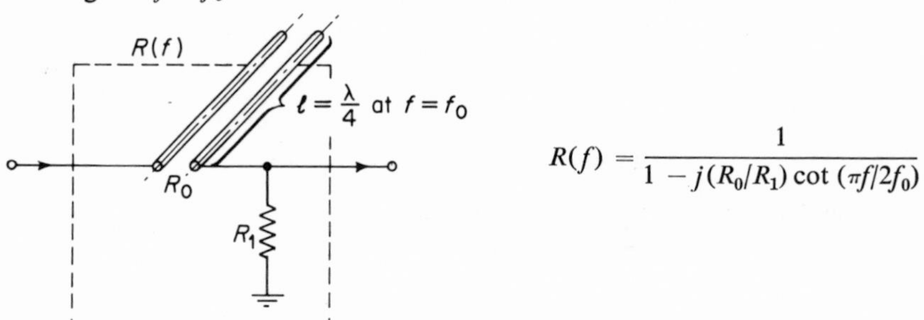

$$R(f) = \frac{1}{1 - j(R_0/R_1)\cot(\pi f/2f_0)}$$

Radar-ambiguity Functions

The radar system designer is concerned with the selection of bandpass signals which afford an adequate range resolution in signals returned from targets. To meet this requirement, the signal should exhibit a large signal space separation for moderately small time shifts. This separation property was characterized by the time-ambiguity function in Example 2.10. A radar signal is relatively narrow-band; i.e., carrier frequencies are typically much higher than the video frequencies involved. The significance of this, for our present purposes, is that inner products of the bandpass signals are simply related to the real part of the inner product of their complex envelopes (see Exercise 4.8). Now consider the distance between $\mathbf{x}$ and $\mathbf{x}_\tau$, where $x_\tau(t) = x(t + \tau)$ and, hence, $\gamma_\tau(t) = \gamma(t + \tau)e^{j2\pi f_0 \tau}$

$$\begin{aligned} \|\mathbf{x} - \mathbf{x}_\tau\|^2 &= 2[\|\mathbf{x}\|^2 - (\mathbf{x}, \mathbf{x}_\tau)] \\ &= \|\boldsymbol{\gamma}\|^2 - \text{Re}\,[\rho_\gamma(\tau)] \end{aligned} \tag{4.62}$$

where we define

$$\begin{aligned} \rho_\gamma(\tau) &= (\boldsymbol{\gamma}, \boldsymbol{\gamma}_\tau) \\ &= e^{-j2\pi f_0 \tau}\int_{-\infty}^{\infty} \gamma(t)\gamma^*(t + \tau)\, d\tau \end{aligned} \tag{4.63}$$

The rapid fluctuation of ρ_γ with τ due to the exponential factor is of no concern in the radar system since it corresponds to range differences smaller than the size of the targets under observation. Also, with an envelope detector, these fluctuations will not be observed. The radar system designer characterizes time ambiguity in terms of the magnitude of ρ_γ (or its square) and tries to select signals for which $|\rho_\gamma|^2$ drops off rapidly with increasing τ.

A radar system is also frequently required to estimate target velocity in terms of the Doppler shift produced on the returned signal by the moving target. For narrow-band signals, the Doppler shift is approximately just a simple frequency translation. The separation between a signal and its frequency-shifted (by an amount ν) version is given by

$$\begin{aligned} \|\mathbf{x} - \mathbf{x}_\nu\|^2 &= 2[\|\mathbf{x}\|^2 - (\mathbf{x}, \mathbf{x}_\nu)] \\ &= \|\boldsymbol{\gamma}\|^2 - \text{Re}\,[\tilde{\rho}_\gamma(\nu)] \end{aligned} \tag{4.64}$$

where we define $\tilde{\rho}_\gamma(\nu) = (\boldsymbol{\gamma}, \boldsymbol{\gamma}_\nu)$ and $\gamma_\nu(t) = \gamma(t)e^{j2\pi\nu t}$. The frequency ambiguity is thus characterized by

$$\tilde{\rho}_\gamma(\nu) = \int_{-\infty}^{\infty} \gamma(t)\gamma^*(t)e^{-j2\pi\nu t}\,dt \tag{4.65}$$

Using Parseval's relation, in the frequency domain (4.65) has a form similar to (4.63).

$$\begin{aligned} \tilde{\rho}_\gamma(\nu) &= \int_{-\infty}^{\infty} \Gamma(f)\Gamma^*(f - \nu)\,df \\ &= \int_{-\infty}^{\infty} \Gamma^*(f)\Gamma(f + \nu)\,df \end{aligned} \tag{4.66}$$

For good velocity resolution, $|\tilde{\rho}_\gamma|$ should drop off rapidly with increasing ν, a requirement more or less incompatible with a good range resolution. To evaluate the performance of a particular radar signal, it has been found very convenient to characterize time ambiguity and frequency ambiguity simultaneously.[8],[10]–[12] To do this, we simply consider the distance between $\mathbf{x}$ and $\mathbf{x}_{\tau,\nu}$, where $\mathbf{x}_{\tau,\nu}$ is a time-shifted and frequency-shifted version of $\mathbf{x}$ characterized by the complex envelope

$$\gamma_{\tau,\nu}(t) = \gamma(t + \tau)e^{j2\pi(f_0\tau + \nu t)} \tag{4.67}$$

Now we have

$$\begin{aligned} \|\mathbf{x} - \mathbf{x}_{\tau,\nu}\|^2 &= 2[\|\mathbf{x}\|^2 - (\mathbf{x}, \mathbf{x}_{\tau,\nu})] \\ &= \|\boldsymbol{\gamma}\|^2 - \text{Re}\,[\rho_\gamma(\tau, \nu)] \end{aligned} \tag{4.68}$$

where we define

$$\begin{aligned} \rho_\gamma(\tau, \nu) &= (\boldsymbol{\gamma}, \boldsymbol{\gamma}_{\tau,\nu}) \\ &= e^{-j2\pi f_0\tau} \int_{-\infty}^{\infty} \gamma(t)\gamma^*(t + \tau)e^{-j2\pi\nu t}\,dt \\ &= e^{-j2\pi f_0\tau} \int_{-\infty}^{\infty} \Gamma^*(f)\Gamma(f + \nu)e^{-j2\pi f\tau}\,df \end{aligned} \tag{4.69}$$

The real, non-negative function $|\rho_\gamma(\tau, \nu)|^2$ is called the *radar-ambiguity function*[8],[10] and is used to characterize the resolution performance of a radar signal in various regions of the $\tau\nu$-plane. Note that, from the Schwarz inequality,

$$|\rho_\gamma(\tau, \nu)|^2 \leqslant |\rho_\gamma(0, 0)|^2 = 4 \, \|\mathbf{x}\|^4 \tag{4.70}$$

It would be most desirable, from the standpoint of simultaneous range and velocity resolution, to have the radar-ambiguity function appear graphically as a narrow spike concentrated about the origin in the $\tau\nu$-plane. There is, however, a fundamental limitation in that

$$\iint_{-\infty}^{\infty} |\rho_\gamma(\tau, \nu)|^2 \, d\tau \, d\nu = |\rho_\gamma(0, 0)|^2 \tag{4.71}$$

for arbitrary $\boldsymbol{\gamma}$. We can interpret (4.71) as a sort of principle of "conservation of ambiguity." The radar designer is concerned with the problem of selecting signals exhibiting a tolerably small ambiguity for various prescribed regions in the $\tau\nu$-plane, letting it "pop up" in other regions. Unfortunately, there are no simple techniques known for synthesizing radar signals from an arbitrary prescribed ambiguity function. Helpful guidelines are provided, however, by considering the ambiguity properties of various types of signals. A comparison of two such signal types is presented in the following example.

Example 4.7. In this example, we consider two types of pulsed-carrier radar signals, each using a single, long-duration, flat-topped modulating pulse sketched graphically in Figure 4.13a. The first signal is straightforward amplitude modulation of a sinusoidal carrier.

$$x_1(t) = w(t) \cos 2\pi f_0 t$$
$$\gamma_1(t) = w(t) + j0$$

The second signal uses the same envelope but has an instantaneous frequency which varies linearly with time; $f_i(t) = f_0 + 2bt$.

$$x_2(t) = w(t) \cos (2\pi f_0 t + 2\pi b t^2)$$
$$\gamma_2(t) = w(t) e^{j2\pi b t^2}$$

This type of signal is referred to as *linear FM* or "*chirp*." [11],[12]

The ambiguity function for the AM signal, from (4.69), is given by

$$|\rho_{\gamma_1}(\tau, \nu)|^2 = \left| \int_{-\infty}^{\infty} w(t) w(t + \tau) e^{-j2\pi\nu t} \, dt \right|^2$$

Some idea of the ambiguity surface over the $\tau\nu$-plane can be obtained by considering constant τ lines in the $\tau\nu$-plane where the ambiguity is given by the

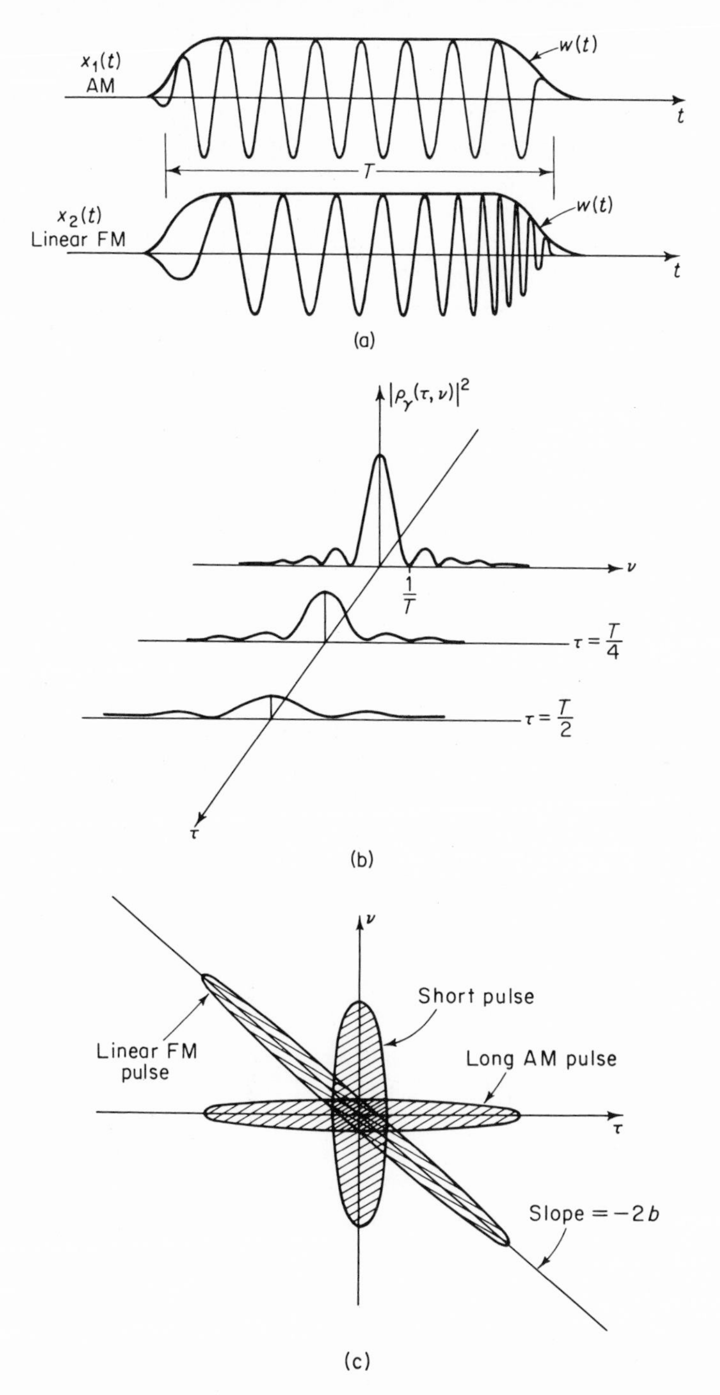

Figure 4.13. Comparison of ambiguity characteristics of two types of radar pulses. (a) Pulsed-carrier radar signals. (b) Ambiguity surface cross sections for AM signal. (c) Regions of large ambiguity in $\tau\nu$ plane.

Fourier transform of the pulse $w(t)w(t + \tau)$. For small values of τ (compared to T), $w(t)w(t + \tau)$ is a long-duration, flat-topped pulse and $|\rho_{\gamma_1}(\tau, \nu)|^2 \cong w^2(0)\,|W(\nu)|^2$. For larger τ, $w(t)w(t + \tau)$ becomes a shorter pulse and its Fourier transform is smaller and broader (see Figure 4.13b). A simpler way to depict ambiguity is to select a particular contour line in the $\tau\nu$-plane and shade in all areas representing higher ambiguity as in Figure 4.13c. As expected, a long-duration AM signal provides desirable ambiguity characteristics in the frequency direction at the expense of a broad time ambiguity. The situation is reversed for very short AM pulses.

For the linear FM signal,

$$|\rho_{\gamma_2}(\tau, \nu)|^2 = \left| \int_{-\infty}^{\infty} w(t)w(t + \tau)e^{j2\pi(bt^2 - b(t+\tau)^2 - \nu t)}\,dt \right|^2$$
$$= \left| \int_{-\infty}^{\infty} w(t)w(t + \tau)e^{-j2\pi(\nu + 2b\tau)t}\,dt \right|^2$$

Hence, for comparison purposes,

$$|\rho_{\gamma_2}(\tau, \nu)|^2 = |\rho_{\gamma_1}(\tau, \nu + 2b\tau)|^2$$

We see that the ridge of peak ambiguity for the linear FM signal lies over the line $\nu = -2b\tau$ instead of over the $\nu = 0$ line as in the case of the AM signal. In this manner, we obtain reasonably good ambiguity performance for range and velocity separately. Ambiguity will be high for certain combinations (depending on the modulation index b) of range and velocity. Physically, it is easy to understand how a Doppler-shifted version of the linear FM signal could be mistaken for an appropriately delayed version of the signal.

Exercise 4.10. Verify that a radar-ambiguity function has the following properties: [10]

a. $|\rho_\gamma(\tau, \nu)|^2 = |\rho_\gamma(-\tau, -\nu)|^2$

b. $|\rho_\gamma(\tau, \nu)|^2 \leqslant |\rho_\gamma(0, 0)|^2 = \|\boldsymbol{\gamma}\|^4 = 4\,\|\mathbf{x}\|^4$

c. $$\iint_{-\infty}^{\infty} |\rho_\gamma(\tau, \nu)|^2\, d\tau\, d\nu = |\rho_\gamma(0, 0)|^2$$

d. $$\iint_{-\infty}^{\infty} |\rho_\gamma(\tau, \nu)|^2\, e^{j2\pi(\nu\xi - \tau\eta)}\, d\tau\, d\nu = |\rho_\gamma(\xi, \eta)|^2$$

REFERENCES

1. L. A. Zadeh, "A General Theory of Linear Signal Transmission Systems," *Jour. Franklin Institute*, Vol. 253, pp. 293–312 (April, 1952).
2. E. C. Titchmarsh, *Introduction to the Theory of Fourier Integrals*, Oxford University Press, 1937.

3. M. J. Lighthill, *Introduction to Fourier Analysis and Generalized Functions*, Cambridge University Press, 1958.

4. H. Bremermann, *Distributions, Complex Variables, and Fourier Transforms*, Addison-Wesley, 1965.

5. I. B. Gel'fand and G. E. Shilov, *Generalized Functions* (Vol. 1, *Properties and Operations*), Academic Press, 1964.

6. I. M. Gel'fand and N. Ya. Vilenkin, *Generalized Functions* (Vol. 4, *Applications of Harmonic Analysis*), Academic Press, 1964.

7. D. Gabor, "Theory of Communication," *Jour. I.E.E.* (Part III), Vol. 93, pp. 429–57 (November, 1946).

8. P. M. Woodward, *Probability and Information Theory, with Applications to Radar*, Pergamon Press, 1953.

9. J. Dugundji, "Envelopes and Pre-Envelopes of Real Waveforms," *Trans. IRE*, Vol. IT-4, No. 1, pp. 53–57 (March, 1958).

10. C. E. Cook and M. Bernfeld, *Radar Signals*, Academic Press, 1967.

11. J. R. Klauder, A. C. Price, S. Darlington, and W. J. Albersheim, "The Theory and Design of Chirp Radars," *Bell Sys. Tech. Jour.*, Vol. 39, No. 4, pp. 745–808 (July, 1960).

12. J. R. Klauder, "The Design of Radar Signals Having Both High Range Resolution and High Velocity Resolution," *Bell Sys. Tech. Jour.*, Vol. 39, No. 4, pp. 809–820 (July, 1960).

REPRESENTATION OF LINEAR OPERATORS

5

5.1 Introduction

In all but the most elementary systems, signals will be subjected to a variety of transformations as they pass from the input to the output of the different system components. Transducers, scanning devices, modulators, amplifiers, filters, coders, transmission channels, etc., all exhibit signal-transforming properties which must be taken into account in an analysis of overall system performance. In view of the wide variety of signal formats encountered, it would be clearly desirable to have general techniques for characterizing the signal-transforming properties of the system components. In addition to characterizing the input-output transformation produced by particular system components, we also have occasion to consider the converse problem of associating physical system components, subsystems, or networks with a particular input-output signal transformation. This problem would arise, for example, in connection with signal processing operations which are intentionally inserted into a system so that it meets specified performance objectives. Examples would be compensating networks that counteract distortion produced by transducers and transmission channels and noise filters that reduce interference from unwanted signals.

From the system theory viewpoint, a signal transformation is regarded as an *input-output relationship* for a system (or network), indicated simply as a block in Figure 5.1. Mathematically, we regard the transformation as a mapping, as described in Section 1.3, from one set of signals into another set of signals. We make no attempt to keep these two viewpoints distinct. On

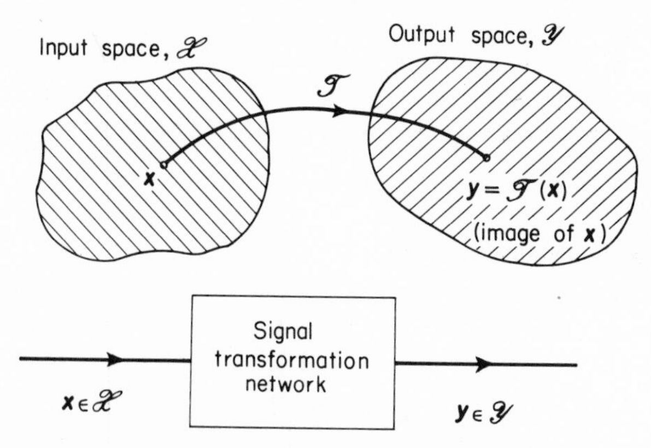

Figure 5.1. Diagrammatic interpretation of signal mappings.

the contrary, it is often conceptually advantageous to interpret mathematical operations in terms of block diagrams. Accordingly, the domain of the mapping will be referred to as the *input space* $\mathscr{X}$, and the *output space* $\mathscr{Y}$ will be selected so that it contains the range of any mapping under consideration. In general, to characterize a mapping $\mathscr{T}$, we must be able to specify all input-output pairs $\{\mathbf{x}, \mathscr{T}(\mathbf{x})\}$, i.e., the *graph* of the mapping. For most transformations of practical interest, the graph is completely unmanageable as a representation because there are simply too many distinct input-output pairs to be enumerated and ordered in any practical way. If, however, we restrict our considerations to *linear transformations*, then we find that the mapping can be completely characterized by only a small subset of all possible input-output pairs. In other words, the image of an arbitrary input signal can be expressed in terms of the images of a particular subset of input signals. This desirable feature is known as the *superposition property* of linear transformations. Fortunately, most systems contain a substantial number of components which are linear, or approximately so, so that a detailed study of linear transformations is found to have a considerable practical utility. In associating linear transformations and linear networks, it should be noted that we always assume that the networks are excited only by the input signal; i.e., the network is at rest, or in the quiescent state, before the input is applied. This ensures that the output will depend only on the input and not on the transient response to other energy sources in the network; otherwise, the output signal will not be linearly related to the input.

Our objective in this chapter will be to develop methods for representing various classes of linear transformations in a manner analogous to that for representing signals. It will be apparent that strong analogies do exist; however, representation of suitably broad classes of linear transformations is a great deal more complex than the comparable problem of signal representation. A satisfyingly complete treatment of these topics is beyond the scope of this text and we shall have to be content with a discussion of the more elementary aspects providing sufficient background for the applications used in later chapters.

5.2 Linear Transformations

A linear transformation $\mathscr{L}$ is a mapping, whose domain $\mathscr{X}$ is a linear space, which possesses the following properties:

$$\begin{aligned}\mathscr{L}(\mathbf{x}_1 + \mathbf{x}_2) &= \mathscr{L}(\mathbf{x}_1) + \mathscr{L}(\mathbf{x}_2) \\ \mathscr{L}(\alpha\mathbf{x}_1) &= \alpha\mathscr{L}(\mathbf{x}_1)\end{aligned} \tag{5.1}$$

or, equivalently,

$$\mathscr{L}(\alpha\mathbf{x}_1 + \beta\mathbf{x}_2) = \alpha\mathscr{L}(\mathbf{x}_1) + \beta\mathscr{L}(\mathbf{x}_2) \tag{5.2}$$

where $\mathbf{x}_1$ and $\mathbf{x}_2$ are arbitrary vectors in $\mathscr{X}$ and α and β are arbitrary scalars. It follows that $\mathscr{L}(\mathbf{0}) = \mathbf{0}$ and $\mathscr{L}(-\mathbf{x}) = -\mathscr{L}(\mathbf{x})$; hence, the range of a linear transformation is a linear space having the same set of scalars as the domain.

Since the set of scalars itself forms a linear space (R or C), it is clear that the linear functionals discussed in Section 2.6 are a particular class of linear transformation (with one-dimensional range). Several of the properties developed in connection with linear functionals can be extended to apply to the more general linear transformation. Also, there are circumstances where a linear transformation is most conveniently expressed in terms of an ordered sequence of linear functionals.

If the input and output linear spaces are normed, then they are metric spaces and we can investigate the continuity properties of a linear transformation. Linear transformations have the property that continuity at a point (say $\mathbf{x} = \mathbf{0}$) is equivalent to continuity (at all points). To show this, we note that continuity at the origin implies that for any $\varepsilon > 0$ there exists a $\delta(\varepsilon) > 0$ such that

$$\|\mathbf{x} - \mathbf{0}\| < \delta \Rightarrow \|\mathscr{L}(\mathbf{x}) - \mathscr{L}(\mathbf{0})\| < \varepsilon \tag{5.3}$$

but since $\mathscr{L}(\mathbf{0}) = \mathbf{0}$,

$$\|\mathbf{x}\| < \delta \Rightarrow \|\mathscr{L}(\mathbf{x})\| < \varepsilon \tag{5.4}$$

if $\mathscr{L}$ is continuous at the origin. Now consider an arbitrary point $\mathbf{x}_0$, replace $\mathbf{x}$ by $\mathbf{x} - \mathbf{x}_0$, and use the same ε and δ as above, giving

$$\|\mathbf{x} - \mathbf{x}_0\| < \delta \Rightarrow \|\mathscr{L}(\mathbf{x} - \mathbf{x}_0)\| = \|\mathscr{L}(\mathbf{x}) - \mathscr{L}(\mathbf{x}_0)\| < \varepsilon \tag{5.5}$$

Hence, continuity and continuity at the origin are equivalent.

Another property is that, as in the case of linear functionals, continuity and boundedness are equivalent. $\mathscr{L}$ is said to be *bounded* if there exists a real constant K such that

$$\|\mathscr{L}(\mathbf{x})\| \leqslant K\|\mathbf{x}\| \qquad \text{for all } \mathbf{x} \in \mathscr{X} \tag{5.6}$$

It is clear that a bounded linear transformation is continuous by using $\delta = \varepsilon/K$ in (5.4). On the other hand, to show boundedness, it is sufficient to show that $\|\mathscr{L}(\mathbf{x})\| \leqslant K$ for all $\mathbf{x}$ such that $\|\mathbf{x}\| = 1$. Let $\mathbf{x} = 2\mathbf{x}_1/\delta$; then

$$\|\mathbf{x}\| = 1 \Rightarrow \|\mathbf{x}_1\| < \delta \Rightarrow \|\mathscr{L}(\mathbf{x}_1)\| < \varepsilon \Rightarrow \|\mathscr{L}(\mathbf{x})\| < \frac{2\varepsilon}{\delta}$$

so continuity implies boundedness and we have shown that boundedness and continuity are equivalent for linear transformations.

For more abstract studies of the theory of linear transformations, it is important to note that the set of all linear transformations, on a particular linear space, is itself a linear space where vector addition and scalar multiplication are defined on a pointwise basis;[1] i.e.,

$$\left.\begin{aligned} \mathscr{L} = \mathscr{L}_1 + \mathscr{L}_2 &\Rightarrow \mathscr{L}(\mathbf{x}) = \mathscr{L}_1(\mathbf{x}) + \mathscr{L}_2(\mathbf{x}) \\ \mathscr{L} = \alpha\mathscr{L}_1 &\Rightarrow \mathscr{L}(\mathbf{x}) = \alpha\mathscr{L}_1(\mathbf{x}) \end{aligned}\right\} \qquad \text{for all } \mathbf{x} \in \mathscr{X} \tag{5.7}$$

These operations have simple system counterparts, as shown in Figure 5.2, with the sum corresponding to a parallel connection of networks and scalar multiplication corresponding to the cascade combination of the network and an ideal amplifier with gain α. For scalar multiplication, the amplifier can be connected either at the input or the output of the network.

The space of linear transformations can be normed in a manner similar to that for linear functionals; i.e.,

$$\|\mathscr{L}\| = \sup \{\|\mathscr{L}\mathbf{x}\|;\ \|\mathbf{x}\| \leqslant 1\} \tag{5.8}$$

or, equivalently,

$$\|\mathscr{L}\| = \inf \{K;\ \|\mathscr{L}\mathbf{x}\| \leqslant K\|\mathbf{x}\|,\ \mathbf{x} \in \mathscr{X}\} \tag{5.9}$$

When the input and output spaces are identical, the linear transformation is called a *linear operator*. Thus a linear operator is a linear transformation which maps its domain into itself. An additional vector operation (called the product) arises in a natural way when considering operators. The product

[1] Henceforth we shall indicate linear transformations by boldface letters. Also we shall drop parentheses so that the image $\mathscr{L}(\mathbf{x})$ will be written as $\mathscr{L}\mathbf{x}$. When $\mathscr{X}$ is a function space, and we have occasion to use the alternative notation $x(t)$ for $\mathbf{x}$, we shall write the image as $\mathscr{L} \cdot x(t)$ to emphasize the fact that $\mathscr{L}$ is a mapping on a time function and not on a scalar.

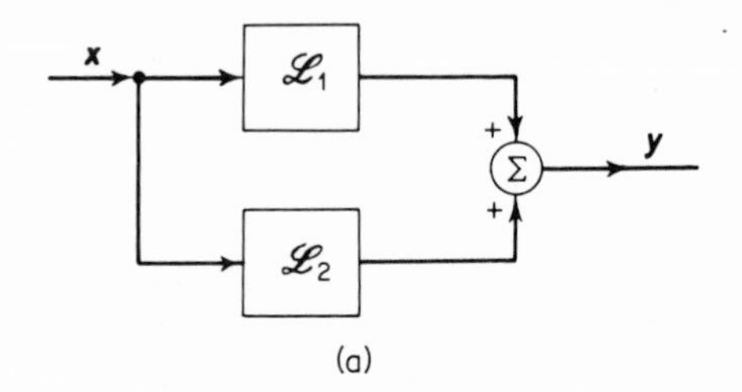

(a)

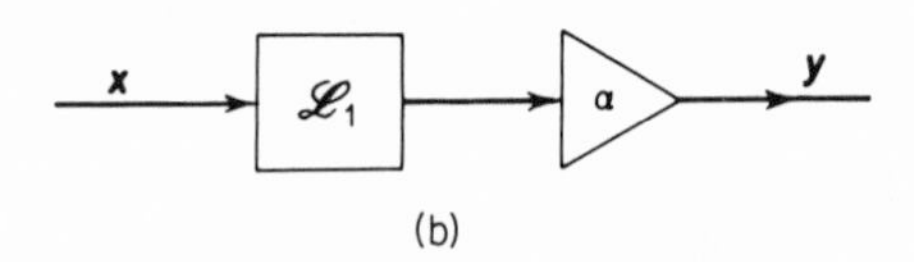

(b)

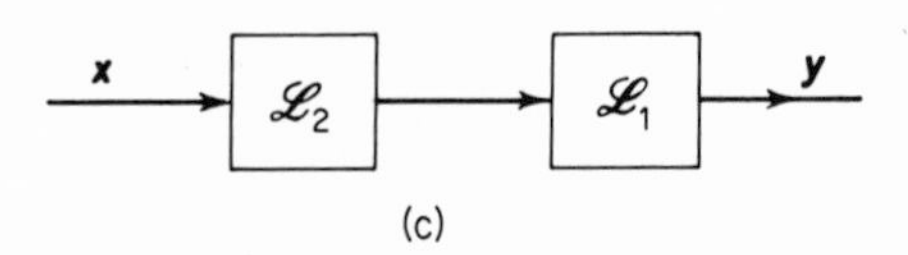

(c)

Figure 5.2. System counterparts of vector and scalar operations with linear transformations. (a) Addition (parallel combination): $\mathscr{L} = \mathscr{L}_1 + \mathscr{L}_2$. (b) Scalar multiplication: $\mathscr{L} = \alpha\mathscr{L}_1$. (c) Vector multiplication (cascade combination): $\mathscr{L} = \mathscr{L}_1\mathscr{L}_2$.

of two operators is the composite mapping defined by

$$\mathscr{L} = \mathscr{L}_1\mathscr{L}_2 \Rightarrow \mathscr{L}(\mathbf{x}) = \mathscr{L}_1[\mathscr{L}_2(\mathbf{x})] \tag{5.10}$$

The physical counterpart of the product operation is simply the cascade combination (in the correct order) of the networks representing the individual operators (Figure 5.2c). Multiplication and addition are distributive with respect to each other in that

$$\begin{aligned} \mathscr{L}_1(\mathscr{L}_2 + \mathscr{L}_3) &= \mathscr{L}_1\mathscr{L}_2 + \mathscr{L}_1\mathscr{L}_3 \\ (\mathscr{L}_1 + \mathscr{L}_2)\mathscr{L}_3 &= \mathscr{L}_1\mathscr{L}_3 + \mathscr{L}_2\mathscr{L}_3 \end{aligned} \tag{5.11}$$

Hence the operators, besides forming a linear space, actually form an algebra.[1] The algebra of operators is non-commutative ($\mathscr{L}_2\mathscr{L}_1 \neq \mathscr{L}_1\mathscr{L}_2$, in general) but it does contain an identity element $\mathscr{I}$, defined by $\mathscr{I}\mathbf{x} = \mathbf{x}$ for all $\mathbf{x}$. If an operator is one-to-one and onto as a mapping, then an inverse mapping exists, which can be shown to be linear also. Such an operator is said to be *non-singular* and it has a multiplicative inverse $\mathscr{L}^{-1}$ such that

$$\mathscr{L}\mathscr{L}^{-1} = \mathscr{L}^{-1}\mathscr{L} = \mathscr{I} \tag{5.12}$$

5.3 Representation of Linear Transformations on a Finite-dimensional Domain

When the domain of a linear transformation is a finite-dimensional inner product space, there are simple and direct methods for obtaining representations for the linear transformations. These methods lead to various forms of representation, each of which has its own merits. Our concern will be with finite-dimensional subspaces of the $L^2(T)$ function space with inner products as defined by (3.3). The primary motivation for examining transformations on the restricted domains is that the results reveal basic approaches that can be extended to the problem of representing transformations on the more extensive domains of interest, such as the entire $L^2(T)$ space.

Representation by Response Vectors

First, we let the input space $\mathcal{X}$ be that which is spanned by the linearly independent set $\{\boldsymbol{\varphi}_i; i = 1, 2, \ldots, n\}$ with reciprocal set $\{\boldsymbol{\theta}_i; i = 1, 2, \ldots, n\}$. Let the output space $\mathcal{Y}$ be an n-dimensional space which contains the range of the transformation $\mathcal{L}$ under consideration. We can always find such a space since the dimension of the range cannot exceed that of the domain. Now, for an arbitrary $\mathbf{x} \in \mathcal{X}$, from (3.2) and (3.7) we have

$$x(t) = \sum_{i=1}^{n} (\mathbf{x}, \boldsymbol{\theta}_i)\varphi_i(t); \qquad t \in T \tag{5.13}$$

Hence, from the linearity of $\mathcal{L}$,

$$y(t) = \mathcal{L} \cdot x(t) = \sum_{i=1}^{n} (\mathbf{x}, \boldsymbol{\theta}_i)\psi_i(t) \tag{5.14}$$

where the set $\{\boldsymbol{\psi}_i; \psi_i(t) = \mathcal{L} \cdot \varphi_i(t), i = 1, 2, \ldots, n\}$ constitutes the "responses" to the various basis functions for $\mathcal{X}$ as signal inputs to $\mathcal{L}$. It is clear that all images of points in $\mathcal{X}$ can be expressed as linear combinations of the $\{\boldsymbol{\psi}_i\}$ using the same n-tuple $\boldsymbol{\alpha} = \{\alpha_i; \alpha_i = (\mathbf{x}, \boldsymbol{\theta}_i), i = 1, 2, \ldots, n\}$ that represents $\mathbf{x} \in \mathcal{X}$. Thus we can consider $\{\boldsymbol{\psi}_i; i = 1, 2, \ldots, n\}$ as a *representation* for $\mathcal{L}$ relative to the basis $\{\boldsymbol{\varphi}_i; i = 1, 2, \ldots, n\}$ for $\mathcal{X}$. In contrast to representation of a signal by an n-tuple, the linear transformation is represented by an ordered sequence of n vectors in $\mathcal{Y}$ relative to a particular basis for $\mathcal{X}$.

It is tempting to think of $\{\boldsymbol{\psi}_i; i = 1, 2, \ldots, n\}$ as forming a basis for the output space $\mathcal{Y}$; however, we cannot rely on this as there is no assurance that the set is linearly independent. If it is not, then (by definition of linear independence) there is some non-zero input $\mathbf{x}$ which maps into zero in $\mathcal{Y}$. If there is at least one vector that maps into zero, then there must be a whole subspace that maps into zero. This subspace (of $\mathscr{X}$) is called the *null space* of the linear transformation. Because of the resulting many-to-one nature of the mapping, there does not exist an inverse mapping and the linear transformation is singular. A singular transformation has a range with dimension less than n, because the $\{\boldsymbol{\psi}_i\}$ are linearly dependent and do not span $\mathscr{Y}$. In fact, it is easily shown that the sum of the dimensions of the range and the null space is equal to n. The dimension of the range space is called the *rank* of the linear transformation.

Representation by a Sequence of Linear Functionals

An alternative method is to represent $\mathscr{L}$ by means of an ordered sequence of n vectors in $\mathscr{X}$. To do this, we let $\{\tilde{\boldsymbol{\varphi}}_i; i = 1, 2, \ldots, n\}$ be a set of linearly independent vectors spanning $\mathscr{Y}$. Let $\{\tilde{\boldsymbol{\theta}}_i; i = 1, 2, \ldots, n\}$ be the corresponding reciprocal basis set. Then any vector in $\mathscr{Y}$ can be expressed as

$$y(t) = \sum_{j=1}^{n} (\mathbf{y}, \tilde{\boldsymbol{\theta}}_j)\tilde{\varphi}_j(t) \tag{5.15}$$

By substitution of (5.14) into (5.15), we obtain

$$y(t) = \mathscr{L} \cdot x(t) = \sum_{j=1}^{n} (\mathbf{x}, \boldsymbol{\omega}_j)\tilde{\varphi}_j(t) \tag{5.16}$$

where

$$\omega_j(t) = \sum_{i=1}^{n} (\tilde{\boldsymbol{\theta}}_j, \boldsymbol{\psi}_i)\theta_i(t) \tag{5.17}$$

Hence, the ordered sequence $\{\boldsymbol{\omega}_j; j = 1, 2, \ldots, n\}$ represents $\mathscr{L}$ relative to a particular basis for $\mathscr{Y}$. The null space is simply the subspace of vectors in $\mathscr{X}$ which are orthogonal to all the $\boldsymbol{\omega}_j$. If the $\{\boldsymbol{\omega}_j; j = 1, 2, \ldots, n\}$ are linearly independent, then the rank of the transformation is n. We can alternatively think of this method of representation as involving an ordered sequence of n linear functionals (transformations of rank 1) of the form $f_j(\mathbf{x}) = (\mathbf{x}, \boldsymbol{\omega}_j); j = 1, 2, \ldots, n$, so that the transformation is expressed by

$$\mathscr{L} \cdot x(t) = \sum_{j=1}^{n} f_j(\mathbf{x})\tilde{\varphi}_j(t) \tag{5.18}$$

Matrix Representation

Since each of the vectors $\boldsymbol{\psi}_i$ in the first method or the vectors $\boldsymbol{\omega}_j$ in the second method could be represented by an n-tuple, it follows that the linear transformation can be represented by an $n \times n$ array of scalars (a matrix). To show this, let

$$\psi_i(t) = \sum_{j=1}^{n} \lambda_{ji}\tilde{\varphi}_j(t) \tag{5.19}$$

where

$$\lambda_{ji} = (\boldsymbol{\psi}_i, \tilde{\boldsymbol{\theta}}_j) = (\mathscr{L}\boldsymbol{\varphi}_i, \tilde{\boldsymbol{\theta}}_j) \tag{5.20}$$

Now substituting (5.19) into (5.14), we have

$$y(t) = \sum_{i=1}^{n}\sum_{j=1}^{n} \lambda_{ji}\alpha_i\tilde{\varphi}_j(t) = \sum_{j=1}^{n}\beta_j\tilde{\varphi}_j(t) \tag{5.21}$$

where

$$\alpha_i = (\mathbf{x}, \boldsymbol{\theta}_i)$$

and

$$\beta_j = (\mathbf{y}, \tilde{\boldsymbol{\theta}}_j) = \sum_{i=1}^{n} \lambda_{ji}\alpha_i$$

or, in standard matrix notation,

$$\boldsymbol{\beta} = \mathbf{L}\boldsymbol{\alpha} \tag{5.22}$$

where $\boldsymbol{\alpha}$ and $\boldsymbol{\beta}$ are n-tuples representing $\mathbf{x}$ and $\mathbf{y}$, respectively, relative to the bases for $\mathscr{X}$ and $\mathscr{Y}$. $\mathbf{L}$ is an $n \times n$ matrix with the scalar $\lambda_{ij} = (\mathscr{L}\boldsymbol{\varphi}_j, \tilde{\boldsymbol{\theta}}_i)$ in the ith row and jth column, representing the transformation $\mathscr{L}$.

Exercise 5.1. Show that the range $\{\mathbf{y}; \mathbf{y} = \mathscr{L}\mathbf{x}, \mathbf{x} \in \mathscr{X}\}$ and the null space $\{\mathbf{x}; \mathscr{L}\mathbf{x} = \mathbf{0}\}$ of a linear transformation $\mathscr{L}$ are linear spaces. If $\mathscr{X}$ is n-dimensional, show that the sum of the dimensions of the range and null space is equal to n.

Exercise 5.2. Let the matrix $\mathbf{L}$ be a representation for $\mathscr{L}$ according to (5.22). Show that the determinant of $\mathbf{L}$ is equal to zero if, and only if, $\mathscr{L}$ has a non-zero null space.

Exercise 5.3. Show that the elements $\lambda_{ij} = (\mathscr{L}\boldsymbol{\varphi}_j, \tilde{\boldsymbol{\theta}}_i)$ of the matrix $\mathbf{L}$ in (5.22) can also be expressed as $\lambda_{ij} = (\boldsymbol{\varphi}_j, \boldsymbol{\omega}_i)$ with $\{\boldsymbol{\omega}_i\}$ as defined in (5.17).

Exercise 5.4. Consider two linear transformations $\mathscr{L}_1: \mathscr{X} \to \mathscr{Y}$ and $\mathscr{L}_2: \mathscr{Y} \to \mathscr{Z}$, where $\mathscr{X}$, $\mathscr{Y}$, and $\mathscr{Z}$ are n-dimensional linear spaces; $\mathscr{Y}$ contains the range of $\mathscr{L}_1$ and $\mathscr{Z}$ contains the range of $\mathscr{L}_2$. Choose arbitrary bases for $\mathscr{X}$, $\mathscr{Y}$, and $\mathscr{Z}$, and show that the composite transformation $\mathscr{L} = \mathscr{L}_2\mathscr{L}_1$ is represented by the ordinary matrix product $\mathbf{L} = \mathbf{L}_2\mathbf{L}_1$, where $\mathbf{L}_1$ and $\mathbf{L}_2$ are the matrix representations for $\mathscr{L}_1$ and $\mathscr{L}_2$, respectively.

5.4 Representation of Operators on the $L^2(T)$ Space

The preceding methods of representation are quite useful when applied to particular linear transformations; but with regard to providing representations for all members of a class of transformations, the methods have some basic limitations. The limitations stem not so much from the restricted domain but from the fact that the ranges of different transformations in the class may be substantially different. The class of transformations whose ranges are contained in a particular n-dimensional output space is too small to be of much practical interest. On the other hand, the class of *bounded* linear transformations whose domain is the entire $L^2(T)$ space represents the major portion of signal transformations of practical interest. We note that a bounded linear transformation on $L^2(-\infty, \infty)$ is actually a *linear operator* since the images of bounded signals are bounded, (5.6). The physical interpretation of bounded operators can be taken as a kind of stability property of the network performing the transformation, in the sense that finite-energy input signals produce only finite-energy output signals.

From an analytical standpoint, linear operators on $L^2(T)$ are most conveniently represented relative to the continuous bases described in Chapter 4. Accordingly, we let the input and output signals $x(t)$ and $y(t)$ be represented by the functions $u(s)$ and $v(s)$, respectively, relative to the basis kernel $\varphi(t, s)$.

$$\begin{aligned} x(t) &= \int_S u(s)\varphi(t, s)\, ds \\ y(t) &= \int_S v(s)\varphi(t, s)\, ds \end{aligned} \tag{5.23}$$

For an operator $\mathscr{L}$ we have

$$y(t) = \mathscr{L} \cdot x(t) = \int_S u(s)\psi(t, s)\, ds \tag{5.24}$$

where $\psi(t, s) = \mathscr{L} \cdot \varphi(t, s)$ is the function resulting from $\mathscr{L}$ operating on $\varphi(t, s)$ considered as a function of t with s a fixed parameter. As mentioned in Chapter 4, we do not require that $\varphi(t, s)$ be an $L^2(T)$ function of t.

Since, from (4.1b),

$$v(s) = \int_T y(t)\theta(s, t)\, dt \tag{5.25}$$

we can combine (5.24) and (5.25) to find

$$\begin{aligned} v(s) &= \int_T \int_S u(\sigma)\psi(t, \sigma)\theta(s, t)\, d\sigma\, dt \\ &= \int_S L(s, \sigma)u(\sigma)\, d\sigma \end{aligned} \tag{5.26}$$

where

$$L(s, \sigma) \triangleq \int_T \theta(s, t)\psi(t, \sigma)\, dt \tag{5.27}$$

The analogy between (5.27) and (5.20) for the discrete case is apparent. The discrete variables j and i in (5.20) are replaced by the continuous variables s and σ in (5.27), and the representations for input and output signals, relative to the basis $\varphi(t, s)$, are related by an integral transform. The integral transform, and hence the operator, is characterized by the kernel function $L(s, \sigma)$, which is analogous to the matrix $\mathbf{L}$ in (5.22). We also see from (5.27) that the kernel function is completely described by the network response (Figure 5.3) to the signal basis $\varphi(t, s)$.

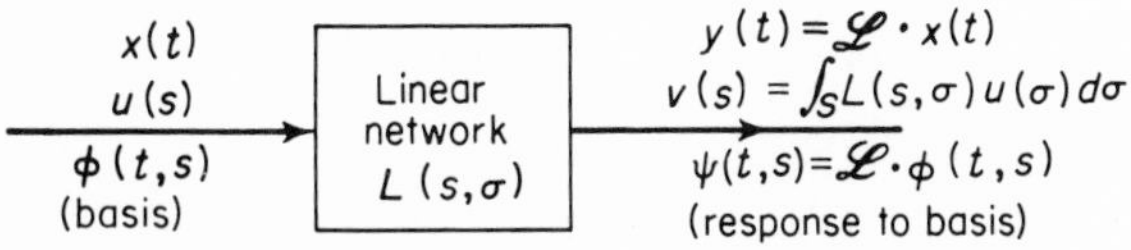

Input	Network	Output
$x(t)$	Linear network	$y(t) = \mathscr{L} \cdot x(t)$
$u(s)$	$L(s, \sigma)$	$v(s) = \int_S L(s, \sigma) u(\sigma) d\sigma$
$\phi(t, s)$ (basis)		$\psi(t, s) = \mathscr{L} \cdot \phi(t, s)$ (response to basis)

Figure 5.3. Representation of signal transforming properties of a linear network.

As a particular example, we consider first the commonly used basis $\varphi(t, s) = \delta(t - s)$ with $T = S = (-\infty, \infty)$. In this case we have simply $\mathbf{u} = \mathbf{x}$ and $\mathbf{v} = \mathbf{y}$, (4.21). The basic response is $\psi(t, s) = h(t, s)$, the *impulse response* of the network. This is the response, as a function of t, to an impulse applied at time s at the input. From (5.27)

$$L(s, \sigma) = \int_{-\infty}^{\infty} \delta(s - t) h(t, \sigma)\, dt = h(s, \sigma) \tag{5.28}$$

Hence, it follows that one way to characterize the input-output relationship for the network is by the integral transform using the impulse response as a kernel.

$$y(t) = \int_{-\infty}^{\infty} h(t, \tau) x(\tau)\, d\tau \tag{5.29}$$

Another commonly used basis is $\varphi(t, s) = e^{j2\pi st}$ with $T = S = (-\infty, \infty)$ leading to frequency-domain descriptions of signals and operators. Here $\mathbf{u} = \mathbf{X}$ and $\mathbf{v} = \mathbf{Y}$ where $\mathbf{X}$ and $\mathbf{Y}$ are Fourier transforms of $\mathbf{x}$ and $\mathbf{y}$. In terms of the impulse response of the network, we can find the response $\psi(t, s)$ to the basis $e^{j2\pi st}$ by means of (5.29).

$$\psi(t, s) = \int_{-\infty}^{\infty} h(t, \tau) e^{j2\pi s\tau}\, d\tau \tag{5.30}$$

Hence, from (5.26) and (5.27),

$$Y(f) = \int_{-\infty}^{\infty} H(f, \nu) X(\nu)\, d\nu$$

where

$$H(f, \nu) = \iint_{-\infty}^{\infty} h(t, \tau) e^{-j2\pi ft} e^{j2\pi\nu\tau}\, dt\, d\tau \tag{5.31}$$

Mixed Bases

For some applications it is useful to represent input and output signals relative to different bases. If $\varphi(t, s)$ is the basis for input signals and $\tilde{\varphi}(t, s)$ is the basis for output signals, then following the preceding development we have

$$v(s) = \int_S L(s, \sigma) u(\sigma)\, d\sigma$$

where

$$L(s, \sigma) = \int_T \tilde{\theta}(s, t) \psi(t, \sigma)\, dt \tag{5.32}$$

and

$$\psi(t, s) = \mathscr{L} \cdot \varphi(t, s)$$

For the analysis of time-variable networks, a frequency-domain description for input signals and a time-domain description for output signals has been used[6]; i.e., $\varphi(t, s) = e^{j2\pi st}$ and $\tilde{\varphi}(t, s) = \delta(t - s)$. Then, using (5.32) for this case, we find

$$y(t) = \int_{-\infty}^{\infty} G(t, f) X(f)\, df \tag{5.33}$$

where, in terms of the impulse response,

$$G(t, f) = \int_{-\infty}^{\infty} h(t, \tau) e^{j2\pi f\tau}\, d\tau \tag{5.34}$$

As another example of representing an operator relative to mixed bases, the dual of the situation above has been found useful[7]; i.e., $\varphi(t, s) = \delta(t - s)$ and $\tilde{\varphi}(t, s) = e^{j2\pi st}$. In this case we have

$$Y(f) = \int_{-\infty}^{\infty} K(f, t) x(t)\, dt \tag{5.35}$$

where

$$K(f, t) = \int_{-\infty}^{\infty} h(\tau, t) e^{-j2\pi f\tau}\, d\tau \tag{5.36}$$

Classification of Operators

Certain properties of linear systems are conveniently expressed in terms of the behavior of the kernel function representing the operator. These properties are useful for classifying various types of operators. A few of the classifications commonly encountered in practical applications are described below. In this discussion we assume that the domain of the operators is $L^2(-\infty, \infty)$.

Time-invariant operator. Many physical system components have signal transforming properties which are essentially independent of the time at which a signal excitation is applied. Time-invariance can be described as an invariance to translations of the time origin in the following sense: If $y(t) = \mathscr{L} \cdot x(t)$, then $y(t - t_0) = \mathscr{L} \cdot x(t - t_0)$ for any $\mathbf{x}$ and any translation t_0 when $\mathscr{L}$ is a time-invariant operator. An operator is time-invariant if, and only if, the impulse response $h(t, \tau)$ depends only on the single variable $t - \tau$. Because of this, it is common practice to write $h(t - \tau)$ for the impulse response or, even more frequently, simply $h(t)$, with the implication that the impulse is applied at the input at time $\tau = 0$. The time-domain description of the operator, from (5.29), is

$$y(t) = \int_{-\infty}^{\infty} h(t - \tau)x(\tau)\, d\tau \tag{5.37}$$

and $\mathbf{y}$ is referred to as the *convolution product* of $\mathbf{h}$ and $\mathbf{x}$, often denoted symbolically by $\mathbf{y} = \mathbf{h} \otimes \mathbf{x}$. By a change of variable of integration in (5.37), we can express the operation as

$$y(t) = \int_{-\infty}^{\infty} k(\sigma)x(\sigma + t)\, d\sigma \tag{5.38}$$

with $k(\sigma) = h(-\sigma)$. This is a natural description of a signal resulting from a *scanning* operation. If a spatially distributed signal $\mathbf{x}$, e.g., an optical image, is "read out" by a scanning device traveling at a constant velocity, then the time function out of the device is given by (5.38) where $k(\sigma)$ can be interpreted as the transmittance across the "window" of the scanning device. As the scanning operation is often encountered in signal processing systems, it is important to remember that these operations are equivalent to time-invariant filtering.

The frequency-domain description of the time-invariant operator is especially convenient. In this case (5.31) becomes

$$\begin{aligned} H(f, \nu) &= \iint\limits_{-\infty}^{\infty} h(t - \tau)e^{-j2\pi ft + j2\pi\nu\tau} dt\, d\tau \\ &= \int_{-\infty}^{\infty} h(\sigma)e^{-j2\pi f\sigma} \int_{-\infty}^{\infty} e^{-j2\pi(f-\nu)\tau}\, d\tau\, d\sigma \\ &= H(f)\, \delta(f - \nu) \end{aligned} \tag{5.39}$$

and hence

$$\begin{aligned} Y(f) &= \int_{-\infty}^{\infty} H(f)\, \delta(f - \nu)X(\nu)\, d\nu \\ &= H(f)X(f) \end{aligned} \tag{5.40}$$

The function $H(f)$ in (5.40) is called the *transfer function* for the time-invariant network. In terms of the weighting function $k(\sigma)$ in the scanning operation, we have

$$Y(f) = K^*(f)X(f) \tag{5.41}$$

Another important property of the class of time-invariant operators is that the multiplication of operators, (5.10), is commutative. This follows from the fact that the transfer function of a cascade combination of time-invariant networks is given by the ordinary numerical product (which is commutative) of the transfer functions; hence the ordering within the cascade is immaterial.

Identity operator. The identity operator $\mathscr{I}$ is described by $\mathbf{x} = \mathscr{I}\mathbf{x}$ for all $\mathbf{x}$. It is clear that this is a time-invariant operator and the corresponding impulse response is $h(t) = \delta(t)$. The corresponding transfer function is $H(f) = 1$.

Delay operator. An operator closely related to the identity operator and of considerable practical interest because of the availability of distrbuted-parameter delay lines is the operator which maps $x(t)$ into $x(t - t_0)$. The corresponding impulse response is $h(t) = \delta(t - t_0)$, and the transfer function is $H(f) = e^{-j2\pi t_0 f}$.

Gating operator (multiplier). There are several physical devices (modulators, signal gates, etc.) which perform the transformation

$$y(t) = w(t)x(t) \tag{5.42}$$

This is clearly not a time-invariant operation, unless $w(t)$ is a constant (in which case, the operation is simply scalar multiplication). The impulse response of the gating operator is given by

$$h(t, \tau) = w(t)\,\delta(t - \tau) \tag{5.43}$$

or, equivalently,

$$h(t, \tau) = w(\tau)\,\delta(t - \tau)$$

In the frequency domain, using (5.31), we find

$$\begin{aligned} H(f, \nu) &= \iint_{-\infty}^{\infty} w(t)\,\delta(t - \tau)e^{-j2\pi ft + j2\pi\nu\tau}\,dt\,d\tau \\ &= \int_{-\infty}^{\infty} w(\tau)e^{-j2\pi(f-\nu)\tau}\,d\tau \\ &= W(f - \nu) \end{aligned} \tag{5.44}$$

Hence the operation is characterized by a convolution product.

$$Y(f) = \int_{-\infty}^{\infty} W(f - \nu)X(\nu)\,d\nu \tag{5.45}$$

Physically realizable operator. A fundamental restriction on operators which characterize a realizable physical device is that they be *non-anticipatory*, or *causal.* This requirement is easily stated in terms of the impulse response. From (5.29) we see that if $y(t)$ is to depend only on past values of the input, then the necessary and sufficient condition on the impulse response is

$$h(t, \tau) = 0 \qquad \text{for all } \tau > t \tag{5.46}$$

Many authors incorporate this condition by expressing the operation as

$$y(t) = \int_{-\infty}^{t} h(t, \tau)x(\tau)\, d\tau$$

with no restriction on the function $h(t, \tau)$. For our purposes, however, it is usually more convenient to retain the fixed integration interval and instead impose an auxillary condition (5.47) to be satisfied by the impulse response.

$$h(t, \tau) = w(t - \tau)h(t, \tau) \tag{5.47}$$

where

$$\begin{aligned} w(t) &= 1 \qquad \text{for } t \geqslant 0 \\ &= 0 \qquad \text{for } t < 0 \end{aligned}$$

Finite-order differential operator. The response characteristics of many physical systems can be approximately described by a finite-order differential equation. The standard approach to the analysis of linear systems is to model the system by a network consisting of an interconnection of lumped elements. There are many well-known techniques for deriving the form of the differential equation from the lumped-element network. It has been shown[8] that for a differential equation of the form

$$\sum_{i=0}^{n} a_i(t) \frac{d^i y(t)}{dt^i} = \sum_{i=0}^{n-1} b_i(t) \frac{d^i x(t)}{dt^i} \tag{5.48}$$

where $a_i(t)$ and $b_i(t)$ are continuous coefficients, the corresponding impulse response of the operation can be expressed as

$$\begin{aligned} h(t, \tau) &= \sum_{k=1}^{n} \psi_k(t)\theta_k(\tau) \qquad \text{for } t \geqslant \tau \\ &= 0 \qquad \text{for } t < \tau \end{aligned} \tag{5.49}$$

In the case that the coefficients are constant, the system is time-invariant and, except for the special case where the polynomial

$$Q(p) = \sum_{i=0}^{n} a_i p^i$$

has multiple roots, the impulse response is given by

$$h(t-\tau) = \sum_{k=1}^{n} \alpha_k e^{p_k(t-\tau)} \qquad \text{for } t \geqslant \tau$$
$$= 0 \qquad \text{for } t < \tau \tag{5.50}$$

where p_k are the roots of $Q(p) = 0$. The transfer function for this time-invariant operator is a rational function of frequency.

$$H(f) = \sum_{k=1}^{n} \frac{\alpha_k}{j2\pi f - p_k} \tag{5.51}$$

or, equivalently,

$$H(f) = \frac{P(j2\pi f)}{Q(j2\pi f)}$$

where the numerator polynomial is

$$P(p) = \sum_{i=0}^{n-1} b_i p^i$$

Degenerate operator. Operators on $L^2(-\infty, \infty)$ which have a finite-dimensional range (finite rank) are said to be degenerate. The kernel functions for degenerate operators have a separability property that is very useful for approximate representations and numerical solution of operator equations. A degenerate operator of rank n has an impulse response given by

$$h(t, \tau) = \sum_{i=1}^{n} \psi_i(t)\theta_i^*(\tau) \tag{5.52}$$

where the $\{\psi_i(t); i = 1, 2, \ldots, n\}$ and $\{\theta_i(t); i = 1, 2, \ldots, n\}$ are linearly independent sets of functions in $L^2(-\infty, \infty)$. The image of this operation on $\mathbf{x}$ can be expressed in terms of n linear functionals on $\mathbf{x}$.

$$y(t) = \sum_{i=1}^{n} (\mathbf{x}, \boldsymbol{\theta}_i)\psi_i(t) \tag{5.53}$$

The impulse response (5.52) looks deceptively like that for the finite-order system (5.49); however, the finite-order system is in general not degenerate because of the different behavior for $t \geqslant \tau$ and $t < \tau$.

Exercise 5.5. Write the expression for the impulse response of the cascade combinations (both ways) of a time-invariant operator and a gating operator. For these two situations, also evaluate the frequency-domain kernel function $H(f, \nu)$ (5.31) and the time-frequency kernels $G(t, f)$ (5.34) and $K(f, t)$ (5.36).

Exercise 5.6. Write the frequency-domain kernel $H(f, \nu)$ for the degenerate operator with impulse response indicated in (5.52). Show that the separable form is maintained for an arbitrary pair of input and output basis kernels.

Exercise 5.7. Describe the range and null space of a degenerate operator with the kernel as shown in (5.52).

5.5 Approximate Representation of Operators on $L^2(T)$

In representing operators on $L^2(T)$ by means of kernel functions in integral transforms, we have sidestepped the important issue of obtaining finite numerical representations for these operators. The problem essentially amounts to finding an operator which can be represented by an $n \times n$ matrix as in Section 5.3 and which, in some sense, is an adequate approximation to the $L^2(T)$ operator under consideration. A useful and straightforward approach is to restrict the domain to a finite-dimensional subspace of $L^2(T)$ on the basis that the set of input signals in a particular situation can be adequately represented by linear combinations of a finite set of basis functions. It might be decided, for example, that the first n terms of one of the complete orthonormal sets described in Section 3.3 provides an adequate representation for all input signals. It would be especially convenient to represent output signals relative to this same basis. Accordingly, we let M_n denote the subspace spanned by $\{\boldsymbol{\varphi}_i; i = 1, 2, \ldots, n\}$. Images of $\mathbf{x} \in M_n$ under the $L^2(T)$ operator $\mathscr{L}$ to be approximated are not, in general, contained in M_n. We select as an approximant, call it $\mathscr{L}_n$, the operator which maps $\mathbf{x}$ into the orthogonal projection of the image $\mathbf{y} = \mathscr{L}\mathbf{x}$ onto M_n. This is illustrated in Figure 5.4.

From (3.9) the orthogonal projection of $\mathbf{y}$ onto M_n is given by

$$\hat{y}(t) = \mathscr{L}_n \cdot x(t) = \sum_{i=1}^{n} (\mathbf{y}, \boldsymbol{\theta}_i)\varphi_i(t) = \sum_{i=1}^{n} (\mathscr{L}\mathbf{x}, \boldsymbol{\theta}_i)\varphi_i(t) \tag{5.54}$$

but $\mathbf{x} \in M_n$ is expressed as

$$x(t) = \sum_{j=1}^{n} \alpha_j \varphi_j(t); \qquad \alpha_j = (\mathbf{x}, \boldsymbol{\theta}_j); \qquad j = 1, 2, \ldots, n \tag{5.55}$$

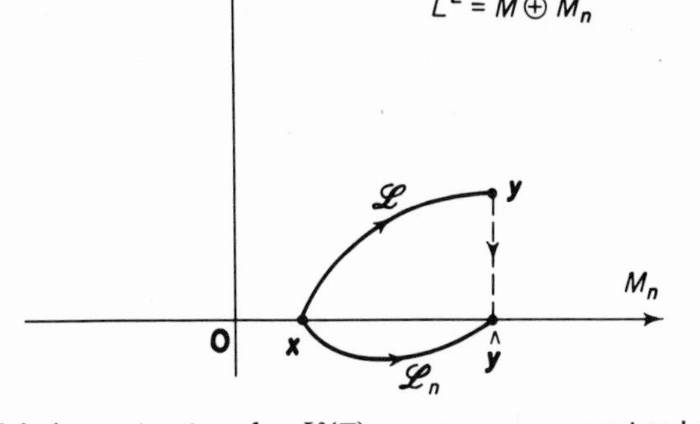

Figure 5.4. Approximation of an $L^2(T)$ operator over a restricted domain.

Hence, substituting (5.55) into (5.54),

$$\hat{y}(t) = \sum_{i=1}^{n}\sum_{j=1}^{n}(\mathscr{L}\boldsymbol{\varphi}_j, \boldsymbol{\theta}_i)\alpha_j\varphi_i(t) = \sum_{i=1}^{n}\beta_i\varphi_i(t)$$

$$\Rightarrow \beta_i = \sum_{j=1}^{n}\lambda_{ij}\alpha_j \Rightarrow \boldsymbol{\beta} = \mathbf{L}\boldsymbol{\alpha} \tag{5.56}$$

In this way, the operator $\mathscr{L}$ is represented by an $n \times n$ matrix $\mathbf{L}$ relative to the basis $\{\boldsymbol{\varphi}_i; i = 1, 2, \ldots, n\}$, where the elements of $\mathbf{L}$ are given by

$$\lambda_{ij} = (\mathscr{L}\boldsymbol{\varphi}_j, \boldsymbol{\theta}_i) \tag{5.57}$$

Example 5.1. To illustrate the application of the matrix representation technique, we consider an example involving finite time-series representations of input and output signals. Let M_n be the subspace spanned by $\{\boldsymbol{\varphi}_i; \varphi_i(t) = \varphi(t - i\tau), i = 0, 1, \ldots, n - 1\}$, where the $\varphi_i(t)$ are orthonormal interpolating functions. By choosing n large enough and τ small enough, we can provide an accurate representation for signals in the interval $0 < t < (n - 1)\tau$. Now suppose that $\mathscr{L}$ is a time-invariant operator corresponding to a network with impulse response $h(t)$. Then

$$\begin{aligned}\mathscr{L} \cdot \varphi_j(t) &= \int_{-\infty}^{\infty} h(t - \sigma)\varphi(\sigma - j\tau)\, d\sigma \\ &\triangleq \psi(t - j\tau)\end{aligned} \tag{5.58}$$

where $\boldsymbol{\psi} = \mathbf{h} \otimes \boldsymbol{\varphi}$ is the network response to the basic interpolating pulse $\boldsymbol{\varphi}$. Knowing this single pulse response, we can completely characterize the operator $\mathscr{L}_n$ since, from (5.57),

$$\begin{aligned}\lambda_{ij} = (\mathscr{L}\boldsymbol{\varphi}_j, \boldsymbol{\varphi}_i) &= \int_{-\infty}^{\infty} \psi(t - j\tau)\varphi(t - i\tau)\, dt \\ &= \int_{-\infty}^{\infty} \psi(\sigma)\varphi(\sigma - (i - j)\tau)\, d\sigma \triangleq h_{i-j}\end{aligned} \tag{5.59}$$

Thus the matrix $\mathbf{L}$ can be constructed from consecutive shifts of the $2n - 1$ element sequence $\{h_i; i = 0, \pm 1, \pm 2, \ldots, \pm(n - 1)\}$.

$$\mathbf{L} = \begin{bmatrix} h_0 & h_{-1} & h_{-2} & \cdots & h_{1-n} \\ h_1 & h_0 & h_{-1} & \cdots & h_{2-n} \\ h_2 & h_1 & h_0 & \cdots & h_{3-n} \\ \cdot & & & \cdot & \cdot \\ \cdot & & & \cdot & \cdot \\ \cdot & & & \cdot & \cdot \\ h_{n-1} & h_{n-2} & & \cdots & h_0 \end{bmatrix} \tag{5.60}$$

The set of n elements $\{h_i; i = 0, 1, 2, \ldots, n-1\}$ form the n-tuple representation for the basic pulse response $\psi(t)$ projected onto M_n.

$$\hat{\psi}(t) = \sum_{i=0}^{n-1} h_i \varphi_i(t) \tag{5.61}$$

since the $\{\boldsymbol{\varphi}_i; i = 0, 1, 2, \cdots, n-1\}$ is an orthonormal set.

If $\mathscr{L}$ is also a physically realizable operator $[h(t) = 0$ for $t < 0]$, then the elements $\{h_i; i = -1, -2, \ldots, 1-n\}$ will tend to be small. If we assume that they are zero, then $\mathbf{L}$ is a triangular matrix (only zeros above the main diagonal) and we have

$$\beta_i = \sum_{j=0}^{n-1} \lambda_{ij}\alpha_j = \sum_{j=0}^{i} h_{i-j}\alpha_j; \qquad i = 0, 1, \ldots, n-1 \tag{5.62}$$

where $\boldsymbol{\alpha}$ and $\boldsymbol{\beta}$ are the n-tuple representations for input and output signals. The form of (5.62) suggests ordinary polynomial multiplication and this has led to a variety of alternate characterizations applicable to time-series analysis of signals and time-invariant, physically realizable networks, e.g., the z-transform characterization.[9] To demonstrate this, we define the $(n-1)$th-degree polynomials

$$A(z) = \sum_{i=0}^{n-1} \alpha_i z^i; \qquad H(z) = \sum_{i=0}^{n-1} h_i z^i \tag{5.63}$$

Then the polynomial

$$B(z) = H(z)A(z) = \sum_{i=0}^{2n-2} \beta_i z^i \tag{5.64}$$

has, for the n lowest-order coefficients, the desired "sample values" of the output signal. The remaining coefficients in $B(z)$ represent parts of the output signal falling outside M_n, but these coefficients will be small if the higher-order coefficients of $A(z)$ and $H(z)$ are sufficiently small.

Figure 5.5 illustrates the case of "staircase" signal approximations resulting from a rectangular interpolating pulse $\boldsymbol{\varphi}$. In this example we let the network be a single section RC low pass filter; and to simplify normalization, we choose $\tau = 1$ and $RC = 1$. Then

$$\begin{aligned}
h(t) &= e^{-t}; \qquad t \geqslant 0 \\
\psi(t) &= 1 - e^{-t} \qquad \text{for } 0 < t \leqslant 1 \\
&= (e-1)e^{-t} \qquad \text{for } t > 1 \\
h_i &= (\boldsymbol{\psi}, \boldsymbol{\varphi}_i) = \int_i^{i+1} \psi(t)\, dt \\
&= 0.368 \qquad \text{for } i = 0 \\
&= 1.085e^{-i} \qquad \text{for } i \geqslant 1
\end{aligned}$$

∎

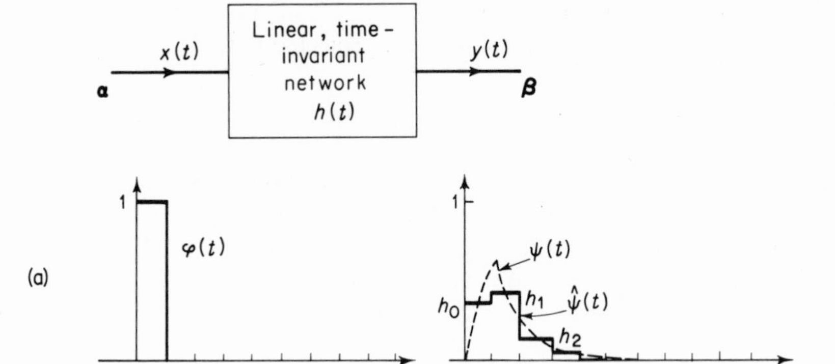

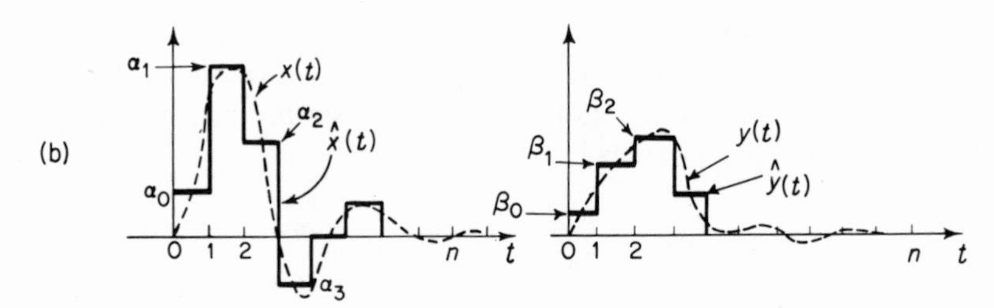

$$B(z) = (1 + 3z + 2z^2 - z^3 + 0 + 0.5z^5 + \cdots)(0.37 + 0.40z + 0.15z^2 + 0.05z^3 + \cdots)$$
$$= 0.37 + 1.51z + 2.09z^2 + 0.97z^3 + 0.05z^4 + 0.14z^5 + \cdots$$

Figure 5.5. Example of time-series representation of time-invariant operator. (a) Basic pulse response. (b) Response to arbitrary signal.

Exercise 5.8. Let $\{\boldsymbol{\xi}_i; i = 1, 2, \ldots, n\}$ be a new basis for the same space M_n spanned by $\{\boldsymbol{\varphi}_i; i = 1, 2, \ldots, n\}$. Show that the matrix representation of (5.57) for $\mathscr{L}$, relative to the new basis, is given by $\widetilde{\mathbf{L}} = \mathbf{\Gamma}^{-1}\mathbf{L}\mathbf{\Gamma}$, where the elements of the $n \times n$ matrix $\mathbf{\Gamma}$ are given by $\gamma_{ij} = (\boldsymbol{\xi}_j, \boldsymbol{\theta}_i)$. $\widetilde{\mathbf{L}}$ is said to be related to $\mathbf{L}$ by a *similarity transformation.*

Exercise 5.9. Show that the operation of orthogonal projection of any $\mathbf{x} \in L^2(T)$ onto M_n can be characterized by the degenerate kernel

$$P(t, s) = \sum_{i=1}^{n} \varphi_i(t)\theta_i^*(s)$$

The corresponding operator $\mathscr{P}$ is called the *orthogonal projection operator*. Show that $\mathscr{P}$ is *idempotent*; i.e., $\mathscr{P}^2 = \mathscr{P}$. What is the norm of $\mathscr{P}$?

Approximation in Norm of Operators

The limitations of (5.57) for representation of arbitrary $L^2(T)$ operators become apparent when we begin to investigate the "closeness" of approximation of $\mathscr{L}$ by $\mathscr{L}_n$. The difficulty is that there are points in M_n for which $\|\mathscr{L}\mathbf{x} - \mathscr{L}_n\mathbf{x}\|$ may be large and, furthermore, this error is not necessarily made small simply by increasing n. It is not immediately clear whether the problem lies in the properties of $\mathscr{L}$ or in the way in which the domain M_n is selected. To avoid the latter problem, we take $L^2(T)$ as the domain of all operators under consideration. Thus $\mathscr{L}_n$ is a degenerate operator described by

$$y(t) = \mathscr{L}_n \cdot x(t) = \int_T L_n(t, s)x(s)\, ds$$

where

$$L_n(t, s) = \sum_{i=1}^{n} \psi_i(t)\theta_i^*(s) \tag{5.65}$$

with

$$\psi_i(t) = \mathscr{L} \cdot \varphi_i(t)$$

Since we have specified a norm for operators (5.8), we can characterize the distance between an operator and its approximant in terms of the induced metric.

$$\begin{aligned} d(\mathscr{L}, \mathscr{L}_n) &= \|\mathscr{L} - \mathscr{L}_n\| \\ &= \sup\{\|\mathscr{L}\mathbf{x} - \mathscr{L}_n\mathbf{x}\|;\ \|\mathbf{x}\| \leqslant 1, \mathbf{x} \in L^2(T)\} \end{aligned} \tag{5.66}$$

At this point we can state the logical counterpart of the problem of approximating signals in $L^2(T)$ by means of complete orthonormal sets. For an arbitrary operator $\mathscr{L}$, we want to find two sets of functions $\{\psi_i(t); i = 1, 2, \ldots\}$ and $\{\theta_i(t); i = 1, 2, \ldots\}$ in $L^2(T)$ such that the corresponding degenerate operator $\mathscr{L}_n$, (5.65), approximates $\mathscr{L}$ in the sense that $\|\mathscr{L} - \mathscr{L}_n\|$ can be made arbitrarily small by choosing n sufficiently large. For the broadest applicability in the following discussion, we shall take the function space to be $L^2(-\infty, \infty)$.

Compact Operators

With only casual thought, it might be suspected that any bounded operator could be approximated by a degenerate operator of rank n to the desired degree by making n sufficiently large. This is not true, and there are simple examples of bounded $\mathscr{L}$ for which no sequence $\{\mathscr{L}_n\}$ converges (in the norm) to $\mathscr{L}$ as $n \to \infty$. The difficulty is that a degenerate operator has only a finite-dimensional range and hence can only approximate those operators whose range is somehow restricted so that all points are close to a finite-dimensional subspace. In particular, we would require that the image set of the unit sphere $\{\mathbf{x};\ \|\mathbf{x}\| = 1\}$ be close to a finite-dimensional subspace.

It is not sufficient that this image set be merely bounded. Consider, for example, the identity operator, which is a bounded operator. The images of the orthonormal sequence $\{\varphi_i(t); i = 1, 2, \ldots\}$ form a bounded set but the distance between any two images is $\sqrt{2}$; hence, for any given finite-dimensional subspace, we can select a $\boldsymbol{\varphi}_i$ which is not close to any point in the subspace. A stronger restriction than boundedness is needed for the image of the unit sphere in order that an operator be approximated by a degenerate operator. A useful restriction is found in the concept of *total boundedness* of a set. A set S with an associated metric d is said to be totally bounded if, for any $\varepsilon > 0$, there exists a finite subset $\{\mathbf{x}_i; i = 1, 2, \ldots, N(\varepsilon)\}$ called an *ε-net*, such that $d(\mathbf{x}, \mathbf{x}_i) < \varepsilon$ for some i and any $\mathbf{x}$ in S. In other words, the set S is sufficiently "compact" so that a finite number $N(\varepsilon)$ of points in the set can be selected in such a way that any point in the set is within a distance ε of one of the selected points. Total boundedness in a metric space is, in fact, equivalent to the more general concept of *compactness* in abstract topological spaces.[1] It is easy to see that a totally bounded set is close to some finite-dimensional subspace, since we could construct a basis from the points in the ε-net and the subspace spanned by this basis would contain points within an ε-distance of any point in the set. Now suppose that $\{\varphi_i(t); i = 1, 2, \ldots, n\}$ spans the subspace M_n which contains an ε-net for the set $\{\mathscr{L}\mathbf{x}; \|\mathbf{x}\| = 1\}$. We choose as an approximant $\mathscr{L}_n$, the operator which maps $\mathbf{x}$ into the orthogonal projection of $\mathscr{L}\mathbf{x}$ onto M_n; i.e.,

$$\mathscr{L}_n \cdot x(t) = \sum_{i=1}^{n} (\mathscr{L}\mathbf{x}, \boldsymbol{\theta}_i)\varphi_i(t) \tag{5.67}$$

Since $\mathscr{L}_n\mathbf{x}$ is the closest point in M_n to $\mathscr{L}\mathbf{x}$ and M_n contains the ε-net for images of the unit sphere, it follows that

$$\|\mathscr{L} - \mathscr{L}_n\| \leqslant \varepsilon \tag{5.68}$$

where ε can be taken arbitrarily small for sufficiently large n. To obtain the kernel function for the degenerate operator $\mathscr{L}_n$, we rewrite (5.67) as

$$\begin{aligned}\mathscr{L}_n \cdot x(t) &= \sum_{i=1}^{n} \left[\iint_{-\infty}^{\infty} L(\sigma, s)x(s)\theta_i^*(\sigma)\, ds\, d\sigma \right] \varphi_i(t) \\ &= \sum_{i=1}^{n} \varphi_i(t) \int_{-\infty}^{\infty} x(s)\omega_i^*(s)\, ds \end{aligned} \tag{5.69}$$

where[2]

$$\omega_i(t) = \mathscr{L}' \cdot \theta_i(t) = \int_{-\infty}^{\infty} L^*(\sigma, t)\theta_i(\sigma)\, d\sigma \tag{5.70}$$

[2] $\mathscr{L}'$ is the *adjoint of* $\mathscr{L}$, defined in Section 5.7.

and the resulting kernel function for $\mathscr{L}_n$ is

$$L_n(t, s) = \sum_{i=1}^{n} \varphi_i(t)\omega_i^*(s) \tag{5.71}$$

Operators which have the property of mapping bounded sets into totally bounded (compact) sets are called *compact operators* or *completely continuous operators.*[2],[3] We see that the approximation problem would be simple except for the difficulty in finding the ε-nets for the operators. For a more restricted class of operators, the spectral representation techniques discussed in Section 5.7 are helpful for the actual determination of the best approximant. In the remainder of this section we shall examine certain types of operators for the property of compactness.

We noted above that the identity operator was not compact. In fact, time-invariant (convolution) operators on $L^2(-\infty, \infty)$ are not compact (see Exercise 5.10), even if the impulse response function is in $L^2(-\infty, \infty)$. Using time-frequency duality, we can show similarly that the gating operator is not compact; however, as shown below, a cascade combination of these two types of operators might be compact.

Hilbert-Schmidt Operators

A useful sufficient condition for a compact operator is that it be of the *Hilbert-Schmidt* type, which is an operator having a square-integrable kernel function.

$$\iint_{-\infty}^{\infty} |L(t, s)|^2 \, dt \, ds < \infty \tag{5.72}$$

It seems reasonable to attempt a function space representation of $L(t, s)$ in terms of kernel functions having the separable property. This can be done and it can be shown[3] that $L_n(t, s)$ converges to $L(t, s)$ in the sense

$$\iint_{-\infty}^{\infty} |L(t, s) - L_n(t, s)|^2 \, dt \, ds \to 0 \qquad \text{as } n \to \infty \tag{5.73}$$

when $L(t, s)$ satisfies (5.72). In (5.73), $L_n(t, s)$ is given by

$$L_n(t, s) = \sum_{i=1}^{n} \psi_i(t)\theta_i^*(s) \tag{5.74}$$

where

$$\psi_i(t) = \int_{-\infty}^{\infty} L(t, s)\varphi_i(s) \, ds$$

and $\{\boldsymbol{\theta}_i\}$ is the reciprocal set for the complete set $\{\boldsymbol{\varphi}_i; i = 1, 2, \ldots\}$. The set $\{\boldsymbol{\psi}_i; i = 1, 2, \ldots\}$ is in $L^2(-\infty, \infty)$ since $\mathscr{L}$ is a bounded operator.

To provide some insight as to why the Hilbert-Schmidt operator maps the unit sphere into a compact set, we let $\{\boldsymbol{\varphi}_i\}$ be an orthonormal set so that $\boldsymbol{\theta}_i = \boldsymbol{\varphi}_i$ and, combining (5.74) and (5.73), we obtain

$$\iint_{-\infty}^{\infty} |L(t, s) - L_n(t, s)|^2 \, dt \, ds = \iint_{-\infty}^{\infty} |L(t, s)|^2 \, dt \, ds - \sum_{i=1}^{n} \|\boldsymbol{\psi}_i\|^2 \geqslant 0 \quad (5.75)$$

Thus

$$\sum_{i=1}^{\infty} \|\boldsymbol{\psi}_i\|^2 < \infty$$

and only a finite number of the transforms $\boldsymbol{\psi}_i$ of $\boldsymbol{\varphi}_i$ can be appreciably different from zero; hence, we tend to think of the range of $\mathscr{L}$ as being "approximately" finite-dimensional.

To show that $\mathscr{L}_n$ converges to $\mathscr{L}$ in this case, we define

$$q_n(t) \triangleq \left[\int_{-\infty}^{\infty} |L(t, s) - L_n(t, s)|^2 \, ds\right]^{\frac{1}{2}} \quad (5.76)$$

Then, from (5.73), $q_n(t)$ is in L^2 and $\|\mathbf{q}_n\| \to 0$ as $n \to \infty$. Using the Schwarz inequality,

$$|\mathscr{L} \cdot x(t) - \mathscr{L}_n \cdot x(t)|^2 \leqslant q_n^2(t) \, \|\mathbf{x}\|^2$$

Hence

$$\|\mathscr{L}\mathbf{x} - \mathscr{L}_n\mathbf{x}\|^2 \leqslant \|\mathbf{q}_n\|^2 \|\mathbf{x}\|^2$$

and it follows that

$$\|\mathscr{L} - \mathscr{L}_n\| \to 0 \qquad \text{as } n \to \infty \quad (5.77)$$

Example 5.2. As an illustrative application of this simple sufficient condition for compactness of an operator, we consider the product operator resulting from the cascade combinations of time-invariant and gating operators shown in Figure 5.6. With $\mathscr{L} = \mathscr{L}_1\mathscr{L}_2$, we have

$$\begin{aligned} L(t, \tau) &= \int_{-\infty}^{\infty} L_1(t, \sigma) L_2(\sigma, \tau) \, d\sigma \\ &= \int_{-\infty}^{\infty} h(t - \sigma) w(\sigma) \, \delta(\sigma - \tau) \, d\sigma = w(\tau) h(t - \tau) \end{aligned} \quad (5.78)$$

and so,

$$\begin{aligned} \iint_{-\infty}^{\infty} |L(t, \tau)|^2 \, dt \, d\tau &= \iint_{-\infty}^{\infty} |w(\tau)|^2 \, |h(t - \tau)|^2 \, dt \, d\tau \\ &= \|\mathbf{w}\|^2 \, \|\mathbf{h}\|^2 \end{aligned} \quad (5.79)$$

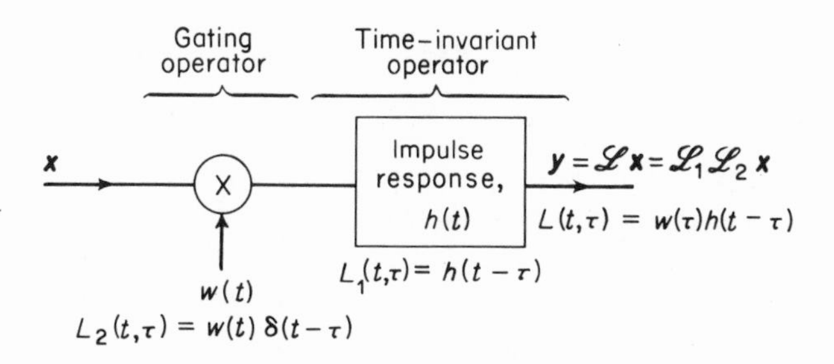

(a)

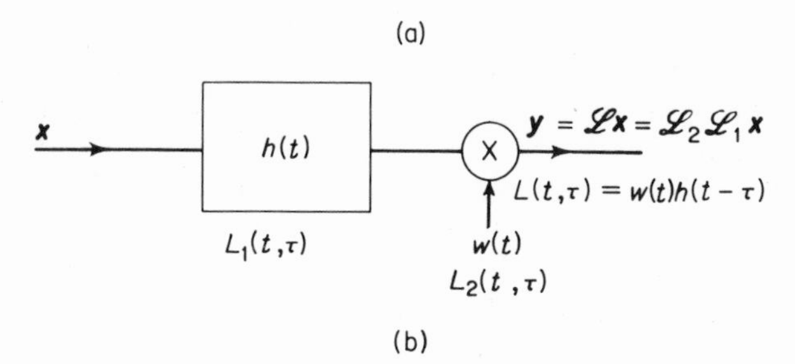

(b)

Figure 5.6. Examples of Hilbert-Schmidt operators with $\|\mathbf{w}\| < \infty$ and $\|\mathbf{h}\| < \infty$.

Hence the product operator is a Hilbert-Schmidt operator (compact) if the functions $w(t)$ and $h(t)$ are in $L^2(-\infty, \infty)$, although individually $\mathscr{L}_1$ and $\mathscr{L}_2$ are not compact. Similarly, for $\mathscr{L} = \mathscr{L}_2\mathscr{L}_1$,

$$L(t, \tau) = w(t)h(t - \tau)$$

and

$$\iint_{-\infty}^{\infty} |L(t, \tau)|^2 \, dt \, d\tau = \|\mathbf{w}\|^2 \, \|\mathbf{h}\|^2 \tag{5.80}$$

also giving a Hilbert-Schmidt operator for $w(t)$ and $h(t)$ in $L^2(-\infty, \infty)$.

For determining whether or not a particular operator corresponding to a physical system is compact, it is often helpful to note that a cascade combination of either network in Figure 5.6 with any (bounded) operator is also a compact operator (see Exercise 5.11). ∎

Exercise 5.10. Show that a time-invariant operator on $L^2(-\infty, \infty)$ is not compact. Show that the gating operator is also not compact. *Hint:* Is $\{\mathbf{x}_i; x_i(t) = x(t - i\tau),\ i = 0, \pm 1, \pm 2, \ldots\}$ a bounded set in $L^2(-\infty, \infty)$? Is $\{\mathbf{y}_i = \mathbf{h} \otimes \mathbf{x}_i; i = 0, \pm 1, \pm 2, \ldots\}$ totally bounded?

Exercise 5.11. Show that the cascade combination (in either order) of a bounded operator and a compact operator is compact.

5.6 Realization of Degenerate Operators

For the special case of time-invariant operators, a great deal of engineering literature has been devoted to techniques for physical realization of networks whose operator approximates the desired time-invariant operator. For the most part, these techniques involve approximation of the desired impulse response function (time-domain approximation) or of the transfer function (frequency-domain approximation). When the approximants are combinations of exponential time functions or rational frequency functions, respectively, then the corresponding network operator is of finite order and classical network synthesis techniques are applicable for finding the particular network realization.

For the more general case of time-variable operators, network realization techniques are not nearly so well developed. There is, however, a variety of canonical network structures which can realize (with some modifications) an arbitrary degenerate operator. Hence, recalling the discussion of the previous section, we can say that there are ways for the approximate physical realization of an arbitrary compact operator.

The required modifications mentioned above stem from the non-anticipatory nature of the response of physical networks. Using the time-domain representation for the rank-n degenerate operator, the impulse response has the separable form

$$L_n(t, \tau) = h(t, \tau) = \sum_{i=1}^{n} \psi_i(t)\theta_i^*(\tau) \tag{5.81}$$

The output signal $\mathbf{y} = \mathscr{L}_n\mathbf{x}$ therefore depends on the inner products $(\mathbf{x}, \boldsymbol{\theta}_i)$ which, unless $\mathbf{x}$ or $\boldsymbol{\theta}_i$ is of finite duration, cannot be evaluated in a finite time. In other words, the network must "see" the entire input signal before it can begin to produce the correct output signal. This gives us the clue as to the circumstances under which an operator of the form of (5.81) can be physically realized. If the input and output signals can be considered to be duration-limited, say to the interval $0 \leqslant t < T$, then the inner products can be evaluated with multipliers and physical integrators,

$$\int_0^t \theta_i^*(\tau)x(\tau)\, d\tau = (\mathbf{x}, \boldsymbol{\theta}_i) \qquad \text{for } t \geqslant T \tag{5.82}$$

Now, assuming that the $\psi_i(t)$ are also duration-limited as above,we can, with a delay of T seconds or more, produce the desired output signal by means of a set of output multipliers as indicated in Figure 5.7. The delay T, which is tolerable in many types of signal processing operations, along with the duration limitation of the signals, is what is required to make the impulse response of the form provided by a finite-order physical system, (5.49).

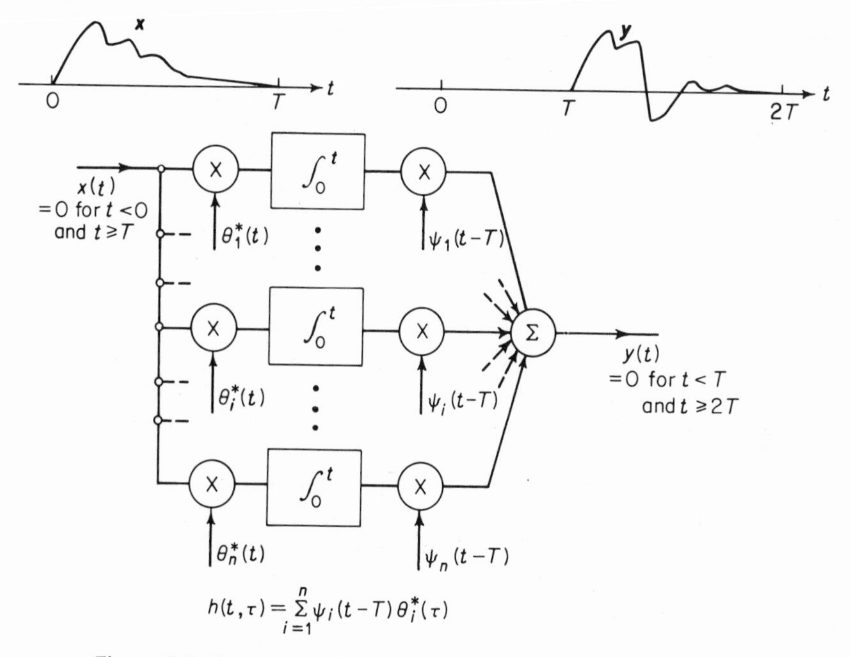

Figure 5.7. Canonical realization of degenerate operator with time delay.

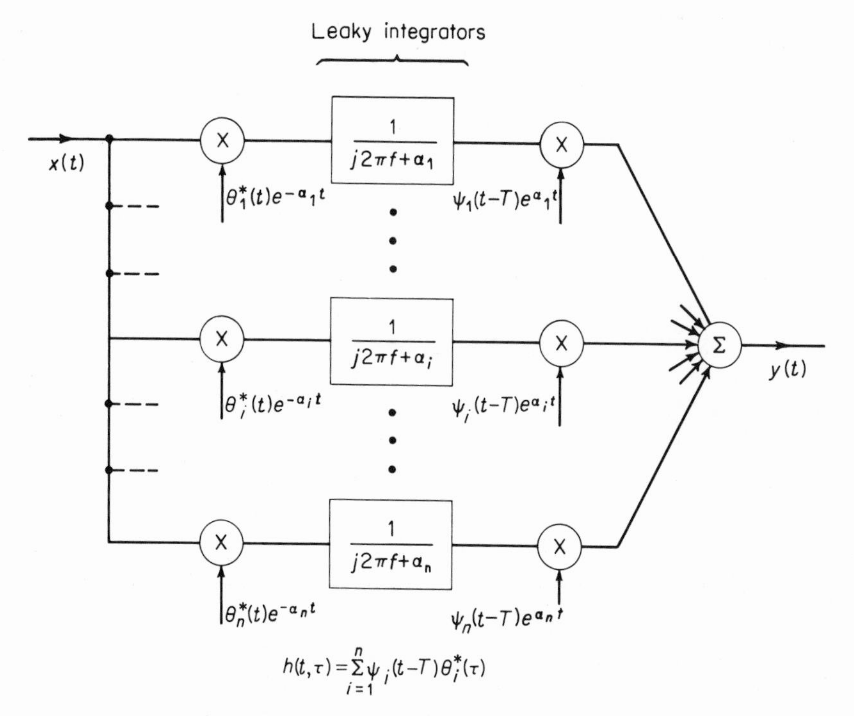

Figure 5.8. Realization equivalent to that of Figure 5.7.

It is not surprising that the realization above for the impulse response (5.81) is not unique. General methods for generating equivalent realizations have been presented.[8],[10] The equivalent realizations can be used to advantage by the network designer to overcome certain practical limitations. For example, suppose the designer has to contend with non-ideal, or "leaky," integrators having an impulse response $e^{-\alpha_i t}$; $t \geqslant 0$, $\alpha_i > 0$. Then the impulse response of the network in Figure 5.7 would be

$$h(t, \tau) = \sum_{i=1}^{n} \psi_i(t - T)e^{-\alpha_i t}\theta_i^*(\tau)e^{\alpha_i \tau} \tag{5.83}$$

Inspection of (5.83) shows that the desired impulse response can be realized in a similar structure using appropriately modified multiplier functions, as shown in Figure 5.8.

Further equivalences involving the basic n-path structure, with n input and output multipliers separated by a time-invariant network, can be developed. These equivalences are readily developed using well-known state-variable techniques for system characterization. The results of such an analysis are briefly presented here. Let the time-invariant part of the structure in Figure

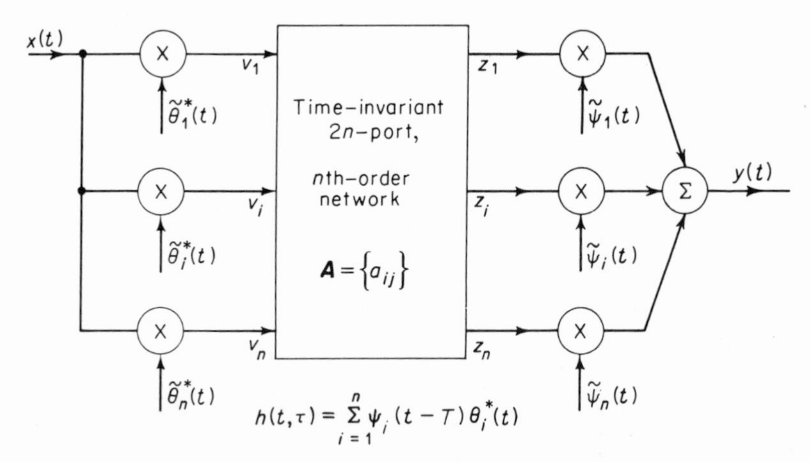

Figure 5.9. Basic n-path structure equivalent to that of Figure 5.7.

5.9 be represented by a $2n$-port, nth-order network described by the following system of n first-order, constant-coefficient equations.

$$\frac{d}{dt} z_i(t) = \sum_{j=1}^{n} a_{ij}z_j(t) + v_i; \qquad i = 1, 2, \ldots, n \tag{5.84}$$

The overall network equivalent to that in Figure 5.7 is obtained by modifying the multiplier functions according to

$$\left.\begin{aligned} \tilde{\theta}_i^*(t) &= \sum_{k=1}^{n} c_{ik}(t)\theta_k^*(t) \\ \tilde{\psi}_i(t) &= \sum_{k=1}^{n} b_{ki}(t)\psi_k(t - T) \end{aligned}\right\} \qquad i = 1, 2, \ldots, n \tag{5.85}$$

where the time-variable transformation matrices $C = \{c_{ik}\}$ and $B = \{b_{ik}\}$ are given by

$$\mathbf{C} = e^{\mathbf{A}t}$$
$$\mathbf{B} = \mathbf{C}^{-1} = e^{-\mathbf{A}t} \tag{5.86}$$

and A is the (possibly complex) $n \times n$ matrix characterizing the time-invariant part of the realization.

In terms of this more general structure, it is seen that the realization of Figure 5.8 results from a diagonal A matrix, while the realization of Figure 5.7 corresponds to a zero A matrix. It is interesting to note that the basic n-path structure has been found useful as a canonical representation for periodically time-varying networks[11] where the restriction to duration-limited signals is dropped. It is also worth mentioning the fact that there are cases, besides the trivial case where the $\boldsymbol{\theta}_i$ and $\boldsymbol{\psi}_i$ are constant, where the n-path structure is equivalent to a time-invariant network (see Exercise 5.12). This has led to some useful alternative realization techniques[12] for transfer functions of time-invariant networks involving band-limiting restrictions on the input and output signals rather than duration-limiting restrictions as employed above.

Exercise 5.12. For the two-path structure with identical time-invariant networks between the input and output multipliers as shown, demonstrate that the input-output relationship is time-invariant. What is the transfer function of the network?

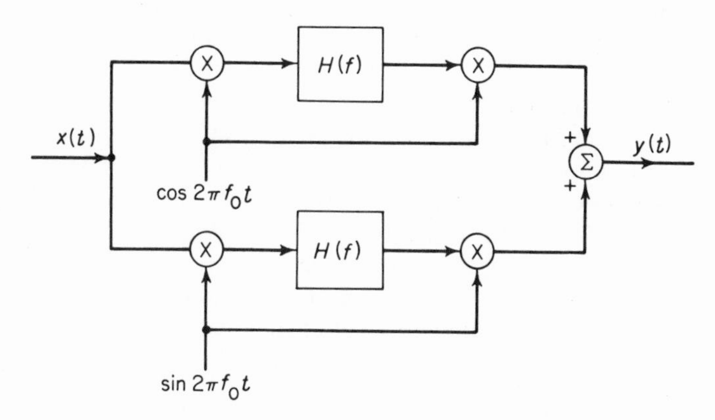

5.7 *Spectral Representation of Operators*

The representation methods we have examined so far can be interpreted as an extension of the finite-dimensional method, described in Section 5.3, for representation in terms of the images of the basis. We now consider extensions of the second method, (5.16), whereby the transformation is represented by a particular set of input vectors $\{\boldsymbol{\omega}_j\}$. First, we define a subset of vectors S which are invariant under the operator $\mathscr{L}$ in the sense that they are simply mapped into scalar multiples of themselves.

$$S = \{\mathbf{x}; \mathscr{L}\mathbf{x} = \lambda\mathbf{x}, \lambda \in C\} \tag{5.87}$$

In many cases the set S, along with the corresponding set of scalar multipliers, can completely characterize the operator and lead to a highly desirable form of representation. To see how this comes about, we re-examine the second method in Section 5.3 for the case of identical input and output spaces. Letting $\tilde{\boldsymbol{\varphi}}_i = \boldsymbol{\varphi}_i$, (5.17) becomes

$$\omega_j(t) = \sum_i (\boldsymbol{\theta}_j, \boldsymbol{\psi}_i)\theta_i(t)$$

where (5.88)

$$\psi_i(t) = \mathscr{L} \cdot \varphi_i(t)$$

Now suppose that the basis functions $\varphi_i(t)$ are selected from S.

$$\boldsymbol{\varphi}_i \in S \Rightarrow \psi_i(t) = \lambda_i \varphi_i(t) \tag{5.89}$$

and (5.88) becomes

$$\omega_j(t) = \lambda_j^* \theta_j(t) \tag{5.90}$$

Using (5.90) in (5.16), the image of any $\mathbf{x}$ in the space spanned by the $\{\boldsymbol{\varphi}_i\}$ is given by

$$y(t) = \mathscr{L} \cdot x(t) = \sum_j \lambda_j(\mathbf{x}, \boldsymbol{\theta}_j)\varphi_j(t)$$

$$= \sum_j \beta_j \varphi_j(t) \Rightarrow \beta_j = \lambda_j \alpha_j \qquad \text{for each } j \tag{5.91}$$

From (5.91) we see that the transformation is simply characterized by a coordinate-by-coordinate scaling of the components of $\mathbf{x}$.

In this section we shall show that for a significantly large class of operators the set S contains enough linearly independent vectors to span the domain of interest and the economy of representation implied by (5.91) is indeed attainable. We shall also see that, in some cases, the problem of best approximation of an operator can be solved in a manner similar to finding the best approximant to a signal in terms of orthogonal projections.

Eigenvalues and Eigenspaces

The particular values of λ for which the equation

$$\mathscr{L}\mathbf{x} = \lambda\mathbf{x} \tag{5.92}$$

has non-trivial solutions ($\mathbf{x} \neq \mathbf{0}$) are called the *eigenvalues* of $\mathscr{L}$. The corresponding solutions (usually normalized) are called *eigenvectors* (or *eigenfunctions* in $L^2(T)$). Equation (5.92) could possibly have a set of linearly independent solutions for one particular eigenvalue. If there are n_i linearly independent solutions for $\lambda = \lambda_i$, then λ_i is said to have a *multiplicity* of n_i. The n_i-dimensional subspace Ω_i spanned by these vectors is said to be an *invariant subspace* under $\mathscr{L}$, in the sense that $\mathbf{x} \in \Omega_i \Rightarrow \mathscr{L}\mathbf{x} \in \Omega_i$. This subspace is called the *eigenspace* for λ_i. In enumerating and ordering the eigenvalues and eigenvectors for $\mathscr{L}$, we can consider only the distinct values of λ satisfying (5.92) and their associated eigenspaces, or we can repeatedly enumerate each λ_i for n_i times and associate a single eigenvector with each eigenvalue. Both of these viewpoints have useful applications.

Operators which have the property that their eigenvectors span the domain have been classified as *simple* operators.[5] We shall consider certain subclasses of simple operators for which a relatively straightforward theory of representation can be developed.

The Adjoint Operator

For defining these subclasses of simple operators, the concept of the *adjoint operator* is very useful. The operator adjoint to $\mathscr{L}$ is defined as the mapping $\mathscr{L}'$ for which

$$(\mathscr{L}\mathbf{x}, \mathbf{y}) = (\mathbf{x}, \mathscr{L}'\mathbf{y}) \tag{5.93}$$

for all $\mathbf{x}$ and $\mathbf{y}$ in the domain. From this definition, the following properties of the adjoint operator are easily established.

$$\begin{aligned}
&\text{a.}\quad \mathscr{L}'(\alpha\mathbf{x} + \beta\mathbf{y}) = \alpha\mathscr{L}'\mathbf{x} + \beta\mathscr{L}'\mathbf{y} \quad \text{(linearity)}\\
&\text{b.}\quad (\mathscr{L}')' = \mathscr{L}\\
&\text{c.}\quad (\alpha\mathscr{L})' = \alpha^*\mathscr{L}'\\
&\text{d.}\quad \|\mathscr{L}'\| = \|\mathscr{L}\|\\
&\text{e.}\quad \|\mathscr{L}'\mathscr{L}\| = \|\mathscr{L}\mathscr{L}'\| = \|\mathscr{L}\|^2\\
&\text{f.}\quad (\mathscr{L}_1\mathscr{L}_2)' = \mathscr{L}_2'\mathscr{L}_1'
\end{aligned} \tag{5.94}$$

Also, if $L(t, s)$ is a kernel function representation for $\mathscr{L}$, then the kernel function $L'(t, s)$ for $\mathscr{L}'$ is given by

$$L'(t, s) = L^*(s, t) \tag{5.95}$$

If $\mathscr{L}$ is a simple operator with eigenvectors $\{\boldsymbol{\varphi}_i\}$, then the eigenvectors of $\mathscr{L}'$ are the reciprocal set $\{\boldsymbol{\theta}_i\}$ and the corresponding eigenvalues of $\mathscr{L}'$ are complex conjugates of the eigenvalues of $\mathscr{L}$. To demonstrate this, we consider, for an arbitrary $\mathbf{x}$, the inner product

$$\begin{aligned}
(\mathscr{L}'\boldsymbol{\theta}_i, \mathbf{x}) &= \left(\mathscr{L}'\boldsymbol{\theta}_i, \sum_j (\mathbf{x}, \boldsymbol{\theta}_j)\boldsymbol{\varphi}_j\right) = \sum_j (\boldsymbol{\theta}_j, \mathbf{x})(\mathscr{L}'\boldsymbol{\theta}_i, \boldsymbol{\varphi}_j)\\
&= \sum_j (\boldsymbol{\theta}_j, \mathbf{x})(\boldsymbol{\theta}_i, \mathscr{L}\boldsymbol{\varphi}_j) = \sum_j \lambda_j^*(\boldsymbol{\theta}_j, \mathbf{x})(\boldsymbol{\theta}_i, \boldsymbol{\varphi}_j) = \lambda_i^*(\boldsymbol{\theta}_i, \mathbf{x})
\end{aligned} \tag{5.96}$$

Since $(\mathscr{L}'\boldsymbol{\theta}_i, \mathbf{x}) = (\lambda_i^*\boldsymbol{\theta}_i, \mathbf{x})$ for arbitrary $\mathbf{x}$, it follows that $\mathscr{L}'\boldsymbol{\theta}_i = \lambda_i^*\boldsymbol{\theta}_i$.

The subclass of simple operators which we shall investigate is the *normal* operators, which are those operators which commute with their adjoints; i.e., if $\mathscr{L}\mathscr{L}' = \mathscr{L}'\mathscr{L}$, then $\mathscr{L}$ is normal. If an operator is normal, its eigenvectors are identical to the eigenvectors of its adjoint; i.e., $\mathscr{L}\mathbf{x} = \lambda\mathbf{x} \Rightarrow \mathscr{L}'\mathbf{x} = \lambda^*\mathbf{x}$. To show this, consider the operator $\mathscr{K} = \mathscr{L} - \lambda\mathscr{I}$, which is normal if $\mathscr{L}$ is, and $\mathscr{K}' = \mathscr{L}' - \lambda^*\mathscr{I}$.

$$\begin{aligned}\|\mathscr{K}\mathbf{x}\|^2 &= (\mathscr{K}\mathbf{x}, \mathscr{K}\mathbf{x}) = (\mathbf{x}, \mathscr{K}'\mathscr{K}\mathbf{x}) \\ &= (\mathbf{x}, \mathscr{K}\mathscr{K}'\mathbf{x}) = (\mathscr{K}'\mathbf{x}, \mathscr{K}'\mathbf{x}) = \|\mathscr{K}'\mathbf{x}\|^2 \end{aligned} \tag{5.97}$$

Hence, $\mathscr{K}\mathbf{x} = \mathbf{0}$ if, and only if, $\mathscr{K}'\mathbf{x} = \mathbf{0}$. Since the eigenvectors of $\mathscr{L}$ and $\mathscr{L}'$ are reciprocal and identical for a normal operator, we would expect that the eigenvectors could be made into an orthonormal set. It is a simple matter to show that eigenvectors for distinct eigenvalues are orthogonal, since

$$\lambda_i(\boldsymbol{\varphi}_i, \boldsymbol{\varphi}_j) = (\mathscr{L}\boldsymbol{\varphi}_i, \boldsymbol{\varphi}_j) = (\boldsymbol{\varphi}_i, \mathscr{L}'\boldsymbol{\varphi}_j) = (\boldsymbol{\varphi}_i, \lambda_j^*\boldsymbol{\varphi}_j) = \lambda_j(\boldsymbol{\varphi}_i, \boldsymbol{\varphi}_j) \tag{5.98}$$

Hence, if $\lambda_i \neq \lambda_j$ in (5.98), then we must have $(\boldsymbol{\varphi}_i, \boldsymbol{\varphi}_j) = 0$. In the case of eigenvalues with a multiplicity greater than one, we can construct an orthonormal set (e.g., by the Gram-Schmidt procedure) which spans the corresponding eigenspace. We can say that the eigenspaces are *pairwise orthogonal*; i.e., $(\mathbf{x}_i, \mathbf{x}_j) = 0$ for $\mathbf{x}_i \in \Omega_i$ and $\mathbf{x}_j \in \Omega_j$, $i \neq j$. Summarizing these results, we have established the important fact that the eigenvectors of a normal operator form a complete orthonormal set.

Exercise 5.13. Verify the properties (5.94) and (5.95) for adjoint operators.

Exercise 5.14. Write the impulse response and show the network realizations for the operators adjoint to the ones shown in Figure 5.6. Under what conditions on $h(t)$ and $w(t)$ are these operators normal?

Exercise 5.15. Show that a time-invariant operator is normal.

Spectral Representation of Normal Degenerate Operators

We now apply these ideas to the representation of degenerate operators on $L^2(-\infty, \infty)$. As before, we express the kernel function of a rank-n operator $\mathscr{L}_n$ by

$$L_n(t, s) = \sum_{i=1}^{n} \psi_i(t)\theta_i^*(s) \tag{5.99}$$

where the $\{\psi_i(t); i = 1, 2, \ldots, n\}$ are linearly independent functions which span a subspace M_n. It is clear that M_n is an invariant subspace under $\mathscr{L}_n$ and that the eigenvectors with non-zero eigenvalues are to be found only in this subspace. Now if $\mathscr{L}_n$ is also a normal operator, from (5.95) we have

$$L'_n(t, s) = \sum_{i=1}^{n} \theta_i(t)\psi_i^*(s) \tag{5.100}$$

Hence the set $\{\theta_i(t); i = 1, 2, \ldots, n\}$ also spans M_n. It follows that the null space of $\mathscr{L}_n$ is the orthogonal complement of M_n since any vector orthogonal to all the $\{\boldsymbol{\theta}_i\}$ must map into zero. This infinite-dimensional subspace is the eigenspace Ω_0 for $\lambda = \lambda_0 = 0$. The problem is thus reduced to a finite-dimensional problem of finding $\mathbf{x} \in M_n$ which are solutions of $\mathscr{L}_n\mathbf{x} = \lambda\mathbf{x}$. This is equivalent to

$$(\mathscr{L}_n\mathbf{x}, \boldsymbol{\theta}_i) - \lambda(\mathbf{x}, \boldsymbol{\theta}_i) = 0; \qquad i = 1, 2, \ldots, n \tag{5.101}$$

With

$$\mathbf{x} = \sum_{j=1}^{n} \alpha_j \boldsymbol{\varphi}_j$$

($\{\boldsymbol{\varphi}_j\}$ reciprocal to $\{\boldsymbol{\theta}_i\}$), (5.101) is equivalent to

$$(\mathbf{L} - \lambda\mathbf{I})\boldsymbol{\alpha} = \mathbf{0} \tag{5.102}$$

where $\mathbf{I}$ is the $n \times n$ identity matrix and $\mathbf{L}$ is an $n \times n$ matrix with elements $\lambda_{ij} = (\boldsymbol{\psi}_j, \boldsymbol{\theta}_i)$. From the theory of linear equations,[4] we know that (5.102) has non-trivial solutions if, and only if, the determinant of $\mathbf{L} - \lambda\mathbf{I}$ is equal to zero. This determinant is an nth-degree polynomial in λ called the *characteristic polynomial* for $\mathbf{L}$. The non-zero eigenvalues of $\mathscr{L}_n$ are thus the roots of the equation

$$P(\lambda) = \det\,[\mathbf{L} - \lambda\mathbf{I}] = 0 \tag{5.103}$$

The corresponding eigenspaces are determined by the solutions of (5.102). The set of eigenvalues, which will provide the desired numerical representation for the operator, is called the *spectrum* of the operator (more accurately, the *point spectrum*[2]).

Since the eigenspaces are pairwise orthogonal, the entire space can be expressed as the direct sum of the eigenspaces and, for arbitrary $\mathbf{x} \in L^2(-\infty, \infty)$, we have

$$\begin{aligned} \mathscr{L}_n\mathbf{x} &= \mathscr{L}_n(\mathbf{x}_0 + \mathbf{x}_1 + \mathbf{x}_2 + \cdots + \mathbf{x}_m) \\ &= \sum_{i=0}^{m} \lambda_i\mathbf{x}_i = \sum_{i=1}^{m} \lambda_i\mathbf{x}_i \qquad (\text{since } \lambda_0 = 0) \end{aligned} \tag{5.104}$$

where $\mathbf{x}_i \in \Omega_i$ is the orthogonal projection of $\mathbf{x}$ onto Ω_i and we have assumed that $P(\lambda)$ has $m \leqslant n$ distinct roots. The orthogonal projections can be characterized by projection operators $\mathscr{P}_i$ (see Exercise 5.9), where $\mathscr{P}_i\mathbf{x} = \mathbf{x}_i$ for any $\mathbf{x}$ in $L^2(-\infty, \infty)$. Finally, $\mathscr{L}_n$ can be expressed as

$$\mathscr{L}_n = \sum_{i=1}^{m} \lambda_i\mathscr{P}_i \tag{5.105}$$

This form of representation is called the *spectral representation* for the operator $\mathscr{L}_n$. Another form of spectral representation is obtained by repeated enumeration of the multiple eigenvalues so that the spectrum is $\{\lambda_i; i = 1, 2, \ldots, n\}$. With $\{\boldsymbol{\omega}_i; i = 1, 2, \ldots, n\}$ as the associated set of orthonormalized eigenvectors, $\mathscr{L}_n$ can be expressed as

$$\mathscr{L}_n = \sum_{i=1}^{n} \lambda_i \mathscr{Q}_i \tag{5.106}$$

where each of the $\mathscr{Q}_i$ are orthogonal projection operators with one-dimensional range; $\mathscr{Q}_i \cdot x(t) = (\mathbf{x}, \boldsymbol{\omega}_i)\omega_i(t)$ for any $\mathbf{x}$ in $L^2(-\infty, \infty)$. The kernel function for the degenerate operator $\mathscr{L}_n$ corresponding to a spectral representation is given simply by

$$L_n(t, s) = \sum_{i=1}^{n} \lambda_i \omega_i(t) \omega_i^*(s) \tag{5.107}$$

To summarize the preceding development, we have shown that for a normal, rank-n operator it is possible to find an orthonormal basis for the domain such that the image of any point in the domain is given simply by the appropriate scaling of its components individually. The scaling factors are the eigenvalues and these are obtained as roots of an nth-degree characteristic polynomial. We note that these roots might be complex, even if $\mathbf{L}$ is real, and this provides a strong motivation for the use of complex signal spaces.

Spectral Representation of Normal Compact Operators

The logical step in generalization from degenerate operators is to go to compact operators on $L^2(-\infty, \infty)$. Since we have shown that compact operators can be approximated by degenerate operators, we would expect to find many similarities in the spectral representation for a compact operator. In fact, the only dissimilarity is that, in general, the compact operator does not have only a finite number of non-zero eigenvalues. We can show, however, that the eigenvalues must tend toward zero in the limit as i approaches infinity. If, on the contrary, the λ_i did not tend to zero, there would be an infinite number of λ_i such that $|\lambda_i| \geqslant \delta > 0$. The linearly independent set $\{\mathbf{y}_i; \mathbf{y}_i = \mathscr{L}\mathbf{x}_i = \lambda \mathbf{x}_i, \|\mathbf{x}_i\| = 1\}$ would therefore not be totally bounded and hence $\mathscr{L}$ would not be a compact operator. As a corollary of this, we can say that for a compact operator the multiplicity of any non-zero eigenvalue is finite. The spectral representation of a normal, compact operator is

$$\mathscr{L} = \sum_{i=1}^{\infty} \lambda_i \mathscr{Q}_i \tag{5.108}$$

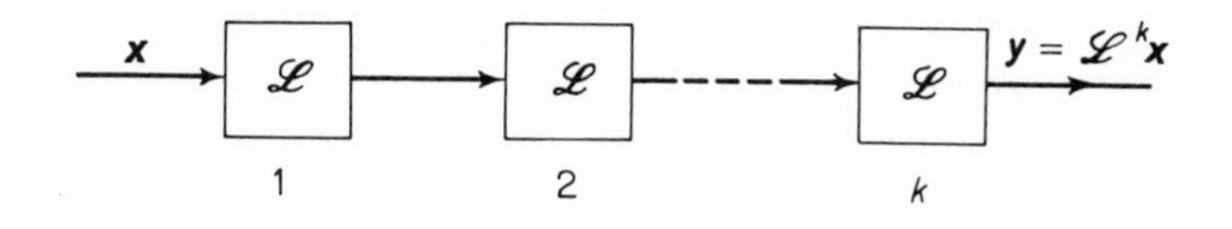

$$\mathscr{L} = \sum_{i=1}^{\infty} \lambda_i \mathscr{Q}_i \Rightarrow \mathscr{L}^k = \sum_{i=1}^{\infty} \lambda_i^k \mathscr{Q}_i$$

Figure 5.10. Spectral representation of cascade of identical networks.

which is a direct extension of (5.106). The kernel function for $\mathscr{L}$, in terms of the eigenvectors, is given by

$$L(t, s) = \sum_{i=1}^{\infty} \lambda_i \omega_i(t) \omega_i^*(s) \tag{5.109}$$

which is a Hilbert-Schmidt kernel provided that

$$\sum_{i=1}^{\infty} |\lambda_i|^2 < \infty$$

It is often useful to order the eigenvalues so that their magnitudes form a non-increasing sequence

$$|\lambda_1| \geqslant |\lambda_2| \geqslant |\lambda_3| \geqslant \cdots \tag{5.110}$$

When this is done, the leading eigenvalue takes on a special significance since it is the norm of the operator. To show this, let

$$x(t) = \sum_{k=1}^{\infty} \alpha_k \varphi_k(t)$$

then

$$\|\mathscr{L}\mathbf{x}\|^2 = (\mathscr{L}\mathbf{x}, \mathscr{L}\mathbf{x}) = \sum_{k=1}^{\infty} |\lambda_k|^2 |\alpha_k|^2 \tag{5.111}$$

Under the constraint $\|\mathbf{x}\| = 1$, which implies that

$$\sum_{k=1}^{\infty} |\alpha_k|^2 = 1 \tag{5.112}$$

the sum in (5.111) is maximized by $|\alpha_1| = 1$, $|\alpha_k| = 0$, $k > 1$. Thus we have

$$\|\mathscr{L}\| = |\lambda_1| \tag{5.113}$$

One of the practical advantages of a spectral representation is readily apparent in problems involving a cascade structure of many identical networks. This is because the k-fold product of $\mathscr{L}$ in (5.108) is simply

$$\mathscr{L}^k = \sum_{i=1}^{\infty} \lambda_i^k \mathscr{Q}_i \tag{5.114}$$

A more general structure than that shown in Figure 5.10 is the *transversal*

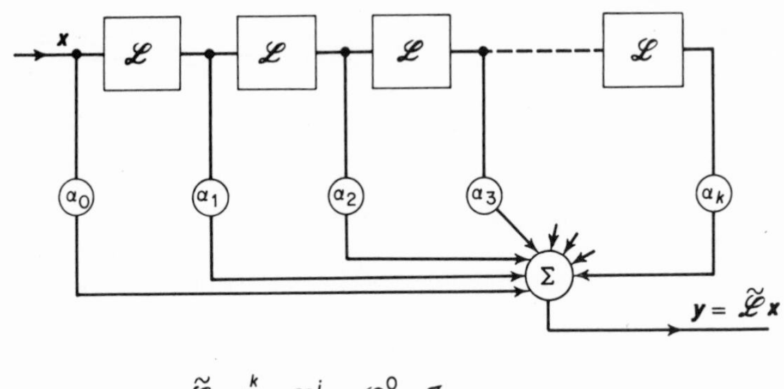

Figure 5.11. Spectral representation for transversal filter.

filter,[13] frequently employed in signal processing operations (Figure 5.11).

Using (5.108) to represent the individual networks in the cascade, we find

$$\tilde{\mathscr{L}} = \sum_{j=0}^{k} \alpha_j \mathscr{L}^j = \sum_{i=1}^{\infty} F(\lambda_i)\mathscr{Q}_i \tag{5.115}$$

where the kth-degree polynomial

$$F(z) = \sum_{j=0}^{k} \alpha_j z^j$$

gives the eigenvalues of the transversal filter in terms of the eigenvalues of the basic element. This provides a convenient structure for the physical realization of a class of operators since it is usually a simple matter to implement the structure with adjustable tap gain coefficients α_j. With this flexibility, any k of the eigenvalues of $\tilde{\mathscr{L}}$ can be arbitrarily set, provided that the corresponding eigenvalues of $\mathscr{L}$ are distinct.

At this point we take up the problem of finding the best approximation to a compact operator in terms of a rank-n degenerate operator. For a certain subclass of normal operators, the spectral representation leads to a neat answer along with a simple expression for the approximation error. This subclass consists of the *self-adjoint* operators, defined by $\mathscr{L}' = \mathscr{L}$. We first note that the eigenvalues of a self-adjoint operator are real; since a self-adjoint operator is normal, $\mathscr{L}\mathbf{x}_i = \lambda_i \mathbf{x}_i \Rightarrow \mathscr{L}'\mathbf{x}_i = \lambda_i^* \mathbf{x}_i$, hence $\lambda_i = \lambda_i^*$. We suppose that the eigenvalues have been ordered in a non-increasing sequence of magnitudes as in (5.110). As an approximant $\mathscr{L}_n$, to a compact, self-adjoint operator $\mathscr{L}$, we simply truncate the spectral representation after the nth term.

$$\mathscr{L} = \sum_{i=1}^{\infty} \lambda_i \mathscr{Q}_i; \qquad \mathscr{L}_n = \sum_{i=1}^{n} \lambda_i \mathscr{Q}_i$$

$$\mathscr{L} - \mathscr{L}_n = \sum_{i=n+1}^{\infty} \lambda_i \mathscr{Q}_i \tag{5.116}$$

It can be shown[3] that this is a best approximant in the sense that

$$\|\mathscr{L} - \mathscr{L}_n\| \leqslant \|\mathscr{L} - \tilde{\mathscr{L}}_n\| \tag{5.117}$$

where $\tilde{\mathscr{L}}_n$ is an arbitrary rank-n operator. Furthermore, from (5.113), we have

$$\|\mathscr{L} - \mathscr{L}_n\| = |\lambda_{n+1}| \tag{5.118}$$

which, since $\{\lambda_i\}$ is a sequence tending to zero, indicates how large the rank of the approximant has to be in order to meet a prescribed approximation error.

As a final comment on spectral representation, we briefly examine some operators which are not compact. A more general spectral theory encounters several difficulties; yet some of the foregoing ideas are conceptually useful in dealing with the more general situations. For example, the gating operator (5.42),

$$\mathscr{L} \cdot x(t) = w(t)x(t) \tag{5.119}$$

would appear, intuitively, to have eigenfunctions given by

$$x_i(t) = \delta(t - t_i) \tag{5.120}$$

with corresponding eigenvalues $\lambda_i = w(t_i)$, although these functions are not contained in $L^2(-\infty, \infty)$. We are tempted to say that the spectrum of $\mathscr{L}$ is the gating function itself and we note that the eigenvalues could be a non-denumerable set, continuously distributed over an interval of the real line.

Similarly, for the time-invariant operator (5.37),

$$\mathscr{L} \cdot x(t) = \int_{-\infty}^{\infty} h(t - \tau)x(\tau)\, d\tau \tag{5.121}$$

the functions $x_i(t) = e^{j2\pi f_i t}$, although not in $L^2(-\infty, \infty)$, are invariant under the operation (5.121).

$$\begin{aligned}\mathscr{L} \cdot x_i(t) &= \int_{-\infty}^{\infty} h(t - \tau)e^{j2\pi f_i \tau}\, d\tau \\ &= \int_{-\infty}^{\infty} h(\sigma)e^{j2\pi f_i (t-\sigma)}\, d\sigma \\ &= H(f_i)x_i(t)\end{aligned} \tag{5.122}$$

Hence the spectrum of $\mathscr{L}$ is the transfer function of the operator. Because of this, we have become quite accustomed to thinking of the frequency domain as providing a spectral representation for time-invariant operators.

Exercise 5.16. State the conditions on the sets of functions $\{\psi_i(t); i = 1, 2, \ldots, n\}$ and $\{\theta_i(t); i = 1, 2, \ldots, n\}$ in (5.99) in order that $\mathscr{L}_n$ be a normal operator.

Exercise 5.17. Show that the eigenvalues of a normal degenerate operator are a "coordinate-free" property of the operator; i.e., the characteristic polynomial is invariant to a change of basis. *Hint:* Use the results of Exercise 5.8 and the following property of determinants: det **AB** = (det **A**)(det **B**).

Exercise 5.18. Show that an orthogonal projection operator is self-adjoint.

REFERENCES

I. General references on linear operators

1. G. F. Simmons, *Introduction to Topology and Modern Analysis*, McGraw-Hill, 1963.
2. N. I. Akhieser and I. M. Glazman, *Theory of Linear Operators in Hilbert Space*, Vol. I, Frederick Ungar, 1961.
3. F. Riesz and B. Sz.-Nagy, *Functional Analysis*, Frederick Ungar, 1955.
4. R. Courant and D. Hilbert, *Methods of Mathematical Physics*, Vol. I, Interscience, 1953.
5. L. A. Zadeh and C. A. Desoer, *Linear System Theory* (Appendix C) McGraw-Hill, 1963.

II. References to specific topics

6. L. A. Zadeh, "Circuit Analysis of Linear Varying-Parameter Networks," *Jour. Appl. Phys.*, Vol. 21, pp. 1171–77 (1950).
7. A. Gersho and N. DeClaris, "Duality Concepts in Time-varying Linear Systems," *1964 IEEE Int'l Conv. Record, Pt. 1*, pp. 344–356.
8. D. C. Youla, "The Synthesis of Linear Dynamical Systems from Prescribed Weighting Patterns," *J. SIAM*, Vol. 14, pp. 527–49 (May, 1966).
9. E. I. Jury, *Theory and Application of the z-Transform Method*, John Wiley & Sons, 1964.
10. L. M. Silverman and H. E. Meadows, "Equivalence and Synthesis of Time-Variable Linear Systems," *Proc. Fourth Annual Allerton Conference on Circuit and System Theory*, pp. 776–84, 1966.
11. H. E. Meadows, L. M. Silverman, and L. E. Franks, "A Canonical Network for Periodically Variable Linear Systems," *Proc. Fourth Annual Allerton Conference on Circuit and System Theory*, pp. 649–58, 1966.
12. L. E. Franks and I. W. Sandberg, "An Alternative Approach to the Realization of Network Transfer Functions: The *N*-Path Filter," *Bell Sys. Tech. Jour.*, Vol. 39, pp. 1321–50 (September, 1960).
13. H. E. Kallmann, "Transversal Filters," *Proc. IRE*, Vol. 38, pp. 302–311 (July, 1940).

CHARACTERIZATION OF SIGNAL PROPERTIES

6

6.1 Introduction

In contrast to representation of a signal, we are often concerned with the problem of characterization of a signal by associating with it a few significant parameters (e.g., bandwidth, energy content, peak value, duration, etc.) rather than obtaining a complete specification of its character. Thus, as with representation, we seek appropriate mappings from the signal space into numerical values; however, the linear functionals, which are so convenient for the representation problem, may not be appropriate for defining the parameters desired in characterizing signal properties. The primary motivation for parameterizing signals is the desire to ascribe performance measures by which various systems can be compared. A suitable characterization of signal properties at various points in the system lays the groundwork for the problem of selection of signals for optimum system performance.

The limitation of linear functionals for signal characterization results from the many-to-one nature of this mapping. We recall that a linear functional can be expressed as an inner product $f_y(\mathbf{x}) = (\mathbf{x}, \mathbf{y})$; hence any signal orthogonal to $\mathbf{y}$ when added to $\mathbf{x}$ gives a signal which maps into the same value as $\mathbf{x}$ does. Signal optimization problems normally involve finding extremal values of a functional. If the signals are characterized only by linear functionals, then signals which assign arbitrary values to the functionals can be found. To avoid these limitations, we want to employ other types of functionals which exhibit definite maximum and minimum values, depending on the signal. In this chapter we shall consider a particular type of functional called the *quadratic functional.* These functionals have several desirable properties. A principal advantage is the high degree of mathematical tractability, particularly in establishing conditions for extremal values and also in the ease with which linear transformations on the signal can be accommodated in the expression of the functional. Another advantage is that quadratic functionals express quantities which have an easily understood physical significance and which have a long history of usage in communication and control system problems.

Duration-bandwidth Product

As an introductory example to illustrate these ideas, we shall consider what is perhaps regarded as the classical problem in signal optimization, that of minimizing the duration-bandwidth product of a signal.[1] Let us borrow from probability theory the concept of measuring the width of a probability density function in terms of its second central moment. Unlike a probability density function, a signal $x(t)$ is not necessarily non-negative and the second moment of $x(t)$ may be meaningless for establishing the width of a signal having a highly oscillatory behavior. To avoid this problem, we consider the second central moment of $x^2(t)$ and since the zeroth moment of $x^2(t)$ is not necessarily unity, we normalize the second moment giving as a measure of signal duration T_x the quantity

$$T_x^2 = \frac{\int_{-\infty}^{\infty} (t - t_0)^2 x^2(t)\, dt}{\int_{-\infty}^{\infty} x^2(t)\, dt} \tag{6.1}$$

where

$$t_0 = \frac{\int_{-\infty}^{\infty} t x^2(t)\, dt}{\int_{-\infty}^{\infty} x^2(t)\, dt}$$

is the normalized first moment of $x^2(t)$.

Another analogy is apparent when we consider the second moment as the moment of inertia of a mass distribution given by $x^2(t)$. For this reason, T_x is often referred to as the "radius of gyration" of the signal $x(t)$. Similarly, we can measure the bandwidth occupied by a signal in terms of the second central moment of the square of the magnitude of its Fourier transform.

$$W_x^2 = \frac{\int_{-\infty}^{\infty} (f - f_0)^2 \, |X(f)|^2\, df}{\int_{-\infty}^{\infty} |X(f)|^2\, df} \tag{6.2}$$

where $f_0 = 0$ because of the symmetry of $X(f)$ about the origin for real signals. Since $|X(f)|$ is independent of a time shift in $x(t)$, we can take $t_0 = 0$ and express the duration-bandwidth product in terms of inner products;[1]

$$(T_x W_x)^2 = \frac{(t\mathbf{x}, t\mathbf{x})(f\mathbf{X}, f\mathbf{X})}{(\mathbf{x}, \mathbf{x})(\mathbf{X}, \mathbf{X})} \tag{6.3}$$

[1] We shall use $t\mathbf{x}$ and $f\mathbf{X}$ to denote the functions $tx(t)$ and $fX(f)$, respectively. Also, the symbol $\mathbf{wx}$, used later, is meant to denote the product of functions, $w(t)x(t)$.

The inner products in (6.3) expressed in the frequency domain can be converted to inner products expressed in the time domain by means of the Parseval relation (4.14): $(\mathbf{x}, \mathbf{y}) = (\mathbf{X}, \mathbf{Y})$. Since $j2\pi f X(f)$ is the Fourier transform of the time derivative of $x(t)$, (6.3) becomes

$$(T_x W_x)^2 = \left(\frac{1}{2\pi}\right)^2 \frac{(t\mathbf{x}, t\mathbf{x})(\dot{\mathbf{x}}, \dot{\mathbf{x}})}{(\mathbf{x}, \mathbf{x})^2} \tag{6.4}$$

Since T_x and W_x are independent of $\|\mathbf{x}\| = (\mathbf{x}, \mathbf{x})^{\frac{1}{2}}$, we can let $\|\mathbf{x}\| = 1$ and apply the Schwarz inequality to (6.4) with the result that

$$2\pi T_x W_x = \|t\mathbf{x}\| \; \|\dot{\mathbf{x}}\| \geqslant |(t\mathbf{x}, \dot{\mathbf{x}})| \tag{6.5}$$

Integration by parts shows that the right-hand term in (6.5) is independent of $x(t)$ under the constraint that $\|\mathbf{x}\| = 1$.

$$\int_{-\infty}^{\infty} tx(t)\dot{x}(t)\, dt = \frac{1}{2}\int_{-\infty}^{\infty} t \frac{d}{dt}[x^2(t)]\, dt = -\frac{1}{2}\int_{-\infty}^{\infty} x^2(t)\, dt = -\tfrac{1}{2}\|\mathbf{x}\|^2 \tag{6.6}$$

Thus, the minimum value of the duration-bandwidth product is given by

$$(T_x W_x)_{\min} = \frac{1}{4\pi} \tag{6.7}$$

and this value is attained by a signal for which the equality in (6.5) holds; i.e.,

$$\dot{x}(t) = c_1 t x(t) \Rightarrow \frac{d}{dt}[\log x(t)] = c_1 t$$

$$\Rightarrow x(t) = c_2 e^{\frac{c_1 t^2}{2}}; \qquad c_1 < 0 \tag{6.8}$$

From this result we conclude that, in the sense specified by (6.1) and (6.2), the *Gaussian pulse* is the signal exhibiting the smallest duration-bandwidth product.

The extremization in this example was especially simple because the Schwarz inequality could be applied directly. In more complicated examples, we might not be so fortunate. In the following sections, we develop a generally applicable method for solving signal optimization problems of this type, retaining only the quadratic nature of the functionals. The communication and control system theory literature contains many interesting examples of this type of problem. A generalized approach allows considerably more insight and simplification of the effort. We shall see that certain necessary conditions to be satisfied by the solution can be written by inspection once the problem is properly formulated. In the following, we focus attention on characterization of signal properties in the time domain and in the frequency domain because of prominent usage of these concepts, but it should be remembered that these techniques are valid for any linear representation of signals.

6.2 Quadratic Functionals

The quadratic functional is derived from the *bilinear functional* which is a mapping of pairs of signals $\mathbf{x}$ and $\mathbf{y}$ into numerical values $f(\mathbf{x}, \mathbf{y})$ and has the following properties:

$$\begin{aligned} f(\alpha_1\mathbf{x}_1 + \alpha_2\mathbf{x}_2, \mathbf{y}) &= \alpha_1 f(\mathbf{x}_1, \mathbf{y}) + \alpha_2 f(\mathbf{x}_2, \mathbf{y}) \\ f(\mathbf{x}, \beta_1\mathbf{y}_1 + \beta_2\mathbf{y}_2) &= \beta_1^* f(\mathbf{x}, \mathbf{y}_1) + \beta_2^* f(\mathbf{x}, \mathbf{y}_2) \end{aligned} \tag{6.9}$$

For example, the inner product $f(\mathbf{x}, \mathbf{y}) = (\mathbf{x}, \mathbf{y})$ is a bilinear functional. For a real space, $\mathbf{f}$ is linear in each of the arguments $\mathbf{x}$ and $\mathbf{y}$. A bilinear functional is said to be bounded if we can find a real number K such that

$$|f(\mathbf{x}, \mathbf{y})| \leqslant K \, \|\mathbf{x}\| \, \|\mathbf{y}\| \tag{6.10}$$

for all $\mathbf{x}$ and $\mathbf{y}$. The greatest lower bound for K provides a norm for the functional.

$$\begin{aligned} \|\mathbf{f}\| &= \inf \{K; |f(\mathbf{x}, \mathbf{y})| \leqslant K \, \|\mathbf{x}\| \, \|\mathbf{y}\|\} \\ &= \sup \{|f(\mathbf{x}, \mathbf{y})|; \|\mathbf{x}\| = 1, \|\mathbf{y}\| = 1\} \end{aligned} \tag{6.11}$$

Note that $f_A(\mathbf{x}, \mathbf{y}) = (\mathscr{A}\mathbf{x}, \mathbf{y})$ also defines a bilinear functional in $\mathbf{x}$ and $\mathbf{y}$. Moreover, it can be shown that any bounded bilinear functional can be written in this form and that

$$\|\mathbf{f}_A\| = \|\mathscr{A}\| \tag{6.12}$$

where $\|\mathscr{A}\|$ is the norm of the linear operator $\mathscr{A}$ as defined previously, (5.8).

We define the quadratic functional, with argument $\mathbf{x}$, simply by replacing $\mathbf{y}$ by $\mathbf{x}$ in $f(\mathbf{x}, \mathbf{y})$.

$$I_A(\mathbf{x}) = f_A(\mathbf{x}, \mathbf{x}) = (\mathscr{A}\mathbf{x}, \mathbf{x}) \tag{6.13}$$

The property of continuity can be easily shown from (6.9) for a bounded quadratic functional.

$$\begin{aligned} |I(\mathbf{x}) - I(\mathbf{x}_0)| &= |f(\mathbf{x}, \mathbf{x}) - f(\mathbf{x}_0, \mathbf{x}_0)| \\ &= |f(\mathbf{x} - \mathbf{x}_0, \mathbf{x}) + f(\mathbf{x}_0, \mathbf{x} - \mathbf{x}_0)| \\ &\leqslant |f(\mathbf{x} - \mathbf{x}_0, \mathbf{x})| + |f(\mathbf{x}_0, \mathbf{x} - \mathbf{x}_0)| \\ &\leqslant \|\mathbf{f}\| \, [\|\mathbf{x}\| + \|\mathbf{x}_0\|] \|\mathbf{x} - \mathbf{x}_0\| \end{aligned} \tag{6.14}$$

so that for any $\varepsilon > 0$, we can find a δ such that

$$\|\mathbf{x} - \mathbf{x}_0\| < \delta \Rightarrow |I(\mathbf{x}) - I(\mathbf{x}_0)| < \varepsilon$$

From this discussion, there is apparent a strong analogy between quadratic and linear functionals. We saw previously that the linear functionals formed a linear space (the conjugate space) and that to each element of this space $\mathbf{f}_y$ there corresponded a vector $\mathbf{y}$ such that $f_y(\mathbf{x}) = (\mathbf{x}, \mathbf{y})$. The analogous situation for quadratic functionals is that to each functional there corresponds a linear operator $\mathscr{A}$ such that $I_A(\mathbf{x}) = (\mathscr{A}\mathbf{x}, \mathbf{x})$. Since we know that the linear operators form a normed linear space, we are not surprised to find that the quadratic functionals also form a normed linear space, and from (6.12) this space is isometric to the space of linear operators.

An extremely useful concept in what follows is that of the *adjoint operator*, (5.93). If $\mathscr{A}'$ is a linear operator, then clearly $(\mathbf{x}, \mathscr{A}'\mathbf{y})$ is a bilinear functional. If

$$(\mathscr{A}\mathbf{x}, \mathbf{y}) = (\mathbf{x}, \mathscr{A}'\mathbf{y}) \tag{6.15}$$

for all $\mathbf{x}$ and $\mathbf{y}$, then $\mathscr{A}'$ is defined as the adjoint of $\mathscr{A}$. The adjoint operator is easily obtained from any representation of the operator. Let $\mathscr{A}$ be represented by the time-domain kernel function $A(t, \tau)$. Then,

$$\mathscr{A}\mathbf{x} = \int_T A(t, \tau)x(\tau)\, d\tau$$

and

$$I_A(\mathbf{x}) = \int_T \int_T A(t, \tau)x(\tau)x^*(t)\, d\tau\, dt \tag{6.16}$$

By interchanging the order of integration in (6.16), we see that the kernel for the adjoint operator is

$$A'(t, \tau) = A^*(\tau, t) \tag{6.17}$$

If $\mathscr{A}' = \mathscr{A} \mid [A(t, \tau) = A^*(\tau, t)]$, then $\mathscr{A}$ is said to be *self-adjoint*. It is also called *Hermitian* (or *symmetric* if $\mathscr{A}$ is real). A broader class, which includes self-adjoint operators, is the class of *normal* operators, which commute with their adjoint.

$$\mathscr{A}'\mathscr{A} = \mathscr{A}\mathscr{A}' \Rightarrow \int_T A^*(\sigma, t)A(\sigma, \tau)\, d\sigma = \int_T A(t, \sigma)A^*(\tau, \sigma)\, d\sigma \tag{6.18}$$

Normal operators have the property that

$$\|\mathscr{A}\mathbf{x}\| = \|\mathscr{A}'\mathbf{x}\|,$$

since

$$\begin{aligned} \|\mathscr{A}\mathbf{x}\|^2 &= (\mathscr{A}\mathbf{x}, \mathscr{A}\mathbf{x}) = (\mathscr{A}'\mathscr{A}\mathbf{x}, \mathbf{x}) = (\mathscr{A}\mathscr{A}'\mathbf{x}, \mathbf{x}) \\ &= (\mathscr{A}'\mathbf{x}, \mathscr{A}'\mathbf{x}) = \|\mathscr{A}'\mathbf{x}\|^2 \end{aligned}$$

An operator is said to be *positive-definite* if

$$(\mathscr{A}\mathbf{x}, \mathbf{x}) > 0 \qquad \text{for all } \mathbf{x} \text{ except } \mathbf{x} = \mathbf{0} \tag{6.19}$$

For example, $\mathscr{A}'\mathscr{A}$ is positive-definite if $\mathscr{A}$ is nonsingular since $(\mathscr{A}'\mathscr{A}\mathbf{x}, \mathbf{x}) = (\mathscr{A}\mathbf{x}, \mathscr{A}\mathbf{x}) = \|\mathscr{A}\mathbf{x}\|^2$.

A linear transformation on the argument function of a quadratic functional can be handled simply by means of the adjoint operator. Let $\mathbf{x} = \mathscr{L}\mathbf{y}$, then

$$(\mathscr{A}\mathbf{x}, \mathbf{x}) = (\mathscr{A}\mathscr{L}\mathbf{y}, \mathscr{L}\mathbf{y}) = (\mathscr{L}'\mathscr{A}\mathscr{L}\mathbf{y}, \mathbf{y}) \tag{6.20}$$

so that the operator $\mathscr{L}'\mathscr{A}\mathscr{L}$ identifies the same quadratic functional with argument **y** instead of **x**.

A *unitary* operator is characterized by the following property:

$$(\mathscr{U}\mathbf{x}, \mathscr{U}\mathbf{y}) = (\mathbf{x}, \mathbf{y}) \tag{6.21}$$

which implies that $\mathscr{U}'\mathscr{U} = \mathscr{U}\mathscr{U}' = \mathscr{I}$. For our purposes, it is important to note that the Fourier transform corresponds to a unitary transformation.

$$X(f) = \mathscr{F} \cdot x(t) = \int_{-\infty}^{\infty} x(\tau)e^{-j2\pi f\tau}\, d\tau$$

$$(\mathbf{X}, \mathbf{Y}) = (\mathscr{F}'\mathscr{F}\mathbf{x}, \mathbf{y}) \tag{6.22}$$

$$\mathscr{F}'\mathscr{F}\mathbf{x} = \iint_{-\infty}^{\infty} e^{j2\pi f(t-\tau)} x(\tau)\, d\tau\, df$$

where

$$\int_{-\infty}^{\infty} e^{j2\pi f(t-\tau)}\, df = \delta(t - \tau)$$

corresponds to the identity operator. From (6.22) we again obtain the *Parseval relation*, (4.14):

$$(\mathbf{x}, \mathbf{y}) = (\mathbf{X}, \mathbf{Y}) \tag{6.23}$$

By application of (6.20) and (6.22), we can express a given quadratic functional in either the time domain or the frequency domain by suitable transformations on the operators. If

$$I_A(\mathbf{x}) = (\mathscr{A}\mathbf{x}, \mathbf{x}) = (\mathscr{B}\mathbf{X}, \mathbf{X})$$

then

$$B(f, \nu) = \iint_{-\infty}^{\infty} A(t, \tau)e^{-j2\pi ft}e^{+j2\pi\nu\tau}\, dt\, d\tau$$

and (6.24)

$$A(t, \tau) = \iint_{-\infty}^{\infty} B(f, \nu)e^{j2\pi ft}e^{-j2\pi\nu\tau}\, df\, d\nu$$

With (6.24) we have the mechanism for shuttling back and forth between the time domain and the frequency domain for the characterization of signal properties. It must be remembered, however, that the inner products in (6.22), (6.23), and (6.24) are valid only for integration intervals of $(-\infty, \infty)$ in both domains. In problems involving characterization over a semi-infinite or finite interval, we prefer to retain the simplicity of (6.23) and (6.24) by using the infinite interval and applying appropriate modifications to the operators to account for the limited intervals. This manipulation will be encountered in several of the examples to follow.

Finally, we mention that the operators $\mathscr{B}$ and $\mathscr{A}$ in (6.24) have many properties in common. In fact, each of the properties described above (self-adjoint, normal, unitary, and positive-definite) is possessed by one operator if, and only if, it is possessed by the other. This similarity results from the unitary nature of the Fourier transform and the relation between $\mathscr{B}$ and $\mathscr{A}$, ($\mathscr{A} = \mathscr{F}'\mathscr{B}\mathscr{F} = \mathscr{F}^{-1}\mathscr{B}\mathscr{F}$) is referred to as a *similarity* transformation.

6.3 Some Specific Time-Domain and Frequency-Domain Quadratic Functionals

For convenience in later discussion, it is worthwhile to examine a few specific quadratic functionals which find frequent application in signal optimization problems.

Let

$$I_1 = \int_{-\infty}^{\infty} |x(t)|^2\, dt = \int_{-\infty}^{\infty} |X(f)|^2\, df$$
$$= (\mathbf{x}, \mathbf{x}) = (\mathbf{X}, \mathbf{X}) = \|\mathbf{x}\|^2 \qquad (6.25)$$

Thus the square of the norm of a signal is a quadratic functional corresponding to the identity operator. As noted previously, this is commonly referred to as the energy content of a signal. In some problems it is important to distinguish this from the physical concept of energy (see Example 6.5).

The functional

$$I_2 = (\mathbf{wx}, \mathbf{x}) = \int_{-\infty}^{\infty} w(t)\, |x(t)|^2\, dt \qquad (6.26)$$

with $w(t)$ real, represents an arbitrarily weighted measure of energy in the time domain. This functional is used to characterize various aspects of concentration of a signal in the time domain. The corresponding operator $A_2(t, \tau)$ has a "diagonal" nature and the weighting can be manipulated to show that the operator is self-adjoint.

$$A_2(t, \tau) = w(t)\, \delta(t - \tau) = w^{\frac{1}{2}}(t) w^{\frac{1}{2}}(\tau)\, \delta(t - \tau) \qquad (6.27)$$

Also, if $w(t) > 0$, for all t, then I_2 is positive-definite. Similarly,

$$I_3 = (\mathbf{VX}, \mathbf{X}) = \int_{-\infty}^{\infty} V(f)\, |X(f)|^2\, df \qquad (6.28)$$

with $V(f)$ real, represents an arbitrarily weighted measure of energy in the frequency domain. I_3 is self-adjoint and if $V(f) > 0$ for all f, it is also positive-definite.

Table 6.1 Some Quadratic Functionals and their Time- and Frequency-Domain Operators

I	$(\mathscr{A}\mathbf{x}, \mathbf{x}) = (\mathscr{B}\mathbf{X}, \mathbf{X})$	$A(t, \tau)$	$B(f, \nu)$
I_1	$(\mathbf{x}, \mathbf{x}) = (\mathbf{X}, \mathbf{X})$	$\delta(t-\tau)$	$\delta(f-\nu)$
I_2	$(\mathbf{wx}, \mathbf{x}) = \int_{-\infty}^{\infty} w(t)\,\lvert x(t)\rvert^2\,dt$	$w(t)\,\delta(t-\tau)$	$W(f-\nu)$
I_3	$(\mathbf{VX}, \mathbf{X}) = \int_{-\infty}^{\infty} V(f)\,\lvert X(f)\rvert^2\,df$	$v(t-\tau)$	$V(f)\,\delta(f-\nu)$
I_4	$(\mathbf{w}[\mathbf{h} \otimes \mathbf{x}], \mathbf{h} \otimes \mathbf{x})$	$\int_{-\infty}^{\infty} w(\sigma)h(\sigma-\tau)h^*(\sigma-t)\,d\sigma$	$H^*(f)H(\nu)W(f-\nu)$
I_5	$(\mathbf{VHX}, \mathbf{HX})$	$\iint_{-\infty}^{\infty} v(\sigma)h(\gamma-\sigma-\tau)h^*(\gamma-t)\,d\gamma\,d\sigma$	$V(f)\,\lvert H(f)\rvert^2\,\delta(f-\nu)$
I_6	$(\mathbf{x}, \mathbf{g})(\mathbf{h}, \mathbf{x}) = (\mathbf{X}, \mathbf{G})(\mathbf{H}, \mathbf{X})$	$h(t)g^*(\tau)$	$H(f)G^*(\nu)$

As additional examples, suppose the arguments in I_2 and I_3 are subjected to a linear, time-invariant transformation according to

$$\tilde{x}(t) = \int_{-\infty}^{\infty} h(t-\tau)x(\tau)\,d\tau$$

then

$$I_4(\mathbf{x}) = I_2(\tilde{\mathbf{x}}) = (\mathbf{w}[\mathbf{h} \otimes \mathbf{x}], \mathbf{h} \otimes \mathbf{x})$$

and

$$I_5(\mathbf{x}) = I_3(\tilde{\mathbf{x}}) = (\mathbf{VHX}, \mathbf{HX}) = (\mathbf{V}\,\lvert\mathbf{H}\rvert^2\mathbf{X}, \mathbf{X}) \tag{6.29}$$

Another useful quadratic functional can be obtained from a pair of linear functionals by taking the product of one with the conjugate of the other.

$$\begin{aligned} I_6(\mathbf{x}) = (\mathbf{x}, \mathbf{g})(\mathbf{x}, \mathbf{h})^* &= (\mathbf{x}, \mathbf{g})(\mathbf{h}, \mathbf{x}) \\ &= (\mathbf{X}, \mathbf{G})(\mathbf{H}, \mathbf{X}) \end{aligned} \tag{6.30}$$

The functionals above, with their corresponding time and frequency-domain operator kernels are tabulated on Table 6.1.

6.4 The Variational Problem

When properly formulated, the signal optimization problem involves finding the signal $x(t)$ which extremizes a functional $I(\mathbf{x})$. This is the single-variable problem without constraints, which will serve to introduce variational techniques for finding necessary conditions to be satisfied by the solution. If $\mathbf{x}$ were a real variable rather than a function, we would simply find the points for which $dI/dx = 0$ and examine I at each of these points to establish which gives a maximum or minimum. When $\mathbf{x}$ is a vector, an arbitrary variation in $\mathbf{x}$ is also a vector and it is not possible to uniquely describe the corresponding variation in I. In other words, we cannot define the derivative of I at the point $\mathbf{x}$. What can be done is to describe the variation in I with respect to variation in $\mathbf{x}$ *along a particular direction.* Let $\mathbf{x}$ and $\mathbf{u}$ be elements of a normed linear space with $\|\mathbf{u}\| = 1$, then with $\varepsilon > 0$ a real number, the *directional derivative* of I at $\mathbf{x}$ along $\mathbf{u}$ is defined as

$$D_u I(\mathbf{x}) = \lim_{\varepsilon \to 0} \frac{I(\mathbf{x} + \varepsilon \mathbf{u}) - I(\mathbf{x})}{\varepsilon} \tag{6.31}$$

assuming that this limit exists when I is a continuous functional.

Now, for a *stationary point* (say $\mathbf{x}_0$) of I we require that $D_u I(\mathbf{x}_0)$ must vanish for all possible values of $\mathbf{u}$. Otherwise, for sufficiently small ε, $I(\mathbf{x}_0 + \varepsilon \mathbf{u}_0)$ would be larger or smaller than $I(\mathbf{x}_0)$ if $D_{u_0} I(\mathbf{x}_0) \neq 0$. Since we shall be using linear and quadratic functionals for signal characterization, we note that

a. For $I(\mathbf{x}) = (\mathbf{x}, \mathbf{g})$

$$D_u I(\mathbf{x}) = (\mathbf{u}, \mathbf{g}); \qquad \|\mathbf{u}\| = 1 \tag{6.32}$$

b. For $I(\mathbf{x}) = (\mathscr{A}\mathbf{x}, \mathbf{x})$

$$D_u I(\mathbf{x}) = (\mathscr{A}\mathbf{u}, \mathbf{x}) + (\mathscr{A}\mathbf{x}, \mathbf{u}); \qquad \|\mathbf{u}\| = 1 \tag{6.33}$$

From (6.32) we see that the directional derivative of a linear functional is independent of $\mathbf{x}$, as we would expect by analogy with functions of a real variable. Extending the analogy, we might expect that the directional derivative of the quadratic functional would be a linear functional on $\mathbf{x}$; however, examination of (6.33) shows that this is not quite true. It is true for real linear spaces, for then $(\mathscr{A}\mathbf{u}, \mathbf{x}) = (\mathbf{x}, \mathscr{A}\mathbf{u})$. Because of this, we are tempted to restrict our consideration to real spaces. On one hand, there is no practical limitation in considering only real functionals on real signals, but on the other hand it is essential that we consider functionals having complex argument functions. For example, $X(f)$ may be complex for real signals. Note, however, that a real space can have complex basis functions provided that all the scalars (inner products) are real. If we restrict the domain of admissible frequency functions such that $X(f) = X^*(-f)$, then all linear functionals and all quadratic functionals corresponding to real time-domain operators will be real, and furthermore $(\mathscr{B}\mathbf{X}, \mathbf{Y}) = (\mathbf{Y}, \mathscr{B}\mathbf{X})$ for all admissible $\mathbf{X}$ and $\mathbf{Y}$. The advantage in restricting consideration to real spaces is primarily conceptual in that all directional derivatives of the functionals can be described by a *gradient vector.*

Gradients of Linear and Quadratic Functionals

Let $\{\boldsymbol{\varphi}_k;\, k = 1, 2, \ldots\}$ be a complete orthonormal set in a real space and let I be a real functional. Let

$$\mathbf{u} = \sum_k \delta_k \boldsymbol{\varphi}_k$$

with

$$\sum_k \delta_k^2 = 1$$

Now, assuming that $D_u I(\mathbf{x})$ is continuous in $\mathbf{x}$ for each $\mathbf{u}$, we can write

$$D_u I(\mathbf{x}) = \sum_k \delta_k D_{\varphi_k} I(\mathbf{x}) = \sum_k (\mathbf{u}, \boldsymbol{\varphi}_k) D_{\varphi_k} I(\mathbf{x}) = (\boldsymbol{\nabla} I, \mathbf{u}) \tag{6.34}$$

where the vector $\boldsymbol{\nabla} I$ may depend on $\mathbf{x}$ but it is independent of $\mathbf{u}$.

$$\boldsymbol{\nabla} I = \sum_k \eta_k \boldsymbol{\varphi}_k$$

with

$$\eta_k = D_{\varphi_k} I(\mathbf{x}) \tag{6.35}$$

The directional derivative at any point, in the direction specified by $\mathbf{u}$, can always be expressed as an inner product of a vector $\boldsymbol{\nabla} I$ with $\mathbf{u}$. $\boldsymbol{\nabla} I$ is called the *gradient* of I.

From the Schwarz inequality,

$$|D_u I(\mathbf{x})| = |(\boldsymbol{\nabla} I, \mathbf{u})| \leqslant \|\boldsymbol{\nabla} I\| \; \|\mathbf{u}\| = \|\boldsymbol{\nabla} I\| \tag{6.36}$$

where the equality holds for $\mathbf{u}$ proportional to $\boldsymbol{\nabla} I$ and the magnitude of the directional derivative has a maximum value equal to the norm of the gradient vector. Some numerical techniques using a step-by-step procedure for locating stationary points of a functional involve successive evaluation of the functional with increments taken in the direction of the gradient vector at each consecutive point. The supposition is that the quickest way to reach an extremum is to take steps in the direction resulting in the maximum change in the value of the functional. These methods are referred to as "methods of steepest ascent (or descent)."

Comparing (6.32) and (6.34), it is evident that the gradient for a real linear functional is given by

$$I(\mathbf{x}) = (\mathbf{x}, \mathbf{g}) \Rightarrow \boldsymbol{\nabla} I = \mathbf{g} \tag{6.37}$$

and from (6.33) the gradient of a real quadratic functional is

$$I(\mathbf{x}) = (\mathscr{A}\mathbf{x}, \mathbf{x}) \quad \boldsymbol{\nabla} I = (\Leftarrow\mathscr{A} + \mathscr{A}')\mathbf{x} \tag{6.38}$$

Suppose that the signal optimization problem involves extremizing a functional composed of a prescribed linear combination of quadratic and linear functionals. Then

$$I = (\mathscr{A}\mathbf{x}, \mathbf{x}) + (\mathbf{x}, \mathbf{g})$$

where

$$\mathscr{A} = \sum_i \alpha_i \mathscr{A}_i; \qquad \mathbf{g} = \sum_j \beta_j \mathbf{g}_j \tag{6.39}$$

A necessary condition for a solution is that it be a stationary point of I, i.e., a point at which ∇I vanishes, since $D_u I(\mathbf{x}) = (\nabla I, \mathbf{u}) = 0$ for all $\mathbf{u}$ if, and only if, $\nabla I = \mathbf{0}$. Hence, the necessary condition is

$$\begin{gathered}(\mathscr{A} + \mathscr{A}')\mathbf{x} + \mathbf{g} = \mathbf{0} \\ \int_{-\infty}^{\infty} [A(t, \tau) + A(\tau, t)]x(\tau)\, d\tau + g(t) = 0\end{gathered} \tag{6.40}$$

or, in the frequency domain,

$$\begin{gathered}(\mathscr{B} + \mathscr{B}')\mathbf{X} + \mathbf{G} = \mathbf{0} \\ \int_{-\infty}^{\infty} [B(f, \nu) + B^*(\nu, f)]X(\nu)\, d\nu + G(f) = 0\end{gathered} \tag{6.41}$$

From these conditions, we can make a few general remarks about the extremization problem. First of all, restricting the functionals involved to quadratic and linear, the necessary conditions are linear equations (in general, Fredholm integral equations) for which there are several known techniques for solution.[2] Note that the kernel in the integral equation is always self-adjoint. If no linear functionals are involved, then the equations are homogeneous, and non-trivial solutions exist only if $\mathscr{A} + \mathscr{A}'$ is singular. Furthermore, if $\mathscr{A}$ is self-adjoint, then $I = 0$ at a stationary point in the homogeneous case.

Problems formulated in the foregoing manner are seldom of any practical significance because of the lack of constraints on the argument functions. Usually practical considerations, such as limited available signal energy, place restrictions on the domain of admissible argument functions. In other words, the functions which extremize the performance functionals must simultaneously satisfy a number of auxiliary conditions. Fortunately, by the method of *Lagrange multipliers*, the necessary conditions for the optimization problem with constraints are obtained in essentially the same manner as in the unconstrained problem. This technique is discussed in the following section.

Exercise 6.1. Show that for a continuous complex-valued functional $I(\mathbf{x})$, defined on an inner product space, the (complex) directional derivative can be expressed as

$$D_u I(\mathbf{x}) = (\mathbf{u}, \mathbf{f}) + (\mathbf{g}, \mathbf{u})$$

where $\mathbf{f}$ and $\mathbf{g}$ may depend on $\mathbf{x}$, but not on $\mathbf{u}$.

Exercise 6.2. If $\mathscr{A}$ is a bounded operator, show that the directional derivative of $(\mathscr{A}\mathbf{x}, \mathbf{x})$ in (6.33) is a continuous functional on $\mathbf{x}$ for each $\mathbf{u}$.

6.5 The Variational Problem with Constraints

Suppose we wish to extremize the functional $I(\mathbf{x})$ subject to the constraint expressed by $J(\mathbf{x}) = c_0$. What might be called the direct approach to the problem would be to first solve for the restricted domain S_0 of admissible argument functions.

$$S_0 = \{\mathbf{x}; J(\mathbf{x}) = c_0\} \tag{6.42}$$

It is helpful to think of the subset S_0 as defining a surface or contour in the unrestricted space. Note that in general S_0 is not a linear subspace. The procedure now is to search S_0 for the points which yield extremal values of I. Necessary conditions for these points can be found by requiring that I be stationary with respect to variations in $\mathbf{x}$ confined to the set S_0. These variations are characterized by $\{\mathbf{u}; D_u J(\mathbf{x}) = 0; \mathbf{x} \in S_0\}$. The stationary points belong to the subset S where

$$S = \{\mathbf{x}; \mathbf{x} \in S_0, D_u I(\mathbf{x}) = 0 \text{ for all } \mathbf{u} \text{ such that } D_u J(\mathbf{x}) = 0\} \tag{6.43}$$

Although the set S is conceptually easy to define, solving for the set S_0 and incorporating the restricted variations often presents a great deal of difficulty.

As an alternative approach, we consider S_0 as one of a family of contours of constant value of the functional $J(\mathbf{x})$. Similarly we consider contours of the functional $I(\mathbf{x})$. The desired solution is the point of intersection of S_0 with the extremal contour of I. At this point, the two contours are tangent, as illustrated in the two-dimensional example in Figure 6.1. A point of tangency is characterized in general by proportionality of the directional derivatives of I and J.

$$D_u I(\mathbf{x}_0) + \lambda D_u J(\mathbf{x}_0) = 0 \qquad \text{for all } \mathbf{u} \tag{6.44}$$

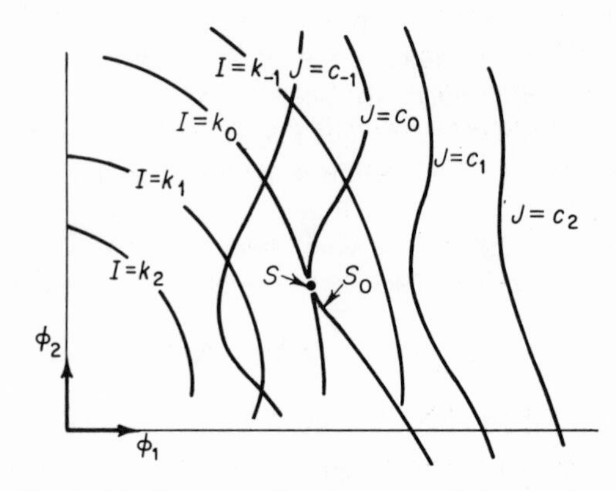

Figure 6.1. Tangency of contours at stationary point.

where λ is a constant and $\mathbf{x}_0$ is the point of tangency. Let S_1 be the subset of points for which (6.44) holds for any λ.

$$S_1 = \{\mathbf{x};\ D_u I(\mathbf{x}) + \lambda D_u J(\mathbf{x}) = 0 \text{ for all } \mathbf{u} \text{ and any } \lambda\} \tag{6.45}$$

The solution set S is the intersection of S_1 and S_0. $S = S_1 \cap S_0$. In practice this intersection is determined by first solving (6.44) for arbitrary λ and then choosing $\lambda = \lambda_0$ to satisfy the constraint $J(\mathbf{x}(\lambda_0)) = c_0$.

The most significant aspect of this approach, evident from inspection of (6.44), is that necessary conditions are obtained from the unconstrained stationary points of a new functional $\tilde{I} = I + \lambda J$. The constant λ is called a *Lagrange multiplier*.

When the functionals involved are real so that the directional derivative can be characterized by a gradient vector, then the tangency condition has a simple geometric interpretation. Since $D_u I(\mathbf{x}) = (\nabla I, \mathbf{u}) = 0$ for any direction $\mathbf{u}$ tangent to the contour at $\mathbf{x}$, it follows that ∇I is a vector normal to the contour at $\mathbf{x}$. If a pair of I and J contours are to be tangent at $\mathbf{x}$, it is necessary that their gradient vectors be colinear at this point; hence (6.44) becomes

$$\nabla I + \lambda \nabla J = \mathbf{0} \tag{6.46}$$

To illustrate these points with a simple, two-dimensional example, consider the minimization of $I(\mathbf{x}) = (\mathscr{A}\mathbf{x}, \mathbf{x}) = 3\xi_1^2 - 2\xi_1\xi_2 + 3\xi_2^2$ subject to $J(\mathbf{x}) = (\mathbf{x}, \mathbf{g}) = \xi_1 - 2\xi_2 = c_0$, where

$$\mathbf{x} = \begin{bmatrix} \xi_1 \\ \xi_2 \end{bmatrix}; \qquad \mathbf{g} = \begin{bmatrix} 1 \\ -2 \end{bmatrix}; \qquad \mathscr{A} = \mathscr{A}' = \begin{bmatrix} 3 & -1 \\ -1 & 3 \end{bmatrix}$$

The gradient condition (6.46) becomes $2\mathscr{A}\mathbf{x} = -\lambda\mathbf{g}$ which is solved by

$$\mathbf{x} = -\frac{\lambda}{2}\mathscr{A}^{-1}\mathbf{g} = -\frac{\lambda}{16}\begin{bmatrix} 3 & 1 \\ 1 & 3 \end{bmatrix}\begin{bmatrix} 1 \\ -2 \end{bmatrix} = -\frac{\lambda}{16}\begin{bmatrix} 1 \\ -5 \end{bmatrix}$$

The intersection of this set S_1 with S_0, the $J = c_0$ contour, is given by $\xi_1 = c_0/11$, $\xi_2 = -5c_0/11$ as shown in Figure 6.2.

Another, perhaps more familiar, example is the extremization of a quadratic functional $(\mathscr{A}\mathbf{x}, \mathbf{x})$ subject to the constraint that the argument function have unit norm $(\mathbf{x}, \mathbf{x}) = 1$. If $\mathscr{A}$ is self-adjoint, then the necessary condition (6.46) becomes the eigenvalue equation for $\mathscr{A}$.

$$\mathscr{A}\mathbf{x} = \lambda\mathbf{x} \tag{6.47}$$

Thus the stationary points are the eigenfunctions of $\mathscr{A}$, and the corresponding eigenvalues are real since $\mathscr{A}$ is self-adjoint. Let λ_M and λ_m be, respectively, the largest and smallest eigenvalues, then

$$\lambda_m \leqslant (\mathscr{A}\mathbf{x}, \mathbf{x}) \leqslant \lambda_M \qquad \text{for all } \mathbf{x};\ \|\mathbf{x}\| = 1 \tag{6.48}$$

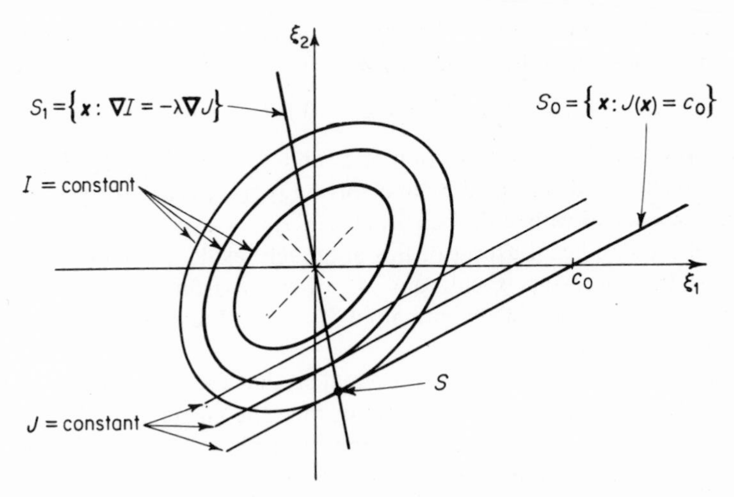

Figure 6.2. Minimization of quadratic functional subject to constraint on linear functional.

When the optimization problem involves a multiplicity of constraints, each expressed as a quadratic or linear functional, the preceding technique is extended by finding the stationary points of a linear combination of all the functionals involved. Hence (6.39) is still valid for the problem with constraints where some of the α_i and β_j are interpreted as Lagrange multipliers rather than initially prescribed multipliers.

Nothing has been said so far about the sufficiency of stationary points as solutions to the extremization problem. Usually, the sufficiency of a particular stationary point is evident from the physical considerations that led to the formulation of the problem. The problem that remains after solving the necessary conditions is the determination of the parameters (Lagrange multipliers) to simultaneously extremize the performance functional and satisfy the constraint equations. Hence another extremization problem with constraints, although with greatly reduced dimensionality, remains to be solved. Unfortunately, the equations resulting are not necessarily linear in these undetermined parameters and often numerical methods are required to obtain the final solution. Rather than pursue a general treatment of these problems, we shall show how they are handled in specific cases by solving several examples drawn from practical signal design considerations.

6.6 Some Examples

Example 6.1 Suppose the linear, time-invariant filter shown in Figure 6.3 has an impulse response $h(t)$. If the input signal $x(t)$ has a fixed amount of energy, what input signal produces maximum energy at the output?

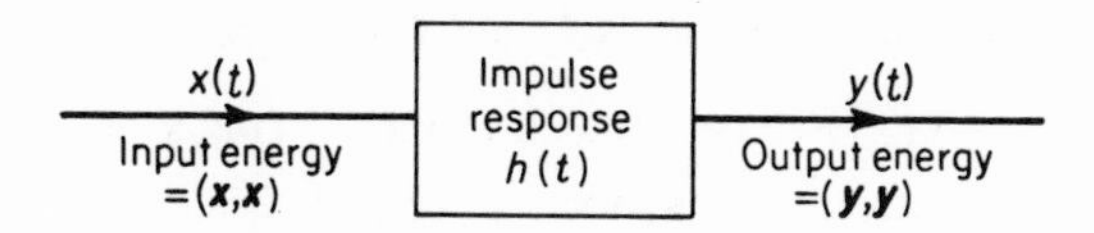

Figure 6.3. Maximum filtered energy.

Let

$$
\begin{aligned}
I_1 &= (\mathbf{y}, \mathbf{y}) = (\mathbf{h} \otimes \mathbf{x}, \mathbf{h} \otimes \mathbf{x}) = (\mathbf{HX}, \mathbf{HX}) \\
&= (\mathbf{H^*HX}, \mathbf{X}) \qquad (6.49) \\
I_2 &= (\mathbf{x}, \mathbf{x}) = (\mathbf{X}, \mathbf{X}) = 1
\end{aligned}
$$

The stationary points of $I_1 + \lambda I_2$ are more simply expressed in the frequency domain.

$$[|H(f)|^2 + \lambda]X(f) = 0 \qquad (6.50)$$

The condition (6.50) requires $X(f) = 0$ for all frequencies except where $|H(f)|^2 = -\lambda$. The corresponding output energy is $I_1 = -\lambda(\mathbf{X}, \mathbf{X}) = -\lambda$. The difficulty with this solution is that we cannot find a signal of unit energy whose Fourier transform is non-zero only at $f = \pm f_0$; however, this condition can be approached arbitrarily closely by the limiting process

$$
\begin{aligned}
x(t) &= \lim_{T\to\infty} T^{-\frac{1}{2}} \cos(2\pi f_0 t + \theta) \qquad \text{in } |t| \leqslant T \\
&= 0 \qquad \text{otherwise} \qquad (6.51)
\end{aligned}
$$

where f_0 is selected so that $|H(f_0)| \geqslant |H(f)|$ for all frequencies in order that the output energy be maximum. This result is intuitively satisfying since it merely says that we want to restrict the excitation of the filter to the frequencies where the gain is maximum. If $|H(f)|$ is constant and also maximum over certain intervals, then other solutions are possible. ∎

Example 6.2 A more interesting example of the preceding type is the problem considered by Chalk,[3] where we solve for the maximum filtered energy with a duration-limited input signal. It turns out that it is easier to incorporate the duration-limiting constraint by the direct method rather than by Lagrange multipliers. In addition to the functionals in Example 6.1, we add the constraint

$$w(t)x(t) = x(t) \qquad \text{for all } t$$

where

$$
\left.
\begin{aligned}
w(t) &= 1 \qquad \text{for } |t| \leqslant T \\
&= 0 \qquad \text{otherwise}
\end{aligned}
\right\} = \tfrac{1}{2}[1 + \operatorname{sgn}(T - |t|)] \qquad (6.52)
$$

Incorporating (6.52), the functionals I_1 and I_2 become

$$\begin{aligned} I_1 &= (\mathbf{y}, \mathbf{y}) = (\mathbf{h} \otimes \mathbf{wx}, \mathbf{h} \otimes \mathbf{wx}) = (\mathscr{A}_1\mathbf{x}, \mathbf{x}) \\ I_2 &= (\mathbf{x}, \mathbf{x}) = (\mathbf{wx}, \mathbf{wx}) = (\mathbf{wx}, \mathbf{x}) = 1 \end{aligned} \tag{6.53}$$

In this problem, the necessary condition is simpler in the time domain.

$$2\mathscr{A}_1\mathbf{x} + 2\lambda\mathbf{wx} = \mathbf{0} \tag{6.54}$$

where $A_1(t, \tau) = w(t)w(\tau)k(t - \tau)$ corresponds to a self-adjoint operator, and the Fourier transform of $k(t)$ is simply

$$K(f) = |H(f)|^2 \tag{6.55}$$

Because of the particular form of $w(t)$, the necessary condition (6.54)

$$w(t)\int_{-\infty}^{\infty} k(t - \tau)w(\tau)x(\tau)\, d\tau + \lambda w(t)x(t) = 0$$

can be simplified to

$$\int_{-T}^{T} k(t - \tau)x(\tau)\, d\tau = \lambda x(t) \qquad \text{for } |t| \leqslant T \tag{6.56}$$

where the sign of λ has been reversed to put (6.56) in the standard form of an eigenvalue equation. A particular case for which the solution of (6.56) was obtained in Reference [3] is the single-section RC lowpass filter with bandwidth $f_0 = 1/2\pi RC$. For this case,

$$K(f) = \frac{1}{1 + (f/f_0)^2} \tag{6.57}$$

so that

$$k(t) = \pi f_0 e^{-2\pi f_0 |t|}$$

and (6.56) becomes

$$\pi f_0 \int_{-T}^{T} e^{-2\pi f_0 |t-\tau|} x(\tau)\, d\tau = \lambda x(t) \qquad \text{for } |t| \leqslant T \tag{6.58}$$

Differentiating (6.58) twice with respect to t,

$$\begin{aligned} \lambda \ddot{x}(t) &= \pi f_0 \int_{-T}^{T} [-4\pi f_0\, \delta(t - \tau) + (2\pi f_0)^2 e^{-2\pi f_0 |t-\tau|}] x(\tau)\, d\tau \\ &= -(2\pi f_0)^2 x(t) + (2\pi f_0)^2 \pi f_0 \int_{-T}^{T} e^{-2\pi f_0 |t-\tau|} x(\tau)\, d\tau \end{aligned} \tag{6.59}$$

and combining (6.58) and (6.59), we have a second-order differential equation applicable to the interval $|t| \leqslant T$.

$$\ddot{x}(t) + (2\pi f_0)^2 \left[\frac{1}{\lambda} - 1\right] x(t) = 0 \tag{6.60}$$

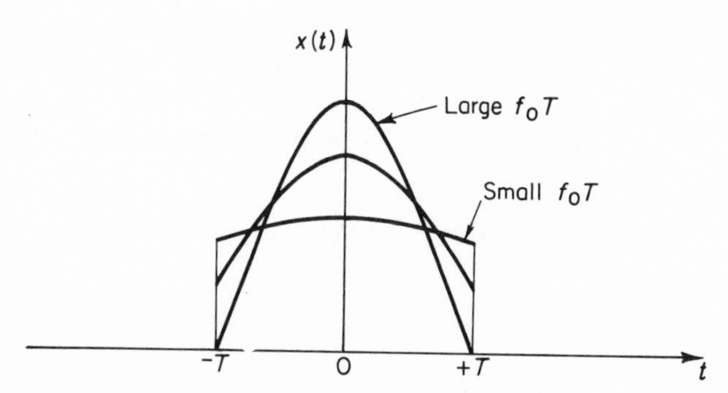

Figure 6.4. Optimal duration-limited input pulses for maximum filtered energy with *RC* lowpass filter.

Substituting the solutions of (6.60) back into (6.58), we find that the eigenvalues can be separated into two sequences:

$$\lambda_n = \frac{(2\pi f_0 T)^2}{a_n^2 + (2\pi f_0 T)^2}; \qquad \lambda_m = \frac{(2\pi f_0 T)^2}{a_m^2 + (2\pi f_0 T)^2} \tag{6.61}$$

where a_n and a_m are given by

$$\tan a_n = \frac{2\pi f_0 T}{a_n}; \qquad \tan a_m = \frac{-a_m}{2\pi f_0 T} \tag{6.62}$$

The corresponding eigenfunctions (not normalized) are

$$x_n(t) = \cos\frac{a_n t}{T}; \qquad x_m(t) = \sin\frac{a_m t}{T} \qquad \text{for } |t| \leqslant T \tag{6.63}$$

When duration-limited versions of these eigenfunctions are used as input signals, we find that the corresponding eigenvalue gives the fraction of energy remaining in the filtered signal. Hence the eigenfunction giving maximum energy at the output of the filter corresponds to the largest eigenvalue λ_1 of the λ_n sequence. The corresponding value of a_1 lies in the range $0 < a_1 < \pi/2$. Small values of f_0T result in small values of a_1 and the optimum input signal is nearly a rectangular pulse. For large values of f_0T, the value of a_1 approaches $\pi/2$ and the optimum signal approaches the "half-cosine" pulse as shown in Figure 6.4. ∎

Example 6.3. Given a time-invariant, linear filter with impulse response $h(t)$, what input signal $x(t)$ with fixed energy maximizes the filter output at $t = t_0$? The peak value of the output signal $y(t_0)$ is a linear functional on $x(t)$. Since $\mathbf{y} = \mathbf{x} \otimes \mathbf{h}$,

$$I_1 = y(t_0) = (\mathbf{x}, \mathbf{g}) = (\mathbf{X}, \mathbf{G})$$

where

$$g(t) = h(t_0 - t) \quad \text{and} \quad G(f) = H^*(f)e^{-j2\pi f t_0} \tag{6.64}$$

and, as before,

$$I_2 = (\mathbf{x}, \mathbf{x}) = (\mathbf{X}, \mathbf{X}) = 1 \tag{6.65}$$

The condition for stationarity of $I_1 + \lambda I_2$ is equally simple in either the time domain or the frequency domain.

$$G(f) + 2\lambda X(f) = 0$$

so that

$$X(f) = -\frac{1}{2\lambda} H^*(f) e^{-j2\pi f t_0} \tag{6.66}$$

and $\lambda = \pm\frac{1}{2}\,\|\mathbf{H}\|$ for $\|\mathbf{X}\| = 1$. The negative sign for λ gives maximum peak output and the positive sign gives the minimum. In the time domain we have, for the optimum input signal,

$$x(t) = \|\mathbf{h}\|^{-1} h(t_0 - t) \tag{6.67}$$

so that, except for the time delay t_0, the input signal is the time reverse of the impulse response of the filter. A signal with this property is said to be *matched* to the filter. The peak output of a filter excited by the matched signal is $y(t_0) = \|\mathbf{x}\| \cdot \|\mathbf{h}\|$. This can be checked using the Schwarz inequality. ∎

Example 6.4. The previous problem could have been motivated by a desire to maximize the detectability of pulses received over a channel having transfer function $H(f)$. In the interest of sharpening the resolution of a time sequence of these pulses at the receiver, we might choose $x(t)$ to minimize the energy of an individual received pulse while maintaining a prescribed peak value (say at $t = t_0$). The effect of minimizing energy for a given peak value is to cause a narrowing of the pulse and a reduction of the overlap of adjacent pulses. Thus we want to minimize

$$I_1 = (\mathbf{y}, \mathbf{y}) = (\mathbf{HX}, \mathbf{HX}) = (|\mathbf{H}|^2\mathbf{X}, \mathbf{X})$$

subject to

$$\begin{aligned} I_2 &= (\mathbf{x}, \mathbf{x}) = (\mathbf{X}, \mathbf{X}) = 1 \\ I_3 &= (\mathbf{x}, \mathbf{g}) = (\mathbf{X}, \mathbf{G}) = y(t_0) \end{aligned} \tag{6.68}$$

where

$$g(t) = h(t_0 - t); \qquad G(f) = H^*(f) e^{-j2\pi f t_0}$$

Stationary points of $I_1 + \lambda_1 I_2 + \lambda_2 I_3$ are given by

$$2\,|H(f)|^2 X(f) + 2\lambda_1 X(f) + \lambda_2 G(f) = 0 \tag{6.69}$$

Hence,

$$X(f) = \frac{-\frac{1}{2}\lambda_2 H^*(f) e^{-j2\pi f t_0}}{|H(f)|^2 + \lambda_1} \tag{6.70}$$

The Lagrange multipliers are obtained from the simultaneous solutions of the constraint equations (note the nonlinearity).

$$I_2 = \frac{\lambda_2^2}{4}\int_{-\infty}^{\infty} \frac{|H(f)|^2\, df}{[|H(f)|^2 + \lambda_1]^2} = 1$$

$$I_3 = -\frac{\lambda_2}{2}\int_{-\infty}^{\infty} \frac{|H(f)|^2\, df}{|H(f)|^2 + \lambda_1} = y(t_0) \tag{6.71}$$

From the general form of the solution (6.70) we see that if λ_1 is large compared to $|H(f)|^2$, then the solution approaches the signal *matched* to the channel. On the other hand, if λ_1 is small, the signal is essentially *inverse* to the channel. These results are intuitively apparent in terms of the severity of the constraint on the peak value of the output pulse. If the peak value can barely be met by a unit energy input signal, then clearly the input signal would have to be close to the matched signal. On the other hand, if the peak value is easily met with unit input energy, then a larger portion of the energy can be discarded by something approaching inverse filtering in order to reduce the width of the output pulse. ∎

Example 6.5. We consider now the problem of finite-time charging of a capacitor with minimum expenditure of energy. In the circuit shown in Figure 6.5, it is required that the output voltage $v_2(t)$ be 1 volt at $t = T$. The

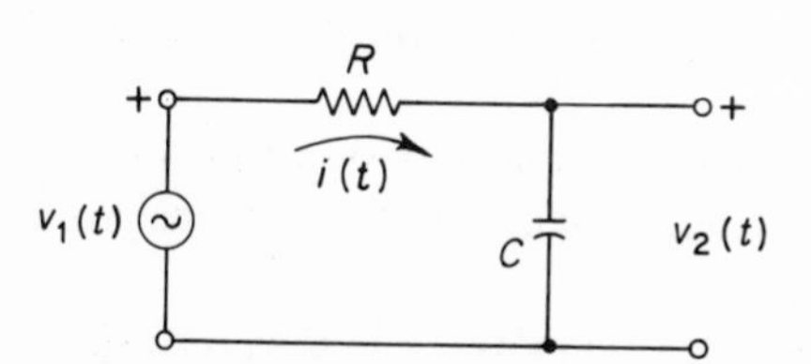

Figure 6.5. Capacitor charging problem.

source voltage $v_1(t)$ is duration-limited to the interval $0 \leqslant t \leqslant T$. What is the source voltage which makes the energy delivered by the source a minimum?

This simple problem illustrates several aspects of the variational problem not covered in the previous examples. First of all, some of the functionals on $v_1(t)$ do not correspond to self-adjoint operators. Secondly, the physical concept of energy is used, not just the square of the norm of $\mathbf{v}_1$. Finally, the choice of argument function has a significant effect on the difficulty of solution. This will be demonstrated by working this problem two ways.

Let $I_1 = v_2(T)$ be the final capacitor voltage and $I_2 = (\mathbf{i}, \mathbf{v}_1)$ be the energy supplied by the source.

Incorporating the duration-limiting constraint

$$v_1(t) = w(t)v_1(t) \tag{6.72}$$

where

$$w(t) = 1 \qquad \text{for } 0 \leqslant t \leqslant T$$
$$= 0 \qquad \text{otherwise}$$

we express the final capacitor voltage as

$$I_1 = (\mathbf{wv}_1, \mathbf{g}) = (\mathbf{v}_1, \mathbf{wg});$$

where

$$g(t) = \frac{1}{RC} e^{-\frac{1}{RC}(T-t)}$$

since

$$v_2(t) = \frac{1}{RC} \int_0^t v_1(\tau) e^{-\frac{1}{RC}(t-\tau)} \, d\tau \tag{6.73}$$

The energy supplied is expressed as

$$I_2 = (\mathbf{i}, \mathbf{wv}_1) = \frac{1}{R}(\mathbf{wv}_1, \mathbf{v}_1) - \frac{1}{R}(\mathscr{A}\mathbf{v}_1, \mathbf{v}_1) \tag{6.74}$$

since

$$\mathbf{i} = \frac{\mathbf{v}_1 - \mathbf{v}_2}{R}$$

The operator $\mathscr{A}$ in (6.74) has a kernel function $A(t, \tau)$ given by

$$A(t, \tau) = \frac{1}{RC} w(t) w(\tau) u(t - \tau) e^{-\frac{1}{RC}(t-\tau)} \tag{6.75}$$

where $u(t) = \frac{1}{2}(1 + \operatorname{sgn} t)$ is the unit step function. Notice that $\mathscr{A}$ is not self-adjoint. The stationary points of $I_2 + \lambda I_1$ are given by

$$\frac{2}{R}\mathbf{wv}_1 - \frac{1}{R}\mathscr{A}\mathbf{v}_1 - \frac{1}{R}\mathscr{A}'\mathbf{v}_1 + \lambda \mathbf{wg} = \mathbf{0} \tag{6.76}$$

Noting the particular form of $w(t)$, (6.76) can be rewritten as

$$\frac{2v_1(t)}{R} - \frac{1}{R^2C}\int_0^t e^{-\frac{1}{RC}(t-\tau)} v_1(\tau)\, d\tau \\ - \frac{1}{R^2C}\int_t^T e^{\frac{1}{RC}(t-\tau)} v_1(\tau)\, d\tau + \frac{\lambda}{RC} e^{-\frac{1}{RC}(T-t)} = 0$$

or

$$\frac{2v_1(t)}{R} - \frac{1}{R^2C}\int_0^T e^{-\frac{1}{RC}|t-\tau|} v_1(\tau)\, d\tau + \frac{\lambda}{RC} e^{-\frac{1}{RC}(T-t)} = 0$$

$$\text{for } 0 \leqslant t \leqslant T \tag{6.77}$$

Note the similarity between (6.77) and the corresponding equation (6.58) for the particular case in Example 6.2. The solution can be obtained in the same manner by twice differentiating (6.77) with respect to t and combining with the undifferentiated equation to give a second-order differential equation applicable to the interval $0 \leqslant t \leqslant T$.

$$\ddot{v}_1(t) = 0 \tag{6.78}$$

Hence,

$$v_1(t) = k_1 + k_2 t \qquad \text{for } 0 \leqslant t \leqslant T \tag{6.79}$$

The constants k_1 and k_2 are found such that $I_1(k_1, k_2) = 1$ and $I_2(k_1, k_2)$ is minimum. When this is done, we have the optimum charging voltage as the sum of a step and a ramp.

$$v_1(t) = \frac{RC}{T} + \frac{t}{T} \qquad \text{for } 0 \leqslant t \leqslant T \tag{6.80}$$

The effort involved in solving this problem is greatly reduced by choosing $\mathbf{i}$ as the argument function rather than $\mathbf{v}_1$.

$$I_1 = v_2(T) = \frac{1}{C}\int_0^T i(t)\,dt = \frac{1}{C}(\mathbf{i}, \mathbf{w}) = 1 \tag{6.81}$$

$$\begin{aligned} I_2 &= \text{energy supplied in 0 to } T \\ &= R\int_0^T i^2(t)\,dt + \frac{C}{2}v_2^2(T) \\ &= R(\mathbf{wi}, \mathbf{i}) + \frac{C}{2}I_1^2 \end{aligned} \tag{6.82}$$

Hence we simply want to minimize $(\mathbf{wi}, \mathbf{i})$ subject to $(\mathbf{i}, \mathbf{w}) = C$, so we require that

$$2w(t)i(t) + \lambda w(t) = 0$$

or

$$i(t) = -\frac{\lambda}{2} \qquad \text{for } 0 \leqslant t \leqslant T \tag{6.83}$$

where $\lambda = -2C/T$ in order to satisfy the constraint. Now

$$v_1(t) = Ri(t) + \frac{1}{C}\int_0^t i(\tau)\,d\tau = \frac{RC}{T} + \frac{t}{T} \qquad \text{for } 0 \leqslant t \leqslant T \tag{6.84}$$

which checks with the result in (6.80).

Frequency-domain techniques for solving problems of this type have also been presented.[4] ∎

Example 6.6 As a final example, we consider a more complex situation typical of actual system design problems. Two new aspects of the variational problem will be introduced in this example: (1) An operator, rather than a signal, can be used as the argument function, and (2) two or more independent system functions can be varied simultaneously to find the optimum performance. In the system shown in Figure 6.6, it is desired that the filtered version of $x(t)$ should be as close as possible to a prescribed signal $g(t)$. As before, we assume that the energy content of $x(t)$ is limited so that $(\mathbf{x}, \mathbf{x}) = c_1$. In addition, we note that most filtering devices have some sort

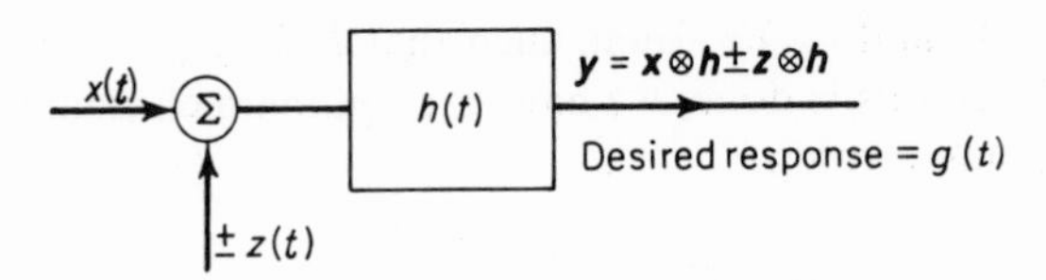

Figure 6.6. Minimum error filtering with interference.

of limitation on *gain-bandwidth product*. This might be expressed by the constraint $(\mathbf{H}, \mathbf{H}) = c_2$. Finally, we suppose that the input signal has an additive component of interference $z(t)$ of either polarity. To control the effect of interference, we could require that the filtered version of $z(t)$ have a tolerably small energy, say c_3. The problem is to find the best combination of input signal $x(t)$ and filter $h(t)$, so that the difference between the filtered signal $\mathbf{x} \otimes \mathbf{h}$ and the desired response $\mathbf{g}$ have minimum norm subject to the constraints above. To summarize, we want to minimize

$$I_1 = \|\mathbf{x} \otimes \mathbf{h} - \mathbf{g}\|^2 = (\mathbf{XH}, \mathbf{XH}) - 2(\mathbf{XH}, \mathbf{G}) + (\mathbf{G}, \mathbf{G})$$

subject to

$$\begin{aligned} I_2 &= (\mathbf{X}, \mathbf{X}) = c_1 \\ I_3 &= (\mathbf{H}, \mathbf{H}) = c_2 \\ I_4 &= \|\mathbf{z} \otimes \mathbf{h}\|^2 = (\mathbf{ZH}, \mathbf{ZH}) = c_3 \end{aligned} \tag{6.85}$$

We proceed by finding the stationary points of

$$\begin{aligned} I &= (|\mathbf{H}|^2\mathbf{X}, \mathbf{X}) - 2(\mathbf{X}, \mathbf{H}^*\mathbf{G}) + (\mathbf{G}, \mathbf{G}) + \lambda_1(\mathbf{X}, \mathbf{X}) \\ &\quad + \lambda_2(\mathbf{H}, \mathbf{H}) + \lambda_3(\mathbf{ZH}, \mathbf{ZH}) \\ &= (|\mathbf{X}|^2\mathbf{H}, \mathbf{H}) - 2(\mathbf{H}, \mathbf{X}^*\mathbf{G}) + (\mathbf{G}, \mathbf{G}) + \lambda_1(\mathbf{X}, \mathbf{X}) \\ &\quad + \lambda_2(\mathbf{H}, \mathbf{H}) + \lambda_3(|\mathbf{Z}|^2\mathbf{H}, \mathbf{H}) \end{aligned} \tag{6.86}$$

with respect to $\mathbf{X}$ and $\mathbf{H}$ simultaneously. The two conditions can be obtained by inspection of the appropriate expression in (6.86).

$$\nabla_X I = 2\,|\mathbf{H}|^2\mathbf{X} - 2\mathbf{H}^*\mathbf{G} + 2\lambda_1\mathbf{X} = \mathbf{0} \tag{6.87}$$

$$\nabla_H I = 2\,|\mathbf{X}|^2\mathbf{H} - 2\mathbf{X}^*\mathbf{G} + 2\lambda_2\mathbf{H} + 2\lambda_3\,|\mathbf{Z}|^2\mathbf{H} = \mathbf{0} \tag{6.88}$$

We want to find $\mathbf{X}$ and $\mathbf{H}$ which simultaneously satisfy (6.87) and (6.88).

Multiply (6.87) by $\mathbf{X}^*$ and (6.88) by $\mathbf{H}^*$ to get

$$2\,|\mathbf{H}|^2|\mathbf{X}|^2 - 2\mathbf{H}^*\mathbf{X}^*\mathbf{G} + 2\lambda_1\,|\mathbf{X}|^2 = \mathbf{0} \tag{6.89}$$

$$2\,|\mathbf{H}|^2|\mathbf{X}|^2 - 2\mathbf{H}^*\mathbf{X}^*\mathbf{G} + 2\lambda_2\,|\mathbf{H}|^2 + 2\lambda_3\,|\mathbf{Z}|^2|\mathbf{H}|^2 = \mathbf{0} \tag{6.90}$$

From (6.89) we see that $\mathbf{H}^*\mathbf{X}^*\mathbf{G}$ is real; hence the phase of $\mathbf{HX}$ is the same as that of $\mathbf{G}$, as would be expected, and these quantities may be replaced by their magnitudes in (6.89) and (6.90). The required phase can be split arbitrarily and assigned to $\mathbf{X}$ and $\mathbf{H}$ since the phase of $\mathbf{X}$ and $\mathbf{H}$ does not affect the constraint equations. Subtracting (6.89) and (6.90), we find

$$|\mathbf{X}|^2 = (k_1\,|\mathbf{Z}|^2 + k_2)\,|\mathbf{H}|^2 \tag{6.91}$$

Assuming $|\mathbf{H}|^2 > 0$, (6.91) can be substituted into (6.89) to obtain

$$|\mathbf{H}|^2 = \frac{|\mathbf{G}|}{[k_1\,|\mathbf{Z}|^2 + k_2]^{\frac{1}{2}}} - k_3 \tag{6.92}$$

and

$$|\mathbf{X}|^2 = (k_1\,|\mathbf{Z}|^2 + k_2)^{\frac{1}{2}}\,|\mathbf{G}| - k_3(k_1\,|\mathbf{Z}|^2 + k_2) \tag{6.93}$$

Nothing much can be said in general about solving for the constants k_1, k_2, and k_3 to satisfy the constraints except that it is likely to be a tedious task unless solved numerically on a digital computer. ▍

Exercise 6.3. In Example 6.2 it was shown that the optimal duration-limited signal, to give maximum energy after filtering by a single-section *RC* network, approaches a rectangular pulse as the parameter f_0T approaches zero (Figure 6.4). Show that for any lowpass filter (and any reasonable measure of bandwidth f_0) the optimum input pulse approaches the rectangular pulse as f_0T approaches zero. Give a physical interpretation of this result.

Exercise 6.4. Write the integral equation that must be satisfied by solutions of the problem "adjoint" to the one in Example 6.2; i.e., what input signal of unit energy, when filtered by a network with transfer function $H(f)$, produces maximum energy at the output in the interval $|t| \leqslant T$?

Exercise 6.5. Show that, except for a time delay, the output signal in Example 6.3, when the input signal is matched to the filter, is proportional to the time-ambiguity function (2.40) for the input signal.

Exercise 6.6. Show that the "time-ambiguity area"

$$r_x^{-2}(0)\int_{-\infty}^{\infty} r_x^2(\tau)\,d\tau$$

for a signal $x(t)$, band-limited to frequencies in the interval $|f| \leqslant W$, cannot be less than $(1/2W)$ seconds. Give an example of a band-limited signal which attains this minimum value.

Exercise 6.7. Find the unit energy input signal, duration-limited to an interval $2T$ seconds wide, which maximizes peak output of a filter with impulse response $h(t)$.

Exercise 6.8. Consider a filter with voltage transfer function $H(f) = V_2(f)/V_1(f)$ and input admittance function $Y(f) = I(f)/V_1(f)$, where $v_1(t)$ and $v_2(t)$ are input and output voltages, respectively, and $i(t)$ is input current. What signal voltage $v_1(t)$ physically delivering unit energy at the input $(\mathbf{v}_1, \mathbf{i}) = 1$ produces maximum output voltage at $t = t_0$? In other words, what is the matched signal for a given filter with a non-resistive input impedance?

Exercise 6.9. A common signal design problem is to find an input signal which, when subjected to a particular time-invariant filtering operation, gives an output pulse of specified shape. This problem is usually approached in the frequency domain. In the absence of other constraints, we simply make $X(f) = H^{-1}(f)G(f)$, where $g(t)$ is the desired output pulse and $h(t)$ is the impulse response of the filter. This solution may require that the input pulse have prohibitively large energy, e.g., if $G(f)$ has appreciable components in regions where $H(f)$ is small. Suppose we place an energy limitation on the input signal and select the signal that minimizes $\|\mathbf{g} - \mathbf{x} \otimes \mathbf{h}\|$. Write the general expression for the Fourier transform of the optimal signal and compare this with the unconstrained result. Give a physical interpretation of the differences for specific types of $G(f)$ and $H(f)$ characteristics. Also, compare the result with that in Example 6.4.

6.7 Duration-bandwidth Occupancy of a Signal

There is a problem of the type discussed in the preceding sections which deserves special attention. It concerns the extent to which the representations for a pulse signal can be simultaneously concentrated into narrow regions in both the time domain and the frequency domain. There are several pulse communication system situations where minimal duration and bandwidth of the pulses lead to desirable performance. Small bandwidth eases channel capacity requirements and also tends to minimize the effects of pulse distortion resulting from transmission over a dispersive channel. Short duration increases the capability of resolution of a time sequence of pulses. Hence, in the radar situation, range resolution is improved with narrow pulses, and in data transmission the information rate can be increased with narrow pulses since pulse rate can be increased while maintaining the detectability of individual pulses.[5],[6] An analogous problem arises in the optimal design of antenna arrays where it is desired to achieve narrow beamwidths with arrays of small dimension.[7]

Generalized Uncertainty Principle

An expression of the fact that there exists a limit to the extent of simultaneous concentration attainable in both domains is popularly called an *uncertainty principle*, after the analogous relations in quantum mechanics. Several authors[8]–[16] have considered this problem and, for the most part, have used some type of energy functional to measure concentration in each domain. In this section we formulate a generalization of the uncertainty problem by measuring the concentrations relevant to duration and bandwidth in terms of arbitrarily weighted energy distributions in the time domain and the frequency domain. For purposes of normalization we constrain the signal to unit total energy. The solution to the uncertainty problem is an expression of the attainable concentrations for a given pair of weighting functions and the corresponding pulse shape which yields the maximum concentration. The variational problem involves the extremization of

$$\mu_1 I_1 + \mu_2 I_2 + I_3$$

where

$$\begin{aligned} I_1 &= (\mathbf{wx}, \mathbf{x}) \\ I_2 &= (\mathbf{VX}, \mathbf{X}) \\ I_3 &= (\mathbf{x}, \mathbf{x}) = (\mathbf{X}, \mathbf{X}) = 1 \end{aligned} \tag{6.94}$$

and $w(t)$ and $V(f)$ are the prescribed real time- and frequency-domain weighting functions. μ_1 and μ_2 are Lagrange multipliers. Only two undetermined multipliers are required and they may be associated with any pair of the three quadratic functionals. Since each of the quadratic functionals corresponds to a self-adjoint operator, the necessary condition for a stationary point of (6.94) can be expressed in the time domain as (see Table 6.1)

$$\mu_1 w(t)x(t) + \mu_2 \int_{-\infty}^{\infty} v(t-\tau)x(\tau)\,d\tau + x(t) = 0 \tag{6.95}$$

or, in the frequency domain, the necessary condition becomes

$$\mu_1 \int_{-\infty}^{\infty} W(f-\nu)X(\nu)\,d\nu + \mu_2 V(f)X(f) + X(f) = 0 \tag{6.96}$$

Hence we see that the necessary condition has the form of a homogeneous Fredholm integral equation in either domain. If the two weighting functions have a different form, the solution may be considerably more convenient in one domain than the other. If the weighting functions are the same, the optimal pulse and its Fourier transform will satisfy the same equation. We also note that with different weightings when one uncertainty problem is solved, we automatically have the solution to a different problem by applying time-frequency duality to (6.95) and (6.96).

The introductory example in Section 6.1 is seen to be an uncertainty problem with $w(t) = t^2$ and $V(f) = f^2$. Instead of minimizing the product of I_1 and I_2 in this case, we could constrain either I_1 or I_2 to a prescribed value and minimize the other (keeping $I_3 = 1$).

Since

$$V(f) = f^2 \Rightarrow v(t) = -\left(\frac{1}{2\pi}\right)^2 \ddot{\delta}(t)$$

the necessary condition (6.95) becomes

$$\mu_1 t^2 x(t) - \left(\frac{1}{2\pi}\right)^2 \mu_2 \ddot{x}(t) + x(t) = 0 \tag{6.97}$$

In the form,

$$\ddot{x}(t) + (a^2 - b^2 t^2)x(t) = 0 \tag{6.98}$$

this second-order differential equation is recognized as the Schrodinger wave equation for a harmonic oscillator with parabolic potential well and its solution has received extensive treatment. It can be shown[17] that for the boundary condition $x(\pm\infty) = 0$, the coefficients in (6.98) must be constrained according to

$$a^2 = (2n + 1)b; \qquad n = 0, 1, 2, \ldots \tag{6.99}$$

and the corresponding orthonormal solutions are

$$\psi_n(\sqrt{b}\, t) = \frac{H_n(\sqrt{b}\, t)e^{-\frac{1}{2}bt^2}}{(2^n n!\, \pi^{\frac{1}{2}})^{\frac{1}{2}}} \tag{6.100}$$

where the H_n are the Hermite polynomials and the ψ_n are the normalized Hermite functions, (3.45). The multiplicity of stationary points typical of homogeneous necessary conditions is evident from (6.100). The absolute minimum of the duration-bandwidth product is easily found by substituting (6.100) into (6.4) with the result that

$$T_{\psi_n} W_{\psi_n} = \frac{1}{4\pi}(2n + 1); \qquad n = 0, 1, 2, \ldots \tag{6.101}$$

Hence, $\psi_0(\sqrt{b}\, t)$, the Gaussian pulse shape, is optimal, in accordance with (6.8).

As another example, it is clear that the problem in Example 6.2 is a type of uncertainty problem with the constraint $I_1 = 1$, where $w(t)$ is the rectangular weighting given by (6.52) and $V(f) = K(f) = |H(f)|^2$ from (6.55). The statement of the uncertainty principle for this case involves finding the largest eigenvalue and corresponding eigenfunction of (6.56) for a given $H(f)$ and T. In particular, for $H(f)$ corresponding to the single-section RC lowpass filter, the minimum value of $f_0 T$ which will yield a prescribed fraction I_2 of filtered energy is found from (6.61) and (6.62) to be given by

$$f_0 T \geqslant \frac{\alpha}{2\pi} \tan^{-1} \alpha \tag{6.102}$$

where

$$\alpha^2 = \frac{I_2}{1 - I_2}$$

In the interest of obtaining a sharper measure of frequency-domain concentration, Chalk[3] also employed rectangular weighting in the frequency domain by assuming an ideal lowpass filter of bandwidth W as indicated in Figure 6.7. The necessary condition for maximizing I_2 with $I_1 = I_3 = 1$,

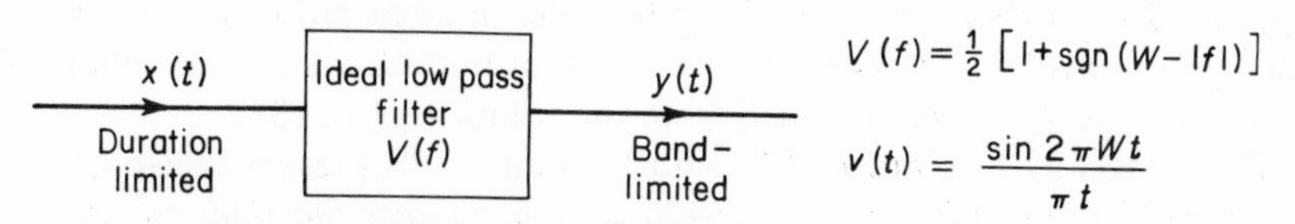

Figure 6.7. Band-limited versions of duration-limited signals.

obtained from (6.56), is given by

$$\int_{-T}^{T} \frac{\sin 2\pi W(t - \tau)}{\pi(t - \tau)} x(\tau)\, d\tau = \lambda x(t) \qquad \text{for } |t| \leqslant T \tag{6.103}$$

We note that the left-hand side of (6.103) is the convolution of the duration-limited pulse with the impulse response of the ideal lowpass filter. This is a result of the special property $K(f) = |H(f)|^2 = H(f)$ in this case.

From this we observe a very curious property of the duration-limited functions which satisfy (6.103); namely, their band-limited versions have the same shape in the interval $|t| \leqslant T$.

$$y(t) = \lambda x(t) \qquad \text{in } |t| \leqslant T \tag{6.104}$$

To obtain the maximum filtered energy, we note that for this case

$$\begin{aligned}(\mathbf{y}, \mathbf{y}) &= (\mathbf{Y}, \mathbf{Y}) = (\mathbf{VX}, \mathbf{Y}) = (\mathbf{x}, \mathbf{y}) = (\mathbf{wx}, \mathbf{y}) \\ &= \lambda(\mathbf{wx}, \mathbf{wx}) = \lambda(\mathbf{x}, \mathbf{x})\end{aligned} \tag{6.105}$$

Hence λ is the fraction of the energy in $x(t)$ which falls in the band $|f| \leqslant W$. This indicates that the eigenvalues of (6.103) are all less than 1, and the maximum filtered energy is equal to the largest eigenvalue, say λ_0.

It is interesting to consider at this point the dual of the problem above; i.e., maximize $I_1 = (\mathbf{wx}, \mathbf{x})$ subject to $I_2 = (\mathbf{VX}, \mathbf{X}) = 1$ and $I_3 = (\mathbf{x}, \mathbf{x}) = 1$. In other words, we want to find the band-limited signal, with bandwidth W, which concentrates the largest portion of its total energy in the interval $|t| \leqslant T$. The constraint $I_2 = 1$ may be incorporated by the direct method $(\mathbf{x} = \mathbf{v} \otimes \mathbf{x})$ by convolving the general expression (6.95) for the necessary condition with $v(t)$, giving for this case

$$\mu_1 \int_{-\infty}^{\infty} \frac{\sin 2\pi W(t - \tau)}{\pi(t - \tau)} w(\tau)x(\tau)\, d\tau + (1 + \mu_2)x(t) = 0 \tag{6.106}$$

which, because of the rectangular form of $w(t)$, can be simplified to

$$\int_{-T}^{T} \frac{\sin 2\pi W(t-\tau)}{\pi(t-\tau)} x(\tau)\, d\tau = \lambda x(t) \qquad \text{for all } t \tag{6.107}$$

This is the same integral equation as (6.103) except that it holds for all t. Thus if $\psi_0(t)$ is the optimal band-limited pulse for this problem, then $w(t)\psi_0(t)$ is the optimal pulse shape for the duration-limited pulse in the dual problem. It is perhaps not surprising that the optimal pulses for the more general case involving rectangular weighting in both domains and arbitrary values for I_1 or I_2 can be obtained from the solution of (6.107).

The papers by Slepian, Pollak, and Landau[10]–[12] present the solutions to (6.107) in terms of normalized *prolate spheroidal wave functions* and derive an interesting set of relations between the attainable values of I_1 and I_2 and the parameter WT. The following is a summary of their results:

1. For any value of $c = 2\pi WT$, there is a countable set of real eigenfunctions $\{\psi_n(c, t); n = 0, 1, 2, \ldots\}$ with corresponding real and positive distinct eigenvalues

$$1 > \lambda_0(c) > \lambda_1(c) > \lambda_2(c) > \cdots \tag{6.108}$$

where the ordering is maintained for all values of c.

2. The $\psi_n(c, t)$ are an orthonormal set of band-limited functions over $(-\infty, \infty)$ and are complete in the $L^2(-\infty, \infty)$ subspace of band-limited functions

$$\int_{-\infty}^{\infty} \psi_i(t)\psi_j(t)\, dt = \delta_{ij} \tag{6.109}$$

3. In the interval $|t| \leqslant T$, the $\psi_n(c, t)$ are orthogonal and form a complete set in $L^2(-T, T)$.

$$\int_{-T}^{T} \psi_i(t)\psi_j(t)\, dt = \lambda_i(c)\, \delta_{ij} \tag{6.110}$$

We now have a partial answer to the uncertainty problem. In the class of duration-limited functions ($I_1 = 1$), the function $w(t)\psi_0(c, t)$ has the largest fraction of its energy inside the band $|f| \leqslant W$. This fraction is $I_2 = \lambda_0(c)$. The graph in Figure 6.8 indicates the variation of λ_0 with c and compares the frequency-domain concentration of the optimal pulse with that of the rectangular pulse of width $2T$.

For band-limited functions ($I_2 = 1$), the function $\psi_0(c, t)$ has the largest fraction of its energy in the time interval $|t| \leqslant T$. The fraction is also $\lambda_0(c)$. Figure 6.8 also serves as a comparison of the time-domain concentrations of the optimal pulse and the $(\sin 2\pi Wt)/\pi t$ pulse.

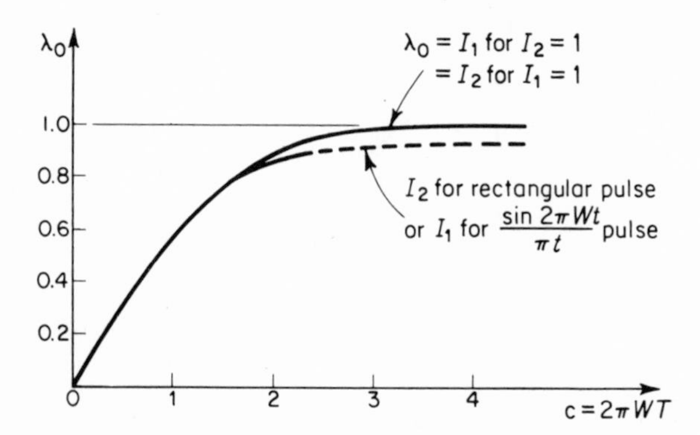

Figure 6.8. Largest eigenvalue of (6.107).

Some idea of the shape of the optimal pulses may be obtained from Figure 6.9, showing the graph of $\psi_0(t)$ for positive values of t (ψ_0 is even) for various values of c.

Finally, the general solution to the uncertainty problem with rectangular weightings where the optimal pulses are neither duration-limited nor band-limited (I_1, $I_2 < 1$) is simply

$$x(t) = p\psi_0(c, t) + qw(t)\psi_0(c, t)$$

where

$$p = \left(\frac{1 - I_1}{1 - \lambda_0}\right)^{\frac{1}{2}}; \qquad q = \left(\frac{I_1}{\lambda_0}\right)^{\frac{1}{2}} - \left(\frac{1 - I_1}{1 - \lambda_0}\right)^{\frac{1}{2}} \tag{6.111}$$

The final result, which is a convenient statement of the uncertainty principle for this problem, gives the maximum attainable value of either I_1 or I_2 when the other is constrained to a particular value. This relationship is

$$\cos^{-1} I_1^{\frac{1}{2}} + \cos^{-1} I_2^{\frac{1}{2}} \geqslant \cos^{-1} \lambda_0^{\frac{1}{2}}(c) \tag{6.112}$$

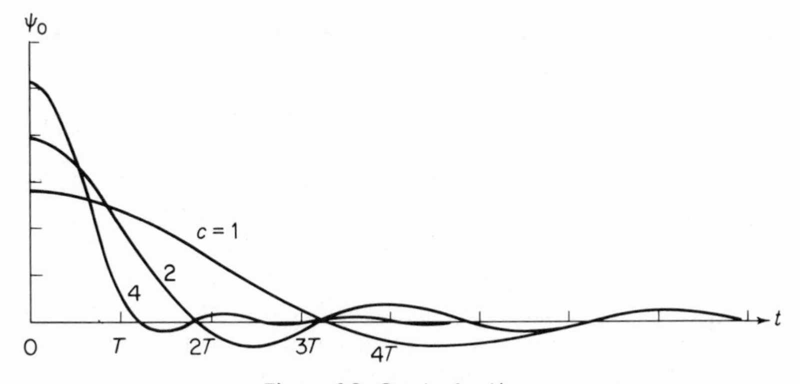

Figure 6.9. Graph of $\psi_0(t)$.

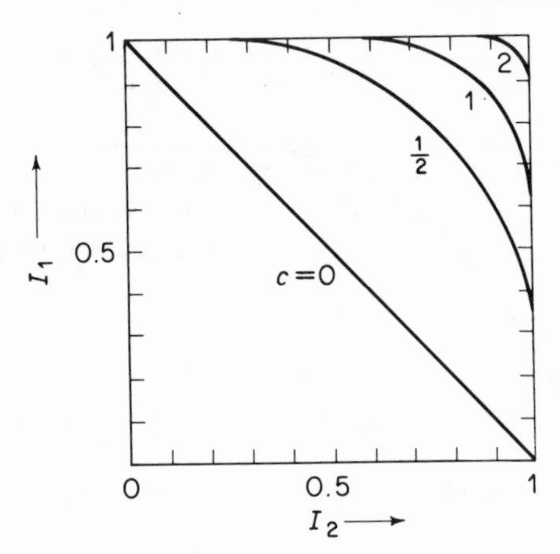

Figure 6.10. Attainable energy concentrations in time and frequency domains.

where the equality is satisfied by (6.111). This result is shown on Figure 6.10 for various values of $c = 2\pi WT$.

The line $I_1 + I_2 = 1$ ($c = 0$) corresponds to the fact that if the sum of the energy in the specified time- and frequency-domain intervals is less than the total energy of the pulse, then there is no restriction on the size of these intervals. Only when $I_1 + I_2$ exceeds the total energy is there a lower bound on the attainable WT product.

Intersymbol Interference

A somewhat different uncertainty problem is motivated by the design of pulses for a communication system which carries information in the amplitudes of a time sequence of pulses. The information is to be recovered from the received signal

$$s(t) = \sum_k a_k x(t - \tau_k) \tag{6.113}$$

by sampling at the appropriate times τ_k, giving sample values $\hat{a}_k$ as estimates of the corresponding a_k.

$$\hat{a}_k = s(\tau_k) = a_k x(0) + \sum_{j;\, j \neq k} a_j x(\tau_k - \tau_j) \tag{6.114}$$

The last term in (6.114) is an error which is called *intersymbol interference* inasmuch as it is produced by adjacent data pulses. For small intersymbol interference we require narrow pulse shapes $x(t)$ so that the values $x(\tau_k - \tau_j)$; $k \neq j$ will all be small. If there are also rather severe bandwidth restrictions on the pulses, it might seem that the preceding problem where pulse energy outside a specified time interval is minimized while maintaining a specified out-of-band energy would be quite appropriate in this case. Because of the asymptotic behavior ($1/t$) of the optimal pulses of the previous example, however, it is possible to choose long pulse sequences which produce an arbitrarily large amount of intersymbol interference at a particular time. In order to counteract this effect, it is desirable to force a faster rate of decay of the "tails" of the individual data pulses by using a different weighting function in the time domain; for example, $w(t) = t^2$. Accordingly, we formulate a new problem by choosing

$$\begin{aligned} w(t) &= t^2 \qquad \text{for all } t \\ V(f) &= \tfrac{1}{2}[1 + \operatorname{sgn}(W - |f|)] \end{aligned} \tag{6.115}$$

and asking, What pulse shape having a specified fraction γ of its total energy outside the frequency interval $|f| \leqslant W$ has the minimum second moment?

The necessary condition (6.96) reduces to a simple second-order differential equation in the frequency domain:

$$\begin{aligned} -(2\pi)^2\mu_1 X''(f) + X(f) &= 0 \qquad \text{in } |f| < W \\ -(2\pi)^2\mu_1 X''(f) + (\mu_2 + 1)X(f) &= 0 \qquad \text{in } |f| > W \end{aligned} \tag{6.116}$$

The solutions to (6.116) which satisfy the boundary conditions $X(\pm\infty) = 0$ and $X(W^+) = X(W^-)$, which are required in order that the functionals be bounded, are given by

$$\begin{aligned} X(f) &= A \cos 2\pi \frac{\alpha f}{W} \qquad \text{in } |f| \leqslant W \\ &= A e^{2\pi\beta} (\cos 2\pi\alpha) e^{-2\pi\beta \left|\frac{f}{W}\right|} \qquad \text{in } |f| > W \end{aligned} \tag{6.117}$$

The constants A, α, and β are determined by substituting (6.117) into the functionals I_1, I_2, and I_3 and making $I_2 = \gamma$ and $I_3 = 1$ while minimizing

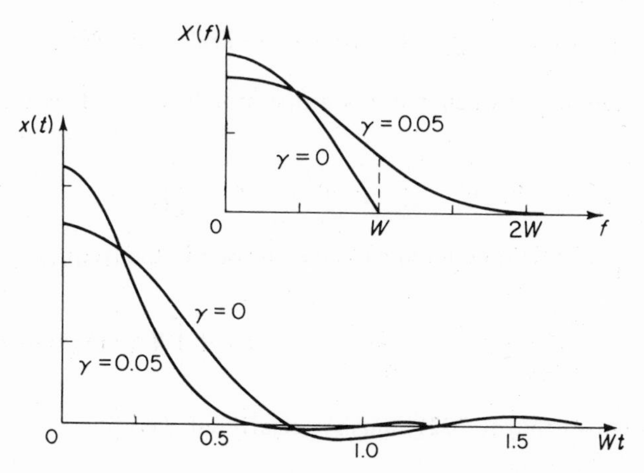

Figure 6.11. Pulses with minimum second moment and specified out-of-band energy.

I_1. This leads to the relations

$$\gamma = \frac{\cos^2 2\pi\alpha}{1 + 2\pi\alpha \tan 2\pi\alpha}$$
$$\beta = \alpha \tan 2\pi\alpha \tag{6.118}$$

In particular, for $\gamma = 0$, we find $\alpha = \frac{1}{4}$ and the Fourier transform of the band-limited pulse of minimum second moment is the "half-cosine" shape. The pulse itself is proportional to the sum of two $(\sin 2\pi Wt)/\pi t$ pulses displaced by $1/2W$ seconds. As shown in Figure 6.11, allowing even a relatively small amount of out-of-band energy causes a marked decrease in the width and magnitude of the tails of the optimal pulses.

Synchronous Pulse Sequences, Nyquist Criterion, Sampling Theorem

If the pulse sequence (6.113) is synchronous, in the sense that $\tau_k = k\tau_0$; $k = 0, \pm 1, \pm 2, \ldots$, then it is possible to select pulses which make the intersymbol interference vanish, even if the pulses are required to be band-limited. Assuming that the a_k can take on arbitrary values in (6.114), we see that intersymbol interference is zero if, and only if, we can make $x(k\tau_0) = 0$ for $k = \pm 1, \pm 2, \ldots$. This is equivalent to the requirement

$$\sum_{k;k\neq 0} x^2(k\tau_0) = \sum_{k=-\infty}^{\infty} x^2(k\tau_0) - x^2(0) = 0 \tag{6.119}$$

which can be expressed as a quadratic functional on $\mathbf{x}$.

$$(\mathscr{A}\mathbf{x}, \mathbf{x}) = 0$$

where

$$A(t, \tau) = \sum_{k=-\infty}^{\infty} \delta(t - k\tau_0)\, \delta(\tau - k\tau_0) - \delta(t)\, \delta(\tau) \tag{6.120}$$

The corresponding frequency-domain operator from (6.24) is given by (see Exercise 4.3).

$$B(f, \nu) = \frac{1}{\tau_0} \sum_{m=-\infty}^{\infty} \delta\left(f - \nu - \frac{m}{\tau_0}\right) - 1 \tag{6.121}$$

Hence the quadratic functional can be expressed in the frequency domain as

$$(\mathscr{B}\mathbf{X}, \mathbf{X}) = \int_{-\infty}^{\infty} \frac{1}{\tau_0}\left[\sum_m X\left(f - \frac{m}{\tau_0}\right)\right] X^*(f)\, df - \iint_{-\infty}^{\infty} X^*(f) X(\nu)\, d\nu\, df \tag{6.122}$$

The requirement that (6.122) vanish is clearly satisfied by

$$\sum_{m=-\infty}^{\infty} X\left(f - \frac{m}{\tau_0}\right) = \tau_0 \int_{-\infty}^{\infty} X(\nu)\, d\nu = \tau_0 x(0) \tag{6.123}$$

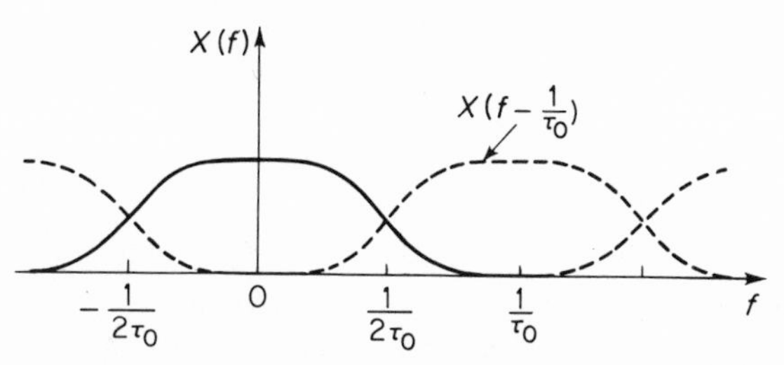

Figure 6.12. Fourier transform of a typical pulse satisfying the Nyquist criterion.

This condition, (6.123), is known as the *Nyquist criterion*[18] for eliminating intersymbol interference in a sequence of uniformly spaced pulses (see also Exercise 1.11). The Fourier transform of a typical pulse which satisfies the criterion is shown in Figure 6.12. It is noticed that $X(f)$ must exhibit a certain type of odd symmetry about the point $f = 1/2\tau_0$ in order that all the translated versions of $X(f)$ will sum up to a constant value as required by (6.123).

If we consider band-limited pulses, with bandwidth W, the criterion can be satisfied with $W \geqslant 1/2\tau_0$. For $W = 1/2\tau_0$, the criterion is satisfied only with $X(f)$ having a rectangular shape.

$$\begin{aligned} X(f) &= \tau_0 x(0) \qquad \text{for } |f| \leqslant \frac{1}{2\tau_0} \\ &= 0 \qquad \text{for } |f| > \frac{1}{2\tau_0} \end{aligned} \tag{6.124}$$

A further insight into the significance of the Nyquist criterion is given by the *Poisson sum formula*[19] which states that, for an arbitrary $x(t)$,

$$\frac{1}{\tau_0}\sum_{\ell=-\infty}^{\infty} X\left(f - \frac{\ell}{\tau_0}\right) = \sum_{k=-\infty}^{\infty} x(k\tau_0)e^{j2\pi k\tau_0 f} \tag{6.125}$$

Hence the sample values $x(k\tau_0)$ are simply the Fourier coefficients for the periodic function

$$\frac{1}{\tau_0}\sum_{\ell} X\left(f - \frac{\ell}{\tau_0}\right)$$

For no intersymbol interference, $x(k\tau_0) = 0$; $k \neq 0$, which requires that the left-hand side of (6.125) be equal to a constant. These ideas are closely related to those of time-series representation for band-limited signals. If $x(t)$ and $s(t)$ are band-limited to $W = 1/2\tau_0$, then we can show that

$$\sum_{k=-\infty}^{\infty} s(k\tau_0)x(t - k\tau_0) = \sum_{k=-\infty}^{\infty} x(k\tau_0)s(t - k\tau_0) \tag{6.126}$$

This result is demonstrated by taking the Fourier transform of each side of (6.126) and using the Poisson sum formula (6.125) to give (see Exercise 4.3)

$$\frac{1}{\tau_0} X(f) \sum_{\ell=-\infty}^{\infty} S\left(f - \frac{\ell}{\tau_0}\right) = \frac{1}{\tau_0} S(f) \sum_{\ell=-\infty}^{\infty} X\left(f - \frac{\ell}{\tau_0}\right) \tag{6.127}$$

which is clearly satisfied if $X(f) = 0$ and $S(f) = 0$ for $|f| > 1/2\tau_0$. Now let $s(t)$ be an arbitrary band-limited signal and let $x(t)$ satisfy the Nyquist condition (6.124). Then, from (6.126), with the additional requirement that $x(0) = 1$, we have

$$\begin{aligned} s(t) &= \sum_{k=-\infty}^{\infty} s(k\tau_0)x(t - k\tau_0) \\ &= \sum_{k=-\infty}^{\infty} s(k\tau_0) \frac{\sin (\pi/\tau_0)(t - k\tau_0)}{(\pi/\tau_0)(t - k\tau_0)} \end{aligned} \tag{6.128}$$

This result is the *sampling theorem*, discussed in Chapter 1, for band-limited signals. We have shown that an arbitrary band-limited signal, of bandwidth W, can be uniquely represented by its sample values taken at intervals of $1/2W$ seconds apart. For a finite-energy signal, only a finite number of these samples, say n, can be substantially different from zero. If the significant samples are concentrated in an interval $2T = n\tau_0$, then the number of significant samples is $n = 4TW$. Because of this relationship, we are tempted to say that the class of signals characterized by an approximate bandwidth W and approximate duration $2T$ is essentially contained in a subspace of $4WT$ dimensions. A more detailed examination of this conjecture is the subject of the following section.

Exercise 6.10. Select a point on or below the $c = 0$ line in Figure 6.10. Sketch the signal that exhibits this amount of energy concentration in time and frequency for very small values of c.

Exercise 6.11. Find the Fourier transform of the real, unit energy, band-limited pulse that has minimum fourth moment; i.e., in (6.94), let $w(t) = t^4$, $V(f) = \frac{1}{2}[1 + \text{sgn}(W - |f|)]$, and $I_2 = I_3 = 1$.

6.8 Approximate Dimensionality of a Signal Space

A question which is intimately related to the uncertainty problem concerns the possibility of classification of subsets of signals by means of constraints on the duration and bandwidth properties of the signals. More specifically, we shall attempt to determine the basis for a finite-dimensional subspace which will be adequate for representation of all signals in a certain restricted class of signals. Let us suppose that we are able to classify signals in terms of a linear operator $\mathscr{A}$, the class being given by

$$\{\mathbf{y};\, \mathbf{y} = \mathscr{A}\mathbf{x} \text{ for all } \mathbf{x} \text{ in } L^2(-\infty, \infty)\} \tag{6.129}$$

It is apparent that classes corresponding to some operators are not in any sense finite-dimensional. Consider, for example, the identity operator; for any finite-dimensional subspace, we can always find an $\mathbf{x}$ in $L^2(-\infty, \infty)$ which is orthogonal to the subspace. On the other hand, there is a broad class of operators producing signals close to a finite-dimensional subspace, in the sense that

$$\left\| \mathbf{y} - \sum_{k=1}^{n} \alpha_k \boldsymbol{\varphi}_k \right\| \leqslant \varepsilon_n \tag{6.130}$$

where $\mathbf{y} = \mathscr{A}\mathbf{x}$; $\|\mathbf{x}\| = 1$ and ε_n can be made arbitrarily small with finite n. These are the compact operators discussed in Section 5.5.

Let us suppose that $\{\varphi_k(t);\, k = 1, 2, \ldots, n\}$ is an orthonormal basis for the subspace in question, then we would naturally pick $\alpha_k = (\mathbf{y}, \boldsymbol{\varphi}_k)$ in (6.130) in order to minimize the norm of the error for any particular $\mathbf{y}$. When this is done, the square of the distance between $\mathbf{y}$ and its orthogonal projection on the subspace is given by

$$\begin{aligned} I(\mathbf{y}) &= (\mathbf{y}, \mathbf{y}) - \sum_{k=1}^{n} |(\mathbf{y}, \boldsymbol{\varphi}_k)|^2 \\ &= (\mathscr{A}\mathbf{x}, \mathscr{A}\mathbf{x}) - \sum_{k=1}^{n} |(\mathscr{A}\mathbf{x}, \boldsymbol{\varphi}_k)|^2 \\ &= (\mathscr{A}'\mathscr{A}\mathbf{x}, \mathbf{x}) - \sum_{k=1}^{n} |(\mathbf{x}, \mathscr{A}'\boldsymbol{\varphi}_k)|^2 \end{aligned} \tag{6.131}$$

We now seek the particular basis $\{\boldsymbol{\varphi}_k;\, k = 1, 2, \ldots, n\}$ such that the maximum value of $I(\mathbf{y})$ is minimized subject to $\|\mathbf{x}\| = 1$. This is equivalent to the problem of approximating an operator by a degenerate operator (of rank n) since

$$\begin{aligned} I(\mathbf{y}) &= (\mathscr{C}\mathbf{x}, \mathbf{x}) - (\mathscr{C}_n\mathbf{x}, \mathbf{x}) = ([\mathscr{C} - \mathscr{C}_n]\mathbf{x}, \mathbf{x}) \\ &\leqslant \|\mathscr{C} - \mathscr{C}_n\|\, \|\mathbf{x}\|^2 = \|\mathscr{C} - \mathscr{C}_n\| \qquad \text{for } \|\mathbf{x}\| = 1 \end{aligned} \tag{6.132}$$

where

$$\mathscr{C} = \mathscr{A}'\mathscr{A}$$

and

$$\begin{aligned} \mathscr{C}_n\mathbf{x} &= \int_{-\infty}^{\infty} C_n(t, \tau)x(\tau)\, d\tau \\ C_n(t, \tau) &= \sum_{k=1}^{n} g_k(t)g_k^*(\tau) \end{aligned} \tag{6.133}$$

From (6.131), we have let $\mathbf{g}_k = \mathscr{A}'\boldsymbol{\varphi}_k$.

We now suppose that the operator $\mathscr{C} = \mathscr{A}'\mathscr{A}$ has a discrete spectral representation given by

$$C(t, \tau) = \sum_{i=1}^{\infty} \lambda_i \psi_i(t) \psi_i^*(\tau) \tag{6.134}$$

where the $\boldsymbol{\psi}_i$ and λ_i are the orthonormalized eigenfunctions and eigenvalues of $\mathscr{C}\mathbf{x} = \lambda\mathbf{x}$. Since $\mathscr{C} = \mathscr{A}'\mathscr{A}$ is a non-negative-definite, self-adjoint operator, its eigenvalues are non-negative and real. We assume that these eigenvalues in (6.134) are indexed so that they form a non-increasing sequence, $\lambda_1 \geqslant \lambda_2 \geqslant \lambda_3 \geqslant \cdots \geqslant 0$. It can be shown[20] that for an arbitrary degenerate kernel $D_n(t, \tau)$ (with n-dimensional range) the norm of the operator $\mathscr{C} - \mathscr{D}_n$ cannot be less than λ_{n+1}.

$$\|\mathscr{C} - \mathscr{D}_n\| \geqslant \|\mathscr{C} - \mathscr{C}_n\| = \lambda_{n+1} \tag{6.135}$$

The right-hand equality in (6.135) holds if we choose

$$\mathscr{A}'\boldsymbol{\varphi}_k = \mathbf{g}_k = \lambda_k^{\frac{1}{2}}\boldsymbol{\psi}_k; \qquad k = 1, 2, \ldots, n \tag{6.136}$$

for then the operator $\mathscr{C} - \mathscr{C}_n$ has a kernel given by

$$C(t, \tau) - C_n(t, \tau) = \sum_{k=n+1}^{\infty} \lambda_i \psi_i(t) \psi_i^*(\tau)$$

and the norm of the operator is given by the largest eigenvalue, (5.113). Now it must be shown that there is an orthonormal basis $\{\boldsymbol{\varphi}_k; k = 1, \ldots, n\}$ which satisfies (6.136). Let

$$\boldsymbol{\varphi}_k = \lambda_k^{-\frac{1}{2}} \mathscr{A} \boldsymbol{\psi}_k \tag{6.137}$$

Then it follows that

$$\mathbf{g}_k = \mathscr{A}'\boldsymbol{\varphi}_k = \lambda_k^{-\frac{1}{2}} \mathscr{C} \boldsymbol{\psi}_k = \lambda_k^{\frac{1}{2}} \boldsymbol{\psi}_k; \qquad k = 1, 2, \ldots, n$$

and

$$\begin{aligned} (\boldsymbol{\varphi}_k, \boldsymbol{\varphi}_j) &= (\lambda_k \lambda_j)^{-\frac{1}{2}} (\mathscr{A}\boldsymbol{\psi}_k, \mathscr{A}\boldsymbol{\psi}_j) \\ &= (\lambda_k \lambda_j)^{-\frac{1}{2}} (\mathscr{C}\boldsymbol{\psi}_k, \boldsymbol{\psi}_j) \\ &= \left(\frac{\lambda_k}{\lambda_j}\right)^{\frac{1}{2}} (\boldsymbol{\psi}_k, \boldsymbol{\psi}_j) = \delta_{kj} \end{aligned}$$

To summarize these results, the norm of the error in an n-dimensional approximation to a signal of the form $\mathbf{y} = \mathscr{A}\mathbf{x}$ with $\|\mathbf{x}\| = 1$ is bounded by

$$\left\| \mathbf{y} - \sum_{k=1}^{n} \alpha_k \boldsymbol{\varphi}_k \right\| \leqslant \lambda_{n+1}^{\frac{1}{2}} \tag{6.138}$$

where the α_k and $\boldsymbol{\varphi}_k$ are chosen according to

$$\alpha_k = (\mathbf{y}, \boldsymbol{\varphi}_k); \qquad \boldsymbol{\varphi}_k = \lambda_k^{-\frac{1}{2}} \mathscr{A} \boldsymbol{\psi}_k$$

and λ_{n+1} is the $n + 1$st eigenvalue of the operator $\mathscr{A}'\mathscr{A}$.

If we now add the requirement that $\mathscr{C}$ have a square-integrable kernel, so that $\mathscr{C}$ is Hilbert-Schmidt, and thus compact, see (5.72),

$$\infty > \int_{-\infty}^{\infty}\!\!\int |C(t, \tau)|^2 \, dt \, d\tau = \sum_{i=1}^{\infty} |\lambda_i|^2 \tag{6.139}$$

then it follows that the λ_i tend to zero for increasing i and the bound on the error in (6.138) can be made arbitrarily small by choosing n large enough.

Gating and Filtering

In order to relate these results to duration and bandwidth properties, we consider the operator $\mathscr{A}_1$, where

$$A_1(t, \tau) = h(t - \tau)w(\tau) \tag{6.140}$$

This operator is interpreted physically as a gating operation followed by a time-invariant filtering operation as shown in Figure 6.13. The gating

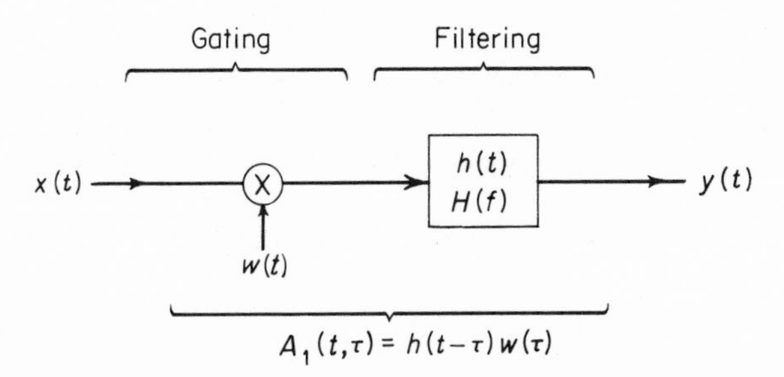

Figure 6.13. Operator providing signals of limited duration and bandwidth.

function $w(t)$ provides, in some sense, a duration limitation of the input signal; while the transfer function $H(f)$ of the filter provides, in some sense, a bandwidth limitation. If **w** and **h** are separately square-integrable, then (6.139) is satisfied and we can proceed to investigate the dimensionality required to represent the class of output signals of $\mathscr{A}_1$ to a prescribed accuracy by examining the size of the eigenvalues of $\mathscr{A}_1'\mathscr{A}_1$. We note here, as we did in Section 5.5, that the gating or filtering operations taken individually do not satisfy (6.139). In fact they do not have discrete spectral representations and there are an infinite number of eigenvalues bounded away from zero. From the discussion at the end of the preceding section we were led to expect that some form of duration and bandwidth limiting is simultaneously required in order to make signals essentially finite-dimensional.

The optimum basis functions for representing $y(t)$ are given, from (6.137), by the gated and then filtered versions of the solutions (corresponding to the n

largest eigenvalues) of

$$\mathscr{C}_1 \cdot x(t) = \int_{-\infty}^{\infty} w^*(t)w(\tau)k(t-\tau)x(\tau)\, d\tau = \lambda x(t) \tag{6.141}$$

where

$$k(t) = \int_{-\infty}^{\infty} h(\sigma)h^*(\sigma - t)\, d\sigma \Rightarrow K(f) = |H(f)|^2 \tag{6.142}$$

If we retain the same gating and filtering functions but reverse their order as shown in Figure 6.14, then

$$A_2(t, \tau) = w(t)h(t-\tau) \tag{6.143}$$

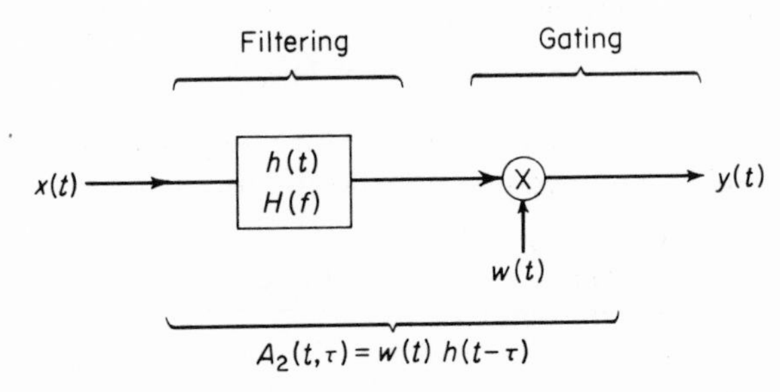

Figure 6.14. Alternate method for limiting duration and bandwidth.

and the eigenvalue equation corresponding to (6.141) is given by

$$\int_{-\infty}^{\infty} C_2(t, \tau)x(\tau)\, d\tau = \lambda x(t) \tag{6.144}$$

where

$$C_2(t, \tau) = \int_{-\infty}^{\infty} |w(\sigma)|^2\, h^*(\sigma - t)h(\sigma - \tau)\, d\sigma$$

In the frequency domain, this equation takes on the same form as (6.141), since (6.144) is equivalent to

$$\int_{-\infty}^{\infty} H^*(f)H(\nu)Z(f-\nu)X(\nu)\, d\nu = \lambda X(f) \tag{6.145}$$

where

$$Z(f) = \int_{-\infty}^{\infty} W^*(\eta - f)W(\eta)\, d\eta \Rightarrow z(t) = |w(t)|^2$$

One thing that we can conclude from this time- and frequency-domain duality is that if the gating and filtering functions employ the same shape; i.e., $H(f) = w(af)$ where a is a real, positive constant, then the integral equations (6.141) and (6.145) have the same eigenvalues and the conclusions regarding the approximate dimensionality are independent of whether the gating precedes the filtering or the filtering precedes the gating. The basis functions for the two cases will be different.

A more general consequence of the duality of (6.141) and (6.145) is that for each operator having the form indicated in Figure 6.13 there is a corresponding operator with suitably modified gating and filtering functions and with filtering preceding the gating (Figure 6.14) such that the eigenvalues of $\mathcal{A}'\mathcal{A}$ are the same in both cases (see Exercise 6.13.)

Example 6.7. If we employ rectangular gating preceding filtering, $w(t) = \frac{1}{2}[1 + \text{sgn}\,(T - |t|)]$, then (6.141) becomes

$$\int_{-T}^{T} k(t - \tau)x(\tau)\, d\tau = \lambda x(t) \qquad \text{for } |t| \leqslant T \tag{6.146}$$

which is identical to (6.56) in the general version of Chalk's problem (Example 6.2). In particular, if the filtering corresponds to that of a single-section RC lowpass filter, then the eigenvalues of (6.146), when properly ordered, are

$$\lambda_i = \frac{(2\pi f_0 T)^2}{a_i^2 + (2\pi f_0 T)^2} \tag{6.147}$$

where

$$\begin{aligned} \tan a_i &= \frac{2\pi f_0 T}{a_i} \qquad \text{for } i = 1, 3, 5, \ldots \\ &= \frac{-a_i}{2\pi f_0 T} \qquad \text{for } i = 2, 4, 6, \ldots \end{aligned}$$

as indicated by (6.61) and (6.62). A graphical solution of (6.147), illustrated in Figure 6.15, shows that

$$(n - 1)\frac{\pi}{2} < a_n < n\frac{\pi}{2}$$

Hence

$$\frac{1}{1 + [(n + 1)/4f_0 T]^2} < \lambda_{n+1} < \frac{1}{1 + (n/4f_0 T)^2} \tag{6.148}$$

As $f_0 T \to 0$, then $\lambda_1 \to 2\pi f_0 T$ is the dominant eigenvalue $[\lambda_2 \to (4f_0 T)^2]$ and a one-dimensional subspace is adequate for representing the filter output. This is intuitively apparent since the filter impulse response is very wide compared to $2T$ in this case and the gated input to the filter appears as an impulsive excitation; thus the filter output is essentially proportional to the impulse response $h(t)$ for any gated input signal.

For large values of $f_0 T$ there will be many eigenvalues near unity, and for $n \simeq 4f_0 T$ we have $\lambda_{n+1} \simeq \frac{1}{2}$. This indicates that for suitable approximation of this class of signals we may require that n be several times larger than $4f_0 T$. The form of the expressions in (6.148) bounding the size of λ_{n+1} suggests that

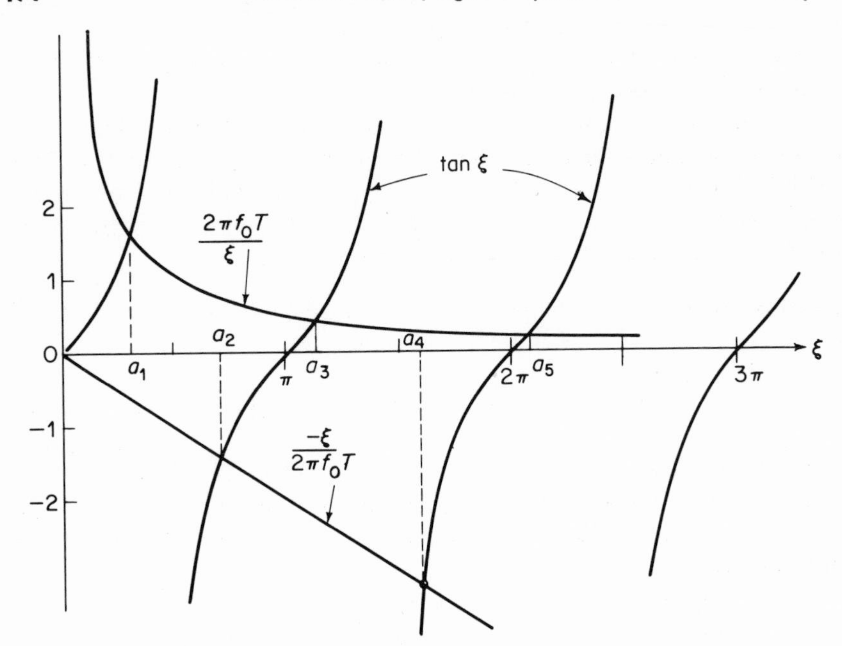

Figure 6.15. Graphical determination of eigenvalues of (6.146).

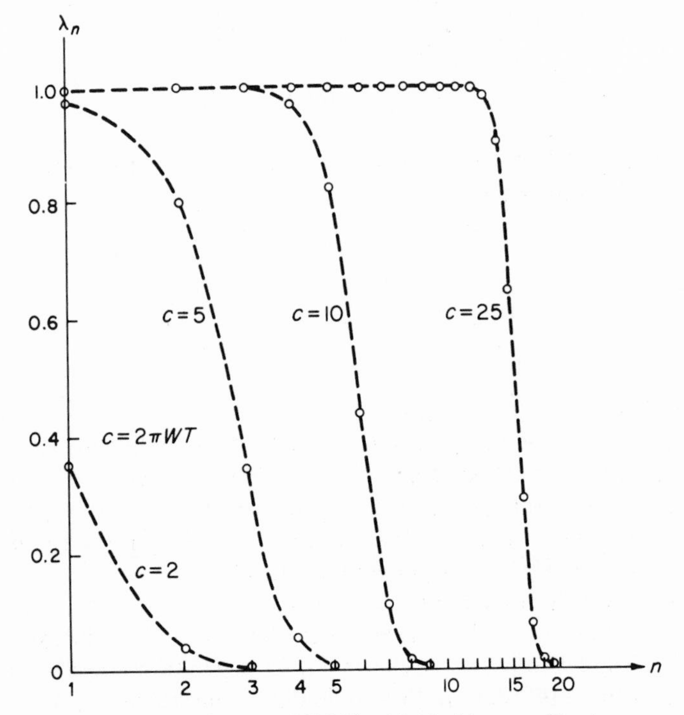

Figure 6.16. Eigenvalues of (6.146) with ideal lowpass filtering.

the relatively slow rate of decrease of λ_n with increasing n is due in part to the gradual decay of $K(f) = |H(f)|^2$ with increasing frequency. If this reasoning is valid, then the λ_n should drop off very sharply after a certain point if ideal lowpass filtering [rectangular $K(f)$] is employed. ▌

Example 6.8. The eigenvalues of (6.146) with

$$K(f) = \tfrac{1}{2}[1 + \operatorname{sgn}(W - |f|)] \tag{6.149}$$

have been evaluated and tabulated[21] for a wide range of values of n and $c = 2\pi WT$. The trends are shown graphically on Figure 6.16 where the sharp drop-off for large c is apparent.

As noted previously, the eigenfunctions are the prolate spheroidal wave functions as are the basis functions in this case [refer to (6.104)]. The problem of approximate dimensionality of a signal space with somewhat less restrictive constraints on duration and bandwidth is presented in detail in Reference [12]. One of the principal results can be summarized as follows. Let $y(t)$ be a signal which satisfies the constraints

$$\begin{aligned} I_1 &= (\mathbf{wy}, \mathbf{y}) = 1 - \varepsilon_T^2 \\ I_2 &= (\mathbf{VY}, \mathbf{Y}) = 1 - \eta_W^2 \\ I_3 &= (\mathbf{y}, \mathbf{y}) = 1 \end{aligned} \tag{6.150}$$

where $w(t)$ and $V(f)$ are the rectangular weighting functions

$$\begin{aligned} w(t) &= \tfrac{1}{2}[1 + \operatorname{sgn}(T - |t|)] \\ V(f) &= \tfrac{1}{2}[1 + \operatorname{sgn}(W - |f|)] \end{aligned} \tag{6.151}$$

Then we can find n constants $\{\alpha_k; k = 0, 1, \ldots, n-1\}$ such that

$$\left\| \mathbf{y} - \sum_{k=0}^{n-1} \alpha_k \boldsymbol{\psi}_k \right\|^2 \leqslant 12(\varepsilon_T + \eta_W)^2 + \eta_W^2 \tag{6.152}$$

where the $\boldsymbol{\psi}_k$ are the prolate spheroidal wave functions and n is the smallest integer greater than $4WT$. ▌

Exercise 6.12. Verify that $\mathscr{C} = \mathscr{A}'\mathscr{A}$ is a self-adjoint, non-negative-definite operator. Show the block diagram network realizations for the operators $\mathscr{C}_1$ and $\mathscr{C}_2$ in (6.141) and (6.144).

Exercise 6.13. For the networks shown below with operators $\mathscr{A}_1$ and $\mathscr{A}_2$, show that the operators $\mathscr{C}_1 = \mathscr{A}_1'\mathscr{A}_1$ and $\mathscr{C}_2 = \mathscr{A}_2'\mathscr{A}_2$ have the same eigenvalues if $H_2(f) = w_1(af)$ and $w_2(t) = H_1(t/a)$, where a is a real, positive constant.

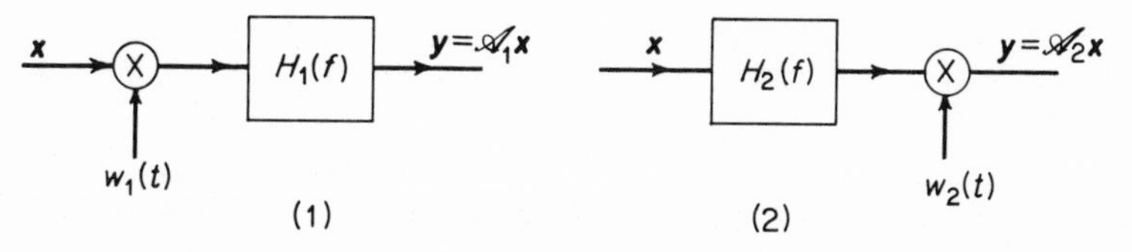

Exercise 6.14. If an operator corresponding to gating followed by filtering is then followed by another gating operation, we would expect a reduction in the "dimensionality" of the output signals. As a particular illustration, show that when rectangular gating and filtering is used, the eigenvalues of $\mathscr{C}_3$ are simply the squares of the eigenvalues of $\mathscr{C}_1$. *Hint:* Show that $\mathscr{A}_3 = \mathscr{C}_1 = \mathscr{A}_1'\mathscr{A}_1$.

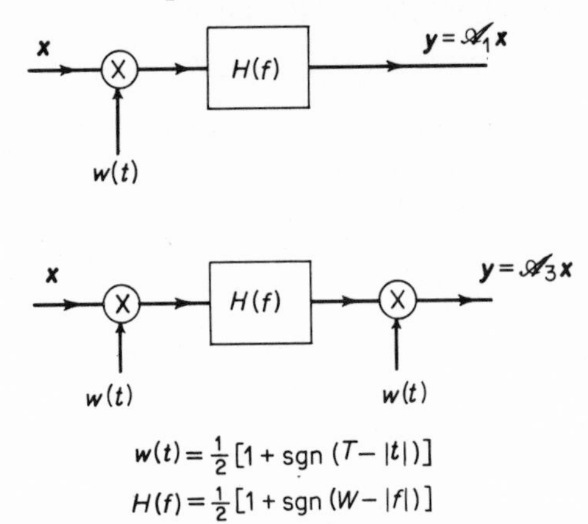

$$w(t) = \tfrac{1}{2}[1 + \text{sgn}\,(T - |t|)]$$
$$H(f) = \tfrac{1}{2}[1 + \text{sgn}\,(W - |f|)]$$

REFERENCES

1. D. Gabor, "Theory of Communication," *Jour. IEE*, Pt. III, Vol. 93, pp. 429–57 (November, 1946).
2. Courant and Hilbert, *Methods of Mathematical Physics*, Vol. 1, Interscience, 1953.
3. J. H. H. Chalk, "The Optimum Pulse-Shape for Pulse Communication," *Proc. IEE*, Pt. III, Vol. 97, pp. 82–92 (March, 1950).
4. D. A. Pierre, "Optimal Time-Limited Energy Constrained Inputs to Linear Systems," Symposium on Signal Transmission and Processing, *IEEE Conference Record 4C9*.
5. A. E. Laemmel, "Optimum Pulse Shape for Digital Transmission," *Proc. Symposium on Modern Network Synthesis*, Polytechnic Institute of Brooklyn, pp. 211–20 (April, 1955).
6. W. R. Bennett and J. R. Davey, *Data Transmission*, Mc-Graw-Hill, 1965.
7. T. Taylor, "Design of Line-Source Antennas for Narrow Beamwidth and Low Side-Lobes," *Trans. IRE-PG on Antennas and Propagation*, Vol. AP-3, No. 1. pp. 16–20 (January, 1955).
8. J. A. Ville and J. Bouzitat, "Note Sur un Signal de Duree Finie et d'energie Filtree Maximum," *Cables et Transmission*, Vol. 11, No. 2, pp. 102–127, April, 1957.
9. M. S. Gurevich, "Signals of Finite Duration Containing Maximum Filtered Energy for a Given Bandwidth" (in Russian), *Radioteknika I Elektronika*, Vol. 1, No. 3, pp. 313–19 (March, 1956).

10. D. Slepian and H. O. Pollak, "Prolate Spheroidal Wave Functions, Fourier Analysis and Uncertainty–I," *Bell Sys. Tech. Jour.*, Vol. 40, No. 1, pp. 43–63 (January, 1961).

11. H. J. Landau and H. O. Pollak, "Prolate Spheroidal Wave Functions, Fourier Analysis and Uncertainty–II," *Bell Sys. Tech. Jour.*, Vol. 40, No. 1, pp. 65–84 (January, 1961).

12. H. J. Landau and H. O. Pollak, "Prolate Spheroidal Wave Functions, Fourier Analysis and Uncertainty–III: The Dimension of the Space of Essentially Time- and Band-Limited Signals," *Bell Sys. Tech. Jour.*, Vol. 41, No. 4 (July, 1962).

13. M. Zakai, "A Class of Definitions of 'Duration' (or 'Uncertainty') and the Associated Uncertainty Relations," *Information and Control*, Vol. 3, No. 2, pp. 101–115 (June, 1960).

14. I. Kay and R. A. Silverman, "On the Uncertainty Relation for Real Signals," *Information and Control*, Vol. 1, No. 1, pp. 64–75 (September, 1957).

15. R. Bourret, "A Note on an Information Theoretic Form of the Uncertainty Principle," *Information and Control*, Vol. 1, No. 4, pp. 398–401 (December 1958).

16. R. Leipnik, "Entropy and the Uncertainty Principle," *Information and Control*, Vol. 2, No. 1, pp. 64–79 (April, 1959).

17. Jeffreys and Jeffreys, *Methods of Mathematical Physics*, Third Edition, Cambridge, 1956, Chapter 23.

18. H. Nyquist, "Certain Topics in Telegraph Transmission Theory," *Trans. AIEE*, Vol. 47, pp. 617–44 (April, 1928).

19. A. Papoulis, *The Fourier Integral and its Applications*, McGraw-Hill, 1962, Chapter 3.

20. F. Riesz and B. Sz.-Nagy, *Functional Analysis*, Frederick Ungar, 1955.

21. D. Slepian and Mrs. E. Sonnenblick, "Eigenvalues Associated with Prolate Spheroidal Wave Functions of Zero Order," *Bell Sys. Tech. Jour.*, Vol. 44, No. 8, pp. 1745–59 (October, 1965).

REPRESENTATION OF RANDOM SIGNAL PROCESSES

7

7.1 Introduction

In a physical system, it often happens that some of the signal sources are capable of producing any one of a large (possibly uncountable) set of time functions. For purposes of analysis, it is convenient to assign a probability law governing the occurrence of each member of the set. The set (along with the probability law) is referred to as the *ensemble* of signals produced by a particular source. We shall say that such a source produces a signal $\mathbf{x}$, called a *random process* (also sometimes called a *stochastic process*). The description of a random signal process must necessarily be quite different from that for the deterministic signals considered previously. It turns out, however, that the signal space concepts (such as distance, norms, inner products, and orthogonality) are equally useful for characterizing random processes. Furthermore, in the most interesting system problems, the two types of description (deterministic and random) appear simultaneously and are strongly interrelated. This will be apparent in the optimum filtering problems discussed in Chapter 9 and also in the problem of finding optimum basis functions for random signals presented later in this chapter.

To be more specific, a random process is defined as a set of *jointly random variables*, each indexed by a time parameter t.

$$\mathbf{x} = \{\mathbf{x}(t);\, t \in T\} \tag{7.1}$$

If T is a countable set, then we say $\mathbf{x}$ is a *discrete-time process*. If T is an interval on the real line, then $\mathbf{x}$ is called a *continuous-time process*. It is important to note that the $\mathbf{x}(t)$ in (7.1) are *not* scalar-valued quantities. For most practical applications, the random variable $\mathbf{x}(t)$ can be viewed as a vector in a Hilbert space.[1],[2] The parameter t simply indexes points in this Hilbert space. Typically, a system designer is interested in characterizing system performance for the entire ensemble of signals rather than for a particular deterministic signal selected from the ensemble. He is able to express this "average" performance in terms of certain statistical averages applied to the random variables. These averages are called *expectations*. Hence, in these problems, the time-domain behavior of a random process is characterized by the expression of various expectations as deterministic functions of time, not by individual signal waveforms. Failure to appreciate this essential difference between descriptions of deterministic and random signals is a common source of confusion.

It is assumed that the reader has a basic background in probability and random variables.[3] The following section serves mainly as a brief summary and as a point to introduce notation that will be employed later.

7.2 Random Variables and Expectations

The *random variable* $\mathbf{x}$ (we drop the index t for the moment)[1] is a *function* defined on a *sample space* Ω. The sample space is a set of *outcomes* $\Omega = \{\omega\}$, and the scalar $x(\omega)$ is called a *realization* of the random variable $\mathbf{x}$. The sample space also has a probability law which assigns probabilities to certain subsets of Ω. The *expected value* $E[\mathbf{x}]$ of the random variable can be interpreted as a linear functional on the function $\mathbf{x}$, expressed heuristically as

$$E[\mathbf{x}] = \sum_{\omega} x(\omega)P(\omega); \qquad \omega \in \Omega \tag{7.2}$$

where $P\colon \Omega \to R$ is a probability function defined on Ω. The *mean-squared value* of $\mathbf{x}$ is a quadratic functional on $\mathbf{x}$ which will be interpreted as the square of the norm of $\mathbf{x}$, expressed by

$$E[\mathbf{x}^2] = \sum_{\omega} x^2(\omega)P(\omega) = \|\mathbf{x}\|^2 \tag{7.3}$$

The difficulty with expressions (7.2) and (7.3) is that the indicated summations may very well be over uncountable sets and, furthermore, it often happens that there is not a clear-cut physical mechanism which relates the random variable to a sample space description. Fortunately, there is an alternative description which completely characterizes random variables in

[1] In this section we index different random variables with subscript integers. When, in later sections, we denote a random process by $\mathbf{x}$, we shall use $\mathbf{x}(t)$ to denote its constituent random variables, (7.1).

terms of *probability density functions.* The probability density description tends to obscure the concept of a Hilbert space as a setting for random variables. Nevertheless, the expectation that defines an inner product (and hence linear and quadratic functionals) is easily expressed in terms of an appropriate probability density function.

A real random variable can be completely characterized by the probability associated with the event $[x(\omega) \leqslant a]$ for all real values of a. The probability density function $p_x(\xi)$, is defined by

$$\Pr [x(\omega) \leqslant a] = \int_{-\infty}^{a} p_x(\xi)\, d\xi \tag{7.4}$$

If $\mathbf{x}$ takes on certain values $a_1, a_2, \ldots,$ with non-zero probabilities $p_1, p_2, \ldots,$ then this feature is reflected in the probability density function as a sequence of δ-functions.[3] These δ-functions may be subtracted out, leaving a remainder density function $\tilde{p}_x(\xi)$ free from singularities.

$$p_x(\xi) = \sum_i p_i\, \delta(\xi - a_i) + \tilde{p}_x(\xi) \tag{7.5}$$

The expected value of a random variable, also called its *mean* value, is defined as

$$E[\mathbf{x}] = \int_{-\infty}^{\infty} \xi p_x(\xi)\, d\xi = \bar{\mathbf{x}} \tag{7.6}$$

The overbar notation, shown in (7.6), is frequently used as a convenient alternative symbolism for the expectation operation. The mean-squared value of $\mathbf{x}$ is given by

$$E[\mathbf{x}^2] = \int_{-\infty}^{\infty} \xi^2 p_x(\xi)\, d\xi = \overline{\mathbf{x}^2} \tag{7.7}$$

Another statistical parameter of importance is the mean-squared value of the zero-mean random variable $\mathbf{x} - \bar{\mathbf{x}}$ called the *variance* of $\mathbf{x}$ and denoted by σ_x^2.

$$\sigma_x^2 = E[(\mathbf{x} - \bar{\mathbf{x}})^2] = E[\mathbf{x}^2] - 2E[\mathbf{x}\bar{\mathbf{x}}] + E[\bar{\mathbf{x}}^2] = \overline{\mathbf{x}^2} - \bar{\mathbf{x}}^2 \tag{7.8}$$

A random variable can also be described by its *characteristic function,* defined by

$$Q_x(\mu) = E[e^{j\mu\mathbf{x}}] = \int_{-\infty}^{\infty} p_x(\xi) e^{j\mu\xi}\, d\xi; \qquad \mu \in R \tag{7.9}$$

The close relationship of the characteristic function to the Fourier transform of the probability density function is apparent in (7.9); i.e.,

$$Q_x(\mu) = P_x(-\mu/2\pi)$$

Two or more real random variables are said to be jointly distributed if they are functions defined on the same sample space and can be described by a *joint probability density function* defined by

$$\Pr[x_1(\omega) \leqslant a_1, x_2(\omega) \leqslant a_2, \ldots, x_n(\omega) \leqslant a_n]$$
$$= \int_{-\infty}^{a_1} \int_{-\infty}^{a_2} \cdots \int_{-\infty}^{a_n} p_{x_1 x_2 \cdots x_n}(\xi_1, \xi_2, \ldots, \xi_n)\, d\xi_1\, d\xi_2 \ldots d\xi_n \qquad (7.10)$$

The jointly distributed real random variables can also be described equivalently by the *joint characteristic function*

$$Q_{x_1 x_2 \cdots x_n}(\mu_1, \mu_2, \ldots, \mu_n) = E[e^{j(\mu_1 \mathbf{x}_1 + \mu_2 \mathbf{x}_2 + \cdots + \mu_n \mathbf{x}_n)}]$$
$$= \int_{-\infty}^{\infty} \int_{-\infty}^{\infty} \cdots \int_{-\infty}^{\infty} p_{x_1 x_2 \cdots x_n}(\xi_1, \xi_2, \ldots, \xi_n)$$
$$\times e^{j(\mu_1 \xi_1 + \mu_2 \xi_2 + \cdots + \mu_n \xi_n)}\, d\xi_1\, d\xi_2 \ldots d\xi_n \qquad (7.11)$$

In addition to the means and variances of the individual random variables, we are most often interested in the expected value of the product of a pair of random variables. This quantity is called the *correlation* of the random variables, and for two random variables $\mathbf{x}_1$ and $\mathbf{x}_2$ it is expressed as

$$E[\mathbf{x}_1\, \mathbf{x}_2] = \int_{-\infty}^{\infty} \int_{-\infty}^{\infty} \xi_1 \xi_2 p_{x_1 x_2}(\xi_1, \xi_2)\, d\xi_1\, d\xi_2$$
$$= -\frac{\partial^2}{\partial \mu_1\, \partial \mu_2} Q_{x_1 x_2}(\mu_1, \mu_2)\bigg|_{\substack{\mu_1 = 0 \\ \mu_2 = 0}} \qquad (7.12)$$

The correlation provides the desired geometric structure for the space of random variables. This is because (7.12) is a valid relation for an inner product of $\mathbf{x}_1$ and $\mathbf{x}_2$ [see (2.28)], and the concepts of norms and distances are induced from this inner product. To show that $E[\mathbf{x}_1 \mathbf{x}_2]$ satisfies properties (a)–(c) of (2.28), we note that property (a) for an inner product requires only symmetry since we are dealing here with real vectors. This symmetry is evident from (7.12). Property (b) is the requirement of linearity which is established by

$$E[(\alpha_1 \mathbf{x}_1 + \alpha_2 \mathbf{x}_2)\mathbf{x}_3]$$
$$= \int_{-\infty}^{\infty} \int_{-\infty}^{\infty} \int_{-\infty}^{\infty} (\alpha_1 \xi_1 + \alpha_2 \xi_2) \xi_3 p_{x_1 x_2 x_3}(\xi_1, \xi_2, \xi_3)\, d\xi_1\, d\xi_2\, d\xi_3$$
$$= \alpha_1 \iint_{-\infty}^{\infty} \xi_1 \xi_3 p_{x_1 x_3}(\xi_1, \xi_3)\, d\xi_1\, d\xi_3 + \alpha_2 \iint_{-\infty}^{\infty} \xi_2 \xi_3 p_{x_2 x_3}(\xi_2, \xi_3)\, d\xi_2\, d\xi_3$$
$$= \alpha_1 E[\mathbf{x}_1 \mathbf{x}_3] + \alpha_2 E[\mathbf{x}_2\, \mathbf{x}_3] \qquad (7.13)$$

In (7.13) we have used a fundamental property of joint density functions which states that[3]

$$\int_{-\infty}^{\infty} p_{x_1 x_2 x_3}(\xi_1, \xi_2, \xi_3)\, d\xi_1 = p_{x_2 x_3}(\xi_2, \xi_3)$$

Finally, for property (c), we note that

$$E[\mathbf{x}_1^2] = \int_{-\infty}^{\infty} \xi^2 p_{x_1}(\xi)\, d\xi \geqslant 0 \tag{7.14}$$

This follows from the fact that $p_{x_1}(\xi)$ is a non-negative function. Also, $E[\mathbf{x}_1^2]$ is zero if, and only if, $\mathbf{x}_1$ assumes non-zero values with zero probability.

Having established an inner product on the space of random variables, we shall interpret the distance between random variables in terms of their *root-mean-squared difference;* i.e.,

$$d(\mathbf{x}_1, \mathbf{x}_2) = \|\mathbf{x}_1 - \mathbf{x}_2\| = \{E[(\mathbf{x}_1 - \mathbf{x}_2)^2]\}^{\frac{1}{2}} \tag{7.15}$$

If the correlation of two random variables vanishes, then they are said to be *orthogonal.* The correlation of the zero-mean random variables $\mathbf{x}_1 - \bar{\mathbf{x}}_1$ and $\mathbf{x}_2 - \bar{\mathbf{x}}_2$ is called the *covariance* m_{12} of $\mathbf{x}_1$ and $\mathbf{x}_2$.

$$\begin{aligned} m_{12} &= E[(\mathbf{x}_1 - \bar{\mathbf{x}}_1)(\mathbf{x}_2 - \bar{\mathbf{x}}_2)] \\ &= E[\mathbf{x}_1\mathbf{x}_2] - \bar{\mathbf{x}}_1\bar{\mathbf{x}}_2 \end{aligned} \tag{7.16}$$

If the covariance vanishes, the random variables are said to be *linearly independent.* From (7.16) we see that orthogonal random variables are linearly independent if either random variable happens to be a zero-mean variable.

A much stronger type of independence is implied by the independence of the events $[x_1(\omega) \leqslant a_1]$ and $[x_2(\omega) \leqslant a_2]$ for all a_1 and a_2. Two random variables with this property are said to be *statistically independent* and their joint density function exhibits a separability property:

$\mathbf{x}_1$ and $\mathbf{x}_2$ statistically independent

$$\Rightarrow p_{x_1x_2}(\xi_1, \xi_2) = p_{x_1}(\xi_1)p_{x_2}(\xi_2); \qquad \text{for all } \xi_1, \xi_2 \tag{7.17}$$

Note that (7.17) implies that $E[\mathbf{x}_1\mathbf{x}_2] = \bar{\mathbf{x}}_1\bar{\mathbf{x}}_2$, and hence statistical independence implies linear independence. The converse is not true in general as statistical independence is a much stronger statement than linear independence.

7.3 Random Processes

In (7.1) we defined a random process as a time sequence of random variables. An alternative viewpoint is that of an ensemble of waveforms (deterministic time functions), each determined by the graph corresponding to a particular point in the sample space on which the jointly distributed random variables are defined. In other words, each waveform is the pairwise collection of scalar values $\{x(t, \omega), t\}$ for all $t \in T$ and a particular $\omega \in \Omega$. Each waveform so produced is called a *realization* of the random process. A typical realization, which we denote by $\mathbf{x}_\omega$, is a vector in the space of time functions, not in the space of random variables. In characterizing the properties of a random process, we normally stick to the first viewpoint, that of a time sequence of random variables. In some problems, however, as in the one described in Section 7.5, both viewpoints must be kept in mind.

A random process can be described by the joint probability density function of certain subsets of its random variables. Many of the continuous-time processes of practical interest have the property that the density functions are invariant to a translation in the time index. Such processes are called *strict-sense stationary* if

$$p_{x_1x_2\cdots x_n}(\xi_1, \xi_2, \ldots, \xi_n) = p_{\tilde{x}_1\tilde{x}_2\cdots\tilde{x}_n}(\xi_1, \xi_2, \ldots, \xi_n) \tag{7.18}$$

where $\mathbf{x}_i = \mathbf{x}(t_i)$ and $\tilde{\mathbf{x}}_i = \mathbf{x}(t_i + \tau)$, for any τ, and any n-element subset $\{t_i\}$ of T, where n is arbitrary.

Autocorrelation and Autocovariance Functions

Whether a process is stationary or not, we are often content to deal merely with the joint properties of pairs of its random variables. In this case, we are dealing with the *second-order statistics* of the random process. In particular for many problems, the correlation between all pairs of random variables may be an adequate description of the process. This correlation will depend jointly on the time indices t_1 and t_2 and is called the *autocorrelation function* $k_{xx}(t_1, t_2)$ for the process $\mathbf{x}$.

$$k_{xx}(t_1, t_2) = E[\mathbf{x}(t_1)\mathbf{x}(t_2)] \tag{7.19}$$

The autocorrelation function of the zero-mean random process created by subtracting out the mean value $E[\mathbf{x}(t)] = \overline{\mathbf{x}(t)}$ is called the *autocovariance function* $m_{xx}(t_1, t_2)$ for the process.

$$\begin{aligned} m_{xx}(t_1, t_2) &= E[\{\mathbf{x}(t_1) - \overline{\mathbf{x}(t_1)}\}\{\mathbf{x}(t_2) - \overline{\mathbf{x}(t_2)}\}] \\ &= k_{xx}(t_1, t_2) - \overline{\mathbf{x}(t_1)}\,\overline{\mathbf{x}(t_2)} \end{aligned} \tag{7.20}$$

Finally, $k_{xx}(t, t)$ is the *mean-squared value* of the process, and $m_{xx}(t, t)$ is the *variance* of the process. Note that both quantities are, in general, time dependent.

For a stationary process, we see from (7.18) that the mean value is independent of time and that the autocorrelation depends only on the time difference $\tau = t_1 - t_2$. It is customary to write the autocorrelation as a function of the single variable τ when it is clear that we are dealing with a stationary process. Thus we have

$$\left.\begin{aligned} k_{xx}(\tau) &= E[\mathbf{x}(t + \tau)\mathbf{x}(t)] \\ \bar{\mathbf{x}} &= E[\mathbf{x}(t)] \end{aligned}\right\} \quad \text{independent of } t \text{ for a stationary process} \tag{7.21}$$

When condition (7.21) is satisfied without regard to the stronger condition (7.18), we refer to the process as being *wide-sense stationary*. From the symmetry and Schwarz inequality properties of an inner product, it is clear that $k_{xx}(\tau) = k_{xx}(-\tau)$ and that $|k_{xx}(\tau)| \leqslant k_{xx}(0) = \overline{\mathbf{x}^2}$.

It seems reasonable that the autocorrelation function can provide at least a partial description of the frequency-domain properties of a random process. Suppose $\mathbf{x}$ is a wide-sense stationary process and we consider the *mean-squared fluctuation* of the process over an interval τ.

$$E[\{\mathbf{x}(t) - \mathbf{x}(t+\tau)\}^2] = E[\mathbf{x}^2(t)] + E[\mathbf{x}^2(t+\tau)] - 2E[\mathbf{x}(t)\mathbf{x}(t+\tau)]$$
$$= 2[k_{xx}(0) - k_{xx}(\tau)] \tag{7.22}$$

Thus, if a process exhibits rapid fluctuations (high frequencies), then samples with relatively small time separation will have a small correlation. The autocorrelation function in this case will be relatively narrow. On the other hand, a process with slow fluctuations (low frequencies) will have a relatively broad autocorrelation function. This is shown graphically in Figure 7.1.

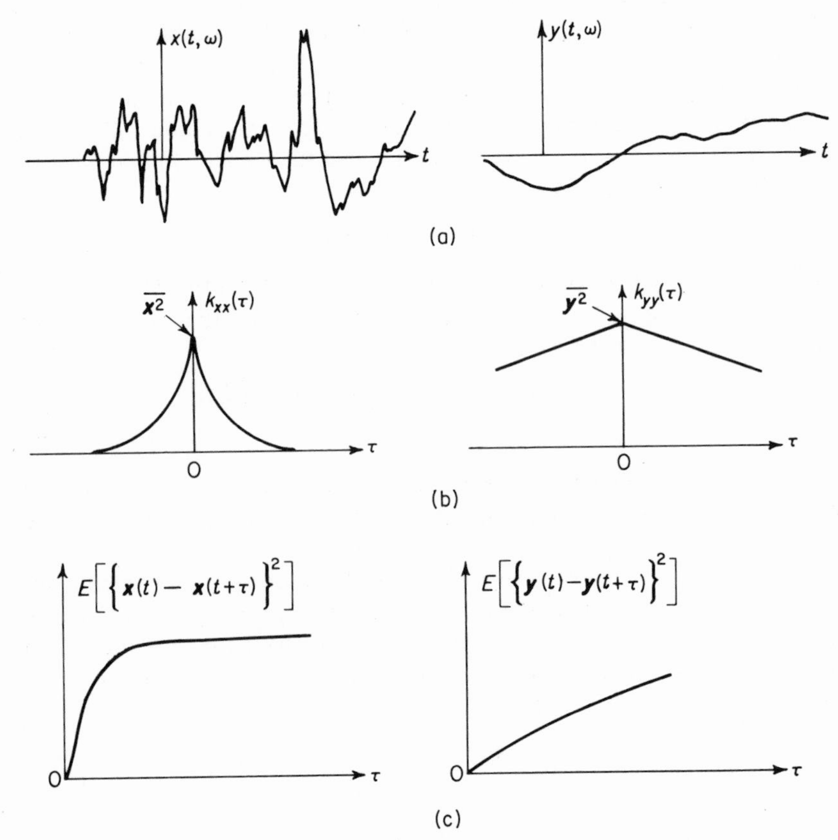

Figure 7.1. Typical realizations (a), autocorrelation functions (b), and mean-squared fluctuations (c), for rapidly and slowly fluctuating processes.

Comparing (7.22) with (2.39), it is clear that the autocorrelation function for a stationary process bears a strong analogy to the time-ambiguity function for a finite-energy, deterministic signal. A more specific frequency-domain description for a stationary process will be presented in the following section.

7.4 Joint Complex Processes

In this section, we shall extend the foregoing concepts to apply to the consideration of the joint properties of a pair of random processes, each of which may be complex. A complex process $\mathbf{x}$ is an ordered pair of real, jointly random processes $\mathbf{x}_R$ and $\mathbf{x}_I$ related by

$$\mathbf{x} = \mathbf{x}_R + j\mathbf{x}_I \tag{7.23}$$

In order that the autocorrelation of $\mathbf{x}$ correspond to an inner product on a complex linear space, it is necessary to use the complex conjugate of one of the random variables. Hence we define

$$k_{xx}(t_1, t_2) = E[\mathbf{x}(t_1)\mathbf{x}^*(t_2)] \tag{7.24}$$

and for a wide-sense stationary complex process,

$$k_{xx}(\tau) = E[\mathbf{x}(t + \tau)\mathbf{x}^*(t)] \tag{7.25}$$

Two processes are jointly random if they are defined on the same sample space. An important joint characteristic of two processes is the correlation between all pairs of random variables with one random variable taken from each process. This correlation is called the *crosscorrelation function* $k_{xy}(t_1, t_2)$ for the two processes.

$$k_{xy}(t_1, t_2) = E[\mathbf{x}(t_1)\mathbf{y}^*(t_2)] \tag{7.26}$$

with the property that

$$k_{yx}(t_2, t_1) = k_{xy}^*(t_1, t_2) \tag{7.27}$$

If $\mathbf{x}$ and $\mathbf{y}$ are jointly wide-sense stationary, then the crosscorrelation depends only on the time difference $\tau = t_1 - t_2$, and in this case we define the crosscorrelation function as

$$k_{xy}(\tau) = E[\mathbf{x}(t + \tau)\mathbf{y}^*(t)] \tag{7.28}$$

Filtered Processes

A situation of particular interest is where $\mathbf{x}$ is related to $\mathbf{y}$ by means of a linear transformation. Suppose $\mathbf{x}$ is a wide-sense stationary process which is filtered by a time-invariant network with impulse response $h(t)$ as shown in

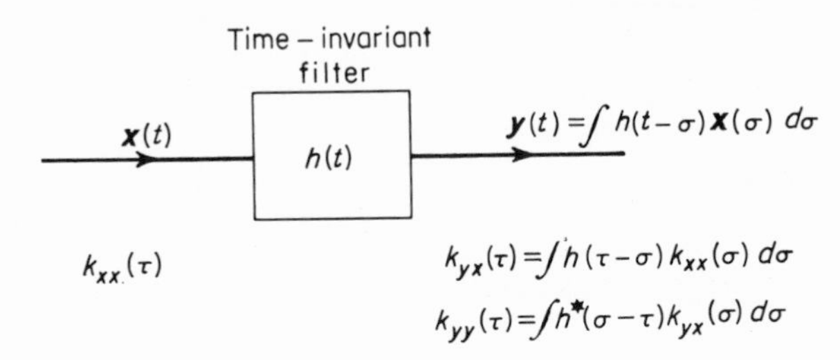

Figure 7.2. Filtered wide-sense stationary random process.

Figure 7.2. In this case, **y** is also a wide-sense stationary process and, assuming interchangeability of integration and expectation operations,

$$\begin{aligned}
k_{yx}(\tau) &= E[\mathbf{y}(t+\tau)\mathbf{x}^*(t)] \\
&= E\left[\int_{-\infty}^{\infty} h(t+\tau-\sigma)\mathbf{x}(\sigma)\mathbf{x}^*(t)\,d\sigma\right] \\
&= \int_{-\infty}^{\infty} h(t+\tau-\sigma)k_{xx}(\sigma-t)\,d\sigma \\
&= \int_{-\infty}^{\infty} h(\tau-\xi)k_{xx}(\xi)\,d\xi
\end{aligned} \tag{7.29}$$

From (7.29) we have the interesting result that the crosscorrelation between input and output of the filter is related to the autocorrelation of the input by the same linear transformation (convolution) that is applied to the input signal process. It also happens that the autocorrelation function of the output process is related to the autocorrelation of the input by means of a convolution integral.

We have

$$\begin{aligned}
k_{yy}(\tau) &= E[\mathbf{y}(t+\tau)\mathbf{y}^*(t)] \\
&= E\left[\iint_{-\infty}^{\infty} h(t+\tau-\sigma)h^*(t-\xi)\mathbf{x}(\sigma)\mathbf{x}^*(\xi)\,d\sigma\,d\xi\right] \\
&= \iint_{-\infty}^{\infty} h(t+\tau-\sigma)h^*(t-\xi)k_{xx}(\sigma-\xi)\,d\sigma\,d\xi
\end{aligned}$$

Hence, by suitable change of integration variables, we find

$$k_{yy}(\tau) = \int_{-\infty}^{\infty} w(\tau-s)k_{xx}(s)\,ds \tag{7.30}$$

where

$$w(\tau) = \int_{-\infty}^{\infty} h(\tau+\eta)h^*(\eta)\,d\eta$$

Since $k_{yy}(\tau)$ and $k_{xx}(\tau)$ are related by a convolution integral, (7.30) is simply expressed in terms of the Fourier transforms of the corresponding quantities.

$$K_{yy}(f) = W(f)K_{xx}(f) = |H(f)|^2 K_{xx}(f) \tag{7.31}$$

where $H(f)$ is the transfer function for the filter network.

Power Spectral Density

The Fourier transform of an autocorrelation function has a special significance in describing the frequency-domain character of the corresponding stationary random process. It is called the *power spectral density* for the process and, as such, it indicates the contribution to mean-squared fluctuation due to signal components in any specified frequency interval. To show this, we let the filter in Figure 7.2 be a (non-physical) narrow-band filter which passes components only in the interval Δf centered about a frequency f as shown in Figure 7.3.

$$\begin{aligned} H(\nu) = W(\nu) &= 1 \qquad \text{for } |\nu - f| \leqslant \tfrac{1}{2}\Delta f \\ &= 0 \qquad \text{otherwise} \end{aligned} \tag{7.32}$$

Now we evaluate the mean-squared value of the filtered process. Using the relation

$$k_{yy}(\tau) = \int_{-\infty}^{\infty} K_{yy}(\nu) e^{j2\pi\nu\tau}\, d\nu$$

we have

$$E[|\mathbf{y}(t)|^2] = k_{yy}(0) = \int_{-\infty}^{\infty} K_{yy}(\nu)\, d\nu \tag{7.33}$$

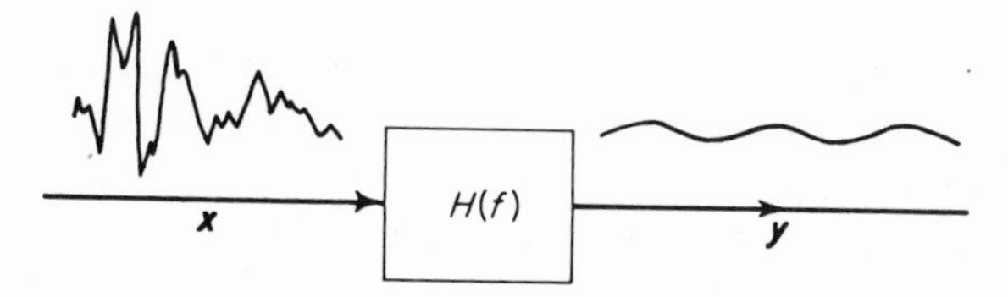

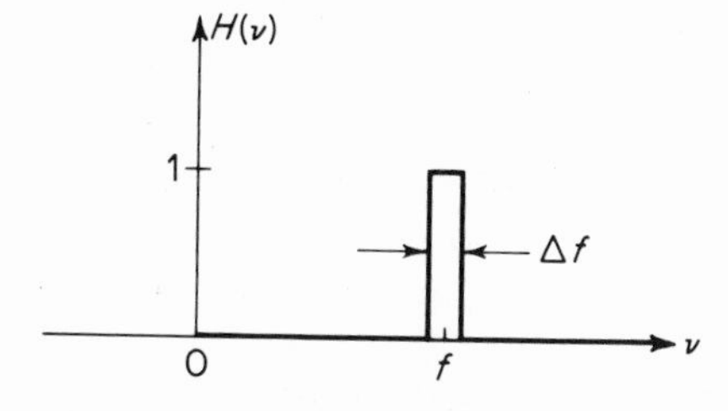

Figure 7.3. Hypothetical filter for determining frequency-domain content of a stationary process in a small region about the frequency f.

Substituting (7.32) into (7.31) and (7.33),

$$E[|\mathbf{y}(t)|^2] = \int_{-\infty}^{\infty} W(\nu) K_{xx}(\nu)\, d\nu$$
$$= \int_{\Delta f} K_{xx}(\nu)\, d\nu \cong K_{xx}(f)\, \Delta f \tag{7.34}$$

for Δf sufficiently small. Thus we can interpret K_{xx} as a density function and

$$\int_{f_1}^{f_2} K_{xx}(\nu)\, d\nu = \begin{cases} \text{contribution to mean squared} \\ \text{value of } \mathbf{x} \text{ due to components} \\ \text{between frequencies } f_1 \text{ and } f_2. \end{cases}$$

The density function defined in this manner is said to be *double-sided* since it considers separately the components at positive and negative frequencies. For a real process, $k_{xx}(\tau)$ is real and even; hence $K_{xx}(f)$ is real and even and the spectral decomposition is equally divided between a band of positive frequencies and the corresponding band of negative frequencies. It now remains to explain why "power" is used as an adjective for this density function. This nomenclature is conventional but somewhat unfortunate. In many cases of practical interest it is possible to attribute to a stationary process the additional property of *ergodicity*. An ergodic process[2] is a stationary process with the property that the "time average" of the square of any realization is equal (with probability 1) to the mean-squared value (ensemble average) of this process.

$$\lim_{T\to\infty} \frac{1}{2T} \int_{-T}^{T} |x(t, \omega)|^2\, dt = E[|\mathbf{x}(t)|^2] = k_{xx}(0) \tag{7.35}$$

for an ergodic process. The left-hand term in (7.35) is commonly called the *power* of the deterministic signal $x(t, \omega)$. In most system problems involving random processes, the concept of ergodicity is not used. An important exception to this is in problems concerned with the use of measured, finite-length time averages as estimates of power spectral density of a process.[4] We shall continue to use the term "power spectral density" in the broader context applicable to any wide-sense stationary process. A particular stationary process that is often employed as an approximate model for a physical process is the *white noise process* characterized by a constant power spectral density; i.e., $K_{xx}(f) = N_0$ (watts/Hz). The autocorrelation function for this process is $k_{xx}(\tau) = N_0\, \delta(\tau)$; hence different samples, $\mathbf{x}(t_1)$ and $\mathbf{x}(t_2)$, however closely spaced in time, will be uncorrelated. Note that the mean-squared value of such an idealized process is not finite.

Exercise 7.1. Show that the time-ambiguity function for any real, finite-energy signal can also be the autocorrelation function for a real, zero-mean, finite-variance, wide-sense stationary random process.

Exercise 7.2. For a finite mean-squared value, wide-sense stationary random process, show that $k_{xx}(\tau)$ is a continuous function of τ if it is continuous at the origin. *Hint:* Apply the Schwarz inequality to the inner product

$$E[\{\mathbf{x}(t+\tau+\varepsilon)-\mathbf{x}(t+\tau)\}\mathbf{x}^*(t)]$$

Exercise 7.3.[3] Let $\mathbf{x}(t)$ be a real, stationary, band-limited, lowpass process [in the sense that $K_{xx}(f) = 0$ for $|f| > W$]. Show that the mean-squared fluctuation $E[\{\mathbf{x}(t) - \mathbf{x}(t+\tau)\}^2]$ is bounded by $(2\pi W\tau)^2 E[\mathbf{x}^2(t)]$. Also show that

$$k_{xx}(\tau) \geqslant k_{xx}(0)\cos 2\pi W\tau \qquad \text{for } |\tau| \leqslant \frac{1}{4W},$$

and that

$$k_{xx}(\tau) \geqslant \tfrac{1}{2}[k_{xx}(0) + k_{xx}(2\tau)] \qquad \text{for } |\tau| \leqslant \frac{1}{4W}$$

Exercise 7.4. Suppose that a time-invariant linear filter, with impulse response $h(t)$, is driven by a wide-sense stationary process at the input. Show that the mean-squared value of the output process is a quadratic functional on $h(t)$. Describe the corresponding operator and show that it is self-adjoint and non-negative-definite.

Exercise 7.5. For the filtering operation indicated in Figure 7.2, show that

$$k_{yy}(\tau) = \int_{-\infty}^{\infty} h^*(\sigma - \tau)k_{yx}(\sigma)\,d\sigma$$

Exercise 7.6. Suppose that a wide-sense stationary process is subjected to a linear transformation (not necessarily time-invariant) characterized by the impulse response $h(t, s)$. Express the output autocorrelation and input-output cross-correlation as linear transformations on the input autocorrelation function $k_{xx}(\tau)$.

Exercise 7.7. In defining the autocorrelation function for a complex, wide-sense stationary process, we might have let

$$k_{xx}(\tau) = E[\mathbf{x}(t)\mathbf{x}^*(t+\tau)]$$

Give a physical interpretation of the Fourier transform of $k_{xx}(\tau)$, thereby showing why the autocorrelation was defined as in (7.25).

7.5 Finite-dimensional Representations for a Random Process

In this section we take up a problem which, besides having considerable practical importance, serves as an excellent tutorial example.[5],[6] This is because it involves simultaneously nearly all the previously discussed concepts of (1) optimization, (2) finite representation by projection onto a subspace of $L^2(T)$, (3) the geometrical concepts (in terms of expectations) associated with a space of random variables, and (4) spectral representation of operators.

The problem is easily stated. We want to find the optimal n-dimensional basis (in $L^2(T)$) for representing particular realizations of a random process such that the $L^2(T)$ norm of the error averaged over the ensemble of realizations is minimized. We shall show that solution of this problem requires only that the autocorrelation function of the process to be represented is specified. Heuristically, there is a strong analogy between this problem and the problem discussed in Section 6.8. There we sought the optimum basis for the class of signals obtained as the image of $L^2(-\infty, \infty)$ under a compact operator. In the present problem, the signals are classified according to the autocorrelation function which, to strengthen the analogy, could be considered to describe a process obtained by linear filtering (time-variable in general) of a white noise process. We would not require that this filtering operator be compact; i.e., it could be a time-invariant operator. We shall instead obtain compact sets of signals by considering representations only over a finite time interval. The reason for this restriction is that many processes of interest, e.g., stationary processes, clearly have typical realizations which are not contained in $L^2(-\infty, \infty)$ and it makes little sense to project these realizations onto finite-dimensional subspaces spanned by functions defined over the entire time axis.

Karhunen-Loève Expansion

The n-dimensional representation for a particular realization $\mathbf{x}_\omega$ will take the form

$$x(t, \omega) \simeq \sum_{i=1}^{n} \alpha_i \varphi_i(t); \qquad |t| \leqslant T \tag{7.36}$$

where we shall assume, for convenience, that the $\{\varphi_i(t)\}$ form an orthonormal basis in $L^2(-T, T)$, i.e.,

$$\int_{-T}^{T} \varphi_i(t)\varphi_j^*(t)\, dt = \delta_{ij} \tag{7.37}$$

In this case, the $L^2(-T, T)$ norm of the error

$$\boldsymbol{\epsilon}_\omega = \mathbf{x}_\omega - \sum_{i=1}^{n} \alpha_i \boldsymbol{\varphi}_i$$

is minimized for each realization by making

$$\alpha_i = (\mathbf{x}_\omega, \boldsymbol{\varphi}_i) = \int_{-T}^{T} x(t, \omega)\varphi_i^*(t)\, dt \tag{7.38}$$

The square of this minimized norm is given by

$$(\boldsymbol{\epsilon}_\omega, \boldsymbol{\epsilon}_\omega) = (\mathbf{x}_\omega, \mathbf{x}_\omega) - \sum_{i=1}^{n} |\alpha_i|^2 \tag{7.39}$$

Notice that $(\boldsymbol{\epsilon}_\omega, \boldsymbol{\epsilon}_\omega)$ and the $\{\alpha_i\}$ are actually realizations of random variables whose statistical properties depend on the choice of the $\{\boldsymbol{\varphi}_i\}$. We shall examine the possibility of selecting n basis functions $\{\boldsymbol{\varphi}_i\}$ such that the expected value of the random variable $(\boldsymbol{\epsilon}, \boldsymbol{\epsilon})$ is minimized. Let I_φ denote the value of this mean-squared norm of error. Then we have

$$\begin{aligned} I_\varphi &= E[(\boldsymbol{\epsilon}, \boldsymbol{\epsilon})] \\ &= E[(\mathbf{x}, \mathbf{x})] - E\left[\sum_{i=1}^{n} |\boldsymbol{\alpha}_i|^2\right] \end{aligned} \tag{7.40}$$

Assuming interchangeability of expectation and the integration implied by the inner products, and using (7.38), we have

$$\begin{aligned} I_\varphi &= \int_{-T}^{T} E[\mathbf{x}(t)\mathbf{x}^*(t)]\, dt - \sum_{i=1}^{n} \iint_{-T}^{T} E[\mathbf{x}(t)\mathbf{x}^*(s)]\varphi_i^*(t)\varphi_i(s)\, dt\, ds \\ &= \int_{-T}^{T} k_{xx}(t, t)\, dt - \sum_{i=1}^{n} \iint_{-T}^{T} k_{xx}(t, s)\varphi_i^*(t)\varphi_i(s)\, dt\, ds \end{aligned} \tag{7.41}$$

Since the first term in (7.41) is independent of the $\{\boldsymbol{\varphi}_i\}$, our problem is to find the set of n orthonormal functions which maximize the quantity

$$\sum_{i=1}^{n} \iint_{-T}^{T} k_{xx}(t, s)\varphi_i^*(t)\varphi_i(s)\, dt\, ds = \sum_{i=1}^{n} (\mathscr{A}_x \boldsymbol{\varphi}_i, \boldsymbol{\varphi}_i) \tag{7.42}$$

In (7.42), we have expressed the quantity to be maximized as a sum of quadratic functionals by interpreting the autocorrelation function as the kernel of an integral operator $\mathscr{A}_x$; i.e.,

$$\mathscr{A}_x \cdot \varphi(t) = \int_{-T}^{T} k_{xx}(t, s)\varphi(s)\, ds \tag{7.43}$$

Because of the fact that the kernel function is an autocorrelation function (of a finite mean-squared value process), there is quite a bit we can say about the properties of the operator. It follows directly that $\mathscr{A}_x$ is (1) Hilbert-Schmidt, (2) self-adjoint, and (3) non-negative-definite. Property (1) is a result of the square integrability, (5.72), of the kernel function, which follows from the boundedness of $k_{xx}(t, s)$ and the finite integration interval. Property (2) is a result of the symmetry $k_{xx}(t, s) = k_{xx}^*(s, t)$, (5.95). To show property (3), we observe that $(\mathscr{A}_x \boldsymbol{\varphi}, \boldsymbol{\varphi}) = E[|(\mathbf{x}, \boldsymbol{\varphi})|^2] \geqslant 0$ for an arbitrary choice of $\boldsymbol{\varphi}$. Taking these properties in combination, we can say that

1. The eigenvalues form a countable, square-summable set. They are real and non-negative, and we shall index them into a non-increasing sequence:

$$\lambda_1 \geqslant \lambda_2 \geqslant \lambda_3 \geqslant \cdots \tag{7.44}$$

2. The kernel function can be represented uniformly as an expansion in the eigenfunctions $\{\boldsymbol{\psi}_i\}$ of $\mathscr{A}_x$:

$$k_{xx}(t, s) = \sum_{i=1}^{\infty} \lambda_i \psi_i(t) \psi_i^*(s) \tag{7.45}$$

3. The eigenfunctions can be orthonormalized and we have

$$(\mathscr{A}_x \boldsymbol{\psi}_i, \boldsymbol{\psi}_j) = (\lambda_i \boldsymbol{\psi}_i, \boldsymbol{\psi}_j) = \lambda_i \, \delta_{ij} \tag{7.46}$$

For an operator $\mathscr{A}_x$ with these properties, it can be shown[7] that, for any orthonormal set $\{\boldsymbol{\varphi}_i\}$,

$$\sum_{i=1}^{n} (\mathscr{A}_x \boldsymbol{\varphi}_i, \boldsymbol{\varphi}_i) \leqslant \sum_{i=1}^{n} (\mathscr{A}_x \boldsymbol{\psi}_i, \boldsymbol{\psi}_i) = \sum_{i=1}^{n} \lambda_i \tag{7.47}$$

This can be demonstrated by a term-by-term maximization of the left-hand side of (7.47). First we select a function $\boldsymbol{\varphi}_1$ of unit norm which maximizes the quadratic functional $(\mathscr{A}_x \boldsymbol{\varphi}, \boldsymbol{\varphi})$. From (6.47) we see that $\boldsymbol{\varphi}_1 = \boldsymbol{\psi}_1$. Next we select $\boldsymbol{\varphi}_2$ of unit norm and orthogonal to $\boldsymbol{\varphi}_1$ to maximize $(\mathscr{A}_x \boldsymbol{\varphi}, \boldsymbol{\varphi})$. This gives $\boldsymbol{\varphi}_2 = \boldsymbol{\psi}_2$ and, continuing the process, we find $\boldsymbol{\varphi}_i = \boldsymbol{\psi}_i$ for $i = 1, 2, \ldots, n$. The performance of this optimal basis, from (7.41), is given by

$$I_\psi = \int_{-T}^{T} k_{xx}(t, t)\, dt - \sum_{i=1}^{n} (\mathscr{A}_x \boldsymbol{\psi}_i, \boldsymbol{\psi}_i) \tag{7.48}$$

Using (7.45) and (7.46) in (7.48), we find

$$\begin{aligned} I_\psi &= \sum_{i=1}^{\infty} \lambda_i (\boldsymbol{\psi}_i, \boldsymbol{\psi}_i) - \sum_{i=1}^{n} \lambda_i \\ &= \sum_{i=n+1}^{\infty} \lambda_i \end{aligned} \tag{7.49}$$

To summarize these results, the optimum n-dimensional subspace of $L^2(-T, T)$ for representing realizations of a random process over the interval $|t| \leqslant T$ is spanned by the n eigenfunctions

$$\int_{-T}^{T} k_{xx}(t, s) \psi_i(s)\, ds = \lambda_i \psi_i(t) \tag{7.50}$$

corresponding to the n largest eigenvalues. The squared $L^2(-T, T)$ norm of error, on the average, is given simply by the sum of the remaining eigenvalues. This relationship is very useful in practice since it simply indicates how much will be gained by adding a few terms to the representation. It is not surprising that, in general, all the eigenvalues will increase as T increases and more terms are required for a satisfactory representation.[8]

Using this optimal basis, the representation

$$\mathbf{x}(t) \simeq \sum_{i=1}^{n} \boldsymbol{\alpha}_i \psi_i(t); \qquad |t| \leqslant T$$

is called the *Karhunen-Loève expansion* for a random process.[3],[5] The coefficients in the expansion are orthogonal random variables, since

$$E[\boldsymbol{\alpha}_i \boldsymbol{\alpha}_j^*] = E[(\mathbf{x}, \boldsymbol{\psi}_i)(\mathbf{x}, \boldsymbol{\psi}_j)^*]$$

$$= \int_{-T}^{T}\!\!\int k_{xx}(t, s)\psi_i^*(t)\psi_j(s)\, dt\, ds = \lambda_j\, \delta_{ij} \tag{7.51}$$

If $\mathbf{x}$ is a zero-mean process, then the coefficients are also zero mean, and they are uncorrelated (linearly independent). If $\mathbf{x}$ is not zero mean, then it is possible to improve the representation, for a given n, if we are willing to add a fixed term $\mu(t)$ to the representation for each realization. We shall show that the best choice for $\mu(t)$ is simply the mean value $\overline{\mathbf{x}(t)}$ for the process, regardless of the orthonormal basis $\{\boldsymbol{\varphi}_i\}$ employed. The representation

$$\mathbf{x}(t) \simeq \sum_{i=1}^{n} \boldsymbol{\alpha}_i \varphi_i(t) + \mu(t); \qquad |t| < T \tag{7.52}$$

is equivalent to that of (7.36) for the random process $\mathbf{y}(t) = \mathbf{x}(t) - \mu(t)$; hence we select $\alpha_i = (\mathbf{x}_\omega - \boldsymbol{\mu}, \boldsymbol{\varphi}_i)$ for each realization and

$$(\boldsymbol{\epsilon}_\omega, \boldsymbol{\epsilon}_\omega) = \left(\mathbf{x}_\omega - \boldsymbol{\mu} - \sum_1^n \alpha_i \boldsymbol{\varphi}_i, \mathbf{x}_\omega - \boldsymbol{\mu} - \sum_1^n \alpha_i \boldsymbol{\varphi}_i\right)$$

$$= (\mathbf{x}_\omega - \boldsymbol{\mu}, \mathbf{x}_\omega - \boldsymbol{\mu}) - \sum_{i=1}^{n} |(\mathbf{x}_\omega - \boldsymbol{\mu}, \boldsymbol{\varphi}_i)|^2 \tag{7.53}$$

The expected value of this squared error norm is

$$I_\varphi = E[(\boldsymbol{\epsilon}, \boldsymbol{\epsilon})]$$

$$= E[(\mathbf{x}, \mathbf{x})] - \sum_{i=1}^{n} E[|(\mathbf{x}, \boldsymbol{\varphi}_i)|^2] - (\bar{\mathbf{x}}, \bar{\mathbf{x}})$$

$$+ \sum_{i=1}^{n} |(\bar{\mathbf{x}}, \boldsymbol{\varphi}_i)|^2 + \|\bar{\mathbf{x}} - \boldsymbol{\mu}\|^2 - \sum_{i=1}^{n} |(\bar{\mathbf{x}} - \boldsymbol{\mu}, \boldsymbol{\varphi}_i)|^2 \tag{7.54}$$

It is only the last two terms in (7.54) which are affected by the choice of $\mu(t)$, and from Bessel's inequality (3.21), we have

$$\|\bar{\mathbf{x}} - \boldsymbol{\mu}\|^2 - \sum_{i=1}^{n} |(\bar{\mathbf{x}} - \boldsymbol{\mu}, \boldsymbol{\varphi}_i)|^2 \geqslant 0 \tag{7.55}$$

for any $\mu(t)$. The obvious choice to minimize I_φ is $\mu(t) = \overline{\mathbf{x}(t)}$. Hence, if we are able to subtract out the mean value of the process, the optimal n-dimensional basis is given by the eigenfunctions (with largest eigenvalues) corresponding to the autocorrelation function of $\mathbf{y} = \mathbf{x} - \bar{\mathbf{x}}$ or, equivalently, by the eigenfunctions corresponding to the autocovariance function for $\mathbf{x}$. In this case the Karhunen-Loève expansion coefficients are zero mean and uncorrelated.

Regarding techniques for solution of (7.50), for stationary processes whose spectral densities are rational functions of frequency, a generalization of the differentiation method used to solve (6.58) is usually employed. This method results in a homogeneous differential equation with constant coefficients, the solution of which contains a number of arbitrary parameters. The parameters must be determined by substitution back into the integral equation. Details of this method are presented in References [9] and [10]. Solutions for particular types of "non-rational" kernels of interest have also been presented.[11]

Exercise 7.8. A stationary, zero-mean, white noise process having power spectral density of $N_0 = 1/\pi f_0$ (watts/Hz) is filtered by a single-section *RC* network with transfer function

$$H(f) = \frac{1}{1 + j(f/f_0)}$$

a. Show that the filtered process has unit variance.
b. Show that a realization of the process has an expected energy of $2T$ in the interval $|t| \leqslant T$.
c. Find the optimum set of basis functions for an n-term representation of realizations of the process over $|t| \leqslant T$.
d. Approximately how many terms (relative to f_0T) are required, using the optimum basis, so that the mean-squared value of the $L^2(-T, T)$ norm of the error $E[(\boldsymbol{\epsilon}, \boldsymbol{\epsilon})]$ should amount to only $0.01(2T)$? *Hint:* Show that the eigenvalues of (7.50), for this case, are approximately

$$\lambda_{n+1} \cong \frac{1}{\pi f_0}\left(\frac{4f_0T}{n}\right)^2$$

for large n.
e. For comparison, suppose that a time-series representation using rectangular pulses is employed. Let n be even and use the orthonormal set

$$\{\boldsymbol{\varphi}_i;\ \varphi_i(t) = \varphi(t - i2T/n + T), i = 1, 2, \ldots, n\}$$

as shown.

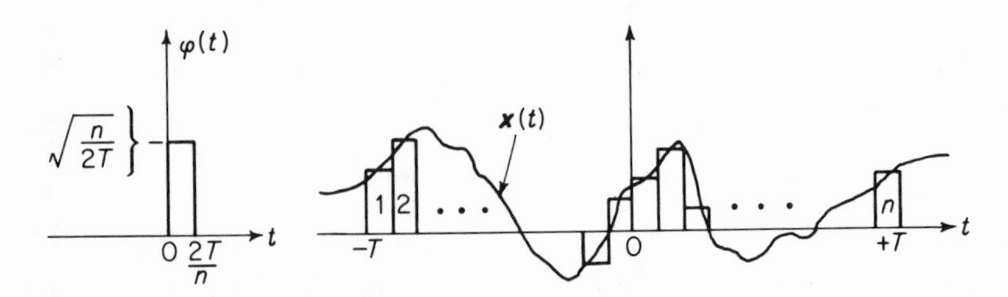

How many terms, relative to the optimum basis, are needed to attain the same accuracy of representation on the average? *Hint:* Show that

$$\sum_{i=1}^{n} \iint_{-T}^{T} k_{xx}(t-s)\varphi_i(s)\varphi_i^*(t)\,dt\,ds = 2n\int_0^{\frac{2T}{n}} e^{-2\pi f_0 t}\left(1 - \frac{nt}{2T}\right)dt$$

for this case, and make a power series expansion in (f_0T/n) to use in (7.41).

7.6 Bandpass Processes

In this section we shall examine some equivalent descriptions for a stationary bandpass process, i.e., one whose power spectral density tends to be concentrated about a frequency removed from the origin. The motivation for this special treatment is exactly the same as for deterministic bandpass signals as discussed in Section 4.4. For random processes also, we want to exploit the advantages inherent in dealing with equivalent lowpass signals and lowpass filtering operations. The starting point for developing the lowpass equivalents is the same as in Section 4.4; i.e., we consider the complex process $\boldsymbol{\psi}$ related to a real bandpass process $\mathbf{x}$ by

$$\boldsymbol{\psi} = \mathbf{x} + j\hat{\mathbf{x}} \tag{7.56}$$

where

$$\hat{\mathbf{x}}(t) = \frac{1}{\pi}\int_{-\infty}^{\infty} \frac{\mathbf{x}(s)}{t-s}\, ds \tag{7.57}$$

is the Hilbert transform of $\mathbf{x}(t)$. We shall assume that $\mathbf{x}$ is a zero-mean, wide-sense stationary process. Then $\hat{\mathbf{x}}$ is a zero-mean, wide-sense stationary process since it is related to $\mathbf{x}$ by a time-invariant transformation (convolution). We can, therefore, write the correlation functions in terms of a single argument.

$$\begin{aligned} k_{\psi\psi}(\tau) &= E[\boldsymbol{\psi}(t+\tau)\boldsymbol{\psi}^*(t)] \\ &= E[\{\mathbf{x}(t+\tau) + j\hat{\mathbf{x}}(t+\tau)\}\{\mathbf{x}(t) - j\hat{\mathbf{x}}(t)\}] \\ &= k_{xx}(\tau) + k_{\hat{x}\hat{x}}(\tau) + j[k_{\hat{x}x}(\tau) - k_{x\hat{x}}(\tau)] \end{aligned} \tag{7.58}$$

Each of the correlation functions in (7.58) can be related to $k_{xx}(\tau)$. For example,

$$\begin{aligned} k_{\hat{x}\hat{x}}(\tau) &= E[\hat{\mathbf{x}}(t+\tau)\hat{\mathbf{x}}(t)] \\ &= \frac{1}{\pi^2}\iint_{-\infty}^{\infty} \frac{E[\mathbf{x}(s)\mathbf{x}(\sigma)]}{(t+\tau-s)(t-\sigma)}\, d\sigma\, ds \\ &= \frac{1}{\pi^2}\int_{-\infty}^{\infty} \frac{1}{t-s+\tau}\int_{-\infty}^{\infty} \frac{k_{xx}(\xi)}{t-s-\xi}\, d\xi \end{aligned}$$

Using (4.24b),

$$k_{\hat{x}\hat{x}}(\tau) = \frac{1}{\pi}\int_{-\infty}^{\infty} \frac{1}{t-s+\tau}\, \hat{k}_{xx}(t-s)\, ds$$

and now using (4.24a) and the fact that the autocorrelation of a real process is even,

$$k_{\hat{x}\hat{x}}(\tau) = k_{xx}(-\tau) = k_{xx}(\tau) \tag{7.59}$$

Following similar steps, we can also show that

$$k_{\hat{x}x}(\tau) = \hat{k}_{xx}(\tau) \tag{7.60a}$$

$$k_{x\hat{x}}(\tau) = \hat{k}_{xx}(-\tau) = -\hat{k}_{xx}(\tau) \tag{7.60b}$$

Substituting (7.59) and (7.60) into (7.58), we see that $k_{\psi\psi}(\tau)$ itself takes on the form of an analytic signal (4.37).

$$k_{\psi\psi}(\tau) = 2k_{xx}(\tau) + j2\hat{k}_{xx}(\tau) \tag{7.61}$$

This result shows that the power spectral density for the complex process is one-sided, as we would have expected in view of the one-sided nature of Fourier transforms of pre-envelope signals discussed in Section 4.4. From (4.28),

$$\begin{aligned} K_{\psi\psi}(f) &= 4K_{xx}(f) \qquad \text{for } f > 0 \\ &= 0 \qquad\qquad\quad \text{for } f < 0 \end{aligned} \tag{7.62}$$

Complex Envelope Processes

We next define a *complex envelope process* $\boldsymbol{\gamma}$ relative to a specified "center" frequency f_0 according to

$$\boldsymbol{\psi}(t) = \boldsymbol{\gamma}(t)e^{j2\pi f_0 t} \tag{7.63}$$

with $\boldsymbol{\gamma} = \mathbf{u} + j\mathbf{v}$. By this means, the bandpass process is expressed in terms of two lowpass processes $\mathbf{u}$ and $\mathbf{v}$ giving in-phase and quadrature components, respectively.

$$\mathbf{x}(t) = \mathbf{u}(t)\cos 2\pi f_0 t - \mathbf{v}(t)\sin 2\pi f_0 t \tag{7.64}$$

In view of the time-dependent multipliers in (7.63) or (7.64), it is not immediately evident that $\mathbf{u}$ or $\mathbf{v}$ are wide-sense stationary processes. To demonstrate the stationarity, we note that

$$\mathbf{u}(t) = \operatorname{Re}\boldsymbol{\gamma}(t) = \tfrac{1}{2}[\boldsymbol{\psi}(t)e^{-j2\pi f_0 t} + \boldsymbol{\psi}^*(t)e^{j2\pi f_0 t}] \tag{7.65}$$

and hence

$$\begin{aligned} E[\mathbf{u}(t+\tau)\mathbf{u}(t)] = \tfrac{1}{4}\{&E[\boldsymbol{\psi}(t+\tau)\boldsymbol{\psi}^*(t)]e^{-j2\pi f_0\tau} + E[\boldsymbol{\psi}^*(t+\tau)\boldsymbol{\psi}(t)]e^{j2\pi f_0\tau} \\ &+ E[\boldsymbol{\psi}(t+\tau)\boldsymbol{\psi}(t)]e^{-j2\pi f_0(2t+\tau)} + E[\boldsymbol{\psi}^*(t+\tau)\boldsymbol{\psi}^*(t)]e^{j2\pi f_0(2t+\tau)}\} \end{aligned} \tag{7.66}$$

The last two terms (which depend on t) in (7.66) vanish since

$$\begin{aligned} E[\boldsymbol{\psi}(t+\tau)\boldsymbol{\psi}(t)] &= E[\{\mathbf{x}(t+\tau) + j\hat{\mathbf{x}}(t+\tau)\}\{\mathbf{x}(t) + j\hat{\mathbf{x}}(t)\}] \\ &= k_{xx}(\tau) - k_{\hat{x}\hat{x}}(\tau) + j[k_{\hat{x}x}(\tau) + k_{x\hat{x}}(\tau)] \\ &= 0 \end{aligned} \tag{7.67}$$

because of the relations (7.59) and (7.60). Thus the autocorrelation for $\mathbf{u}$ is independent of t, and $\mathbf{u}$ is zero mean since we assumed $\mathbf{x}$ was zero mean.

This shows that **u** is wide-sense stationary and (7.66) becomes

$$k_{uu}(\tau) = \tfrac{1}{4}[k_{\psi\psi}(\tau)e^{-j2\pi f_0\tau} + k^*_{\psi\psi}(\tau)e^{j2\pi f_0\tau}]$$
$$= k_{xx}(\tau)\cos 2\pi f_0\tau + \hat{k}_{xx}(\tau)\sin 2\pi f_0\tau \tag{7.68}$$

Following a similar sequence of steps, we can show that

$$E[\mathbf{v}(t+\tau)\mathbf{v}(t)] = k_{vv}(\tau) = k_{uu}(\tau) \tag{7.69}$$

and that

$$E[\mathbf{v}(t+\tau)\mathbf{u}(t)] = k_{vu}(\tau) = -k_{uv}(\tau)$$
$$= \hat{k}_{xx}(\tau)\cos 2\pi f_0\tau - k_{xx}(\tau)\sin 2\pi f_0\tau \tag{7.70}$$

Finally, as a result of these relations, we have

$$k_{\gamma\gamma}(\tau) = k_{\psi\psi}(\tau)e^{-j2\pi f_0\tau}$$
$$= 2k_{uu}(\tau) + j2k_{vu}(\tau) \tag{7.71}$$

and the power spectral density of the complex envelope process is the desired translated version of the one-sided spectral density for ψ, shown graphically in Figure 7.4.

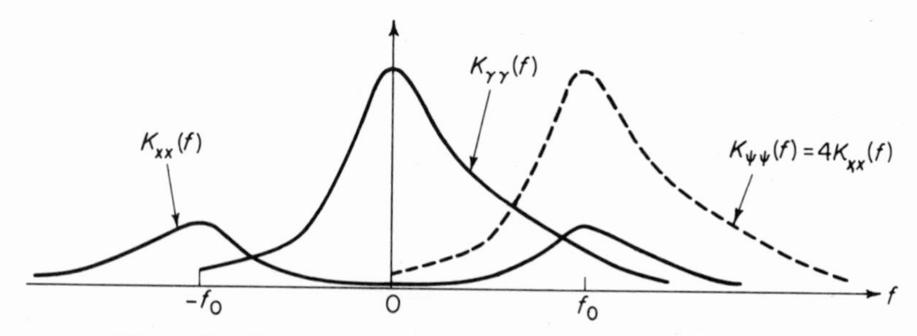

Figure 7.4. Power spectral density of complex envelope process.

$$K_{\gamma\gamma}(f) = K_{\psi\psi}(f+f_0)$$
$$= 4K_{xx}(f+f_0) \qquad \text{for } f > -f_0$$
$$= 0 \qquad \text{for } f < -f_0 \tag{7.72}$$

Note that if $K_{\gamma\gamma}(f)$ is even ($K_{xx}(f)$ symmetrical about f_0), then $k_{\gamma\gamma}(\tau)$ is real and, from (7.71), the in-phase and quadrature components are uncorrelated.

Bandpass Filtering

For characterizing the effects of a time-invariant bandpass filtering operation on the process in terms of lowpass equivalents, we shall, of course, use the same lattice arrangements developed in Section 4.4. Of additional interest, in the case of random processes, is the linear transformation which relates the output autocorrelation function to the input autocorrelation function, (7.30) and (7.31). We shall develop here the lowpass equivalent of this transformation. As in Section 4.4, let the bandpass filter have a transfer function $R(f)$ with lowpass equivalent $\Lambda(f)$. From (7.31), the linear transformation relating input and output correlation functions is given by

$$K_{x_2x_2}(f) = |R(f)|^2 K_{x_1x_1}(f) \tag{7.73}$$

Using (4.57), the corresponding relation in terms of complex envelope processes is

$$K_{\gamma_2\gamma_2}(f) = |\tfrac{1}{2}\Lambda(f)|^2 K_{\gamma_1\gamma_1}(f) \tag{7.74}$$

In order to express the correlation functions for the real lowpass processes, we need to identify the real and imaginary parts of the autocorrelation of the complex process (7.71). These parts, however, correspond to the even and odd parts, respectively, of the power spectral density of the complex process. Making this separation in (7.74), we find

$$\begin{aligned} K_{u_2u_2}(f) &= \text{Ev}\,|\tfrac{1}{2}\Lambda(f)|^2 K_{u_1u_1}(f) + j\,\text{Od}\,|\tfrac{1}{2}\Lambda(f)|^2 K_{v_1u_1}(f) \\ K_{v_2u_2}(f) &= -j\,\text{Od}\,|\tfrac{1}{2}\Lambda(f)|^2 K_{u_1u_1}(f) + \text{Ev}\,|\tfrac{1}{2}\Lambda(f)|^2 K_{v_1u_1}(f) \end{aligned} \tag{7.75}$$

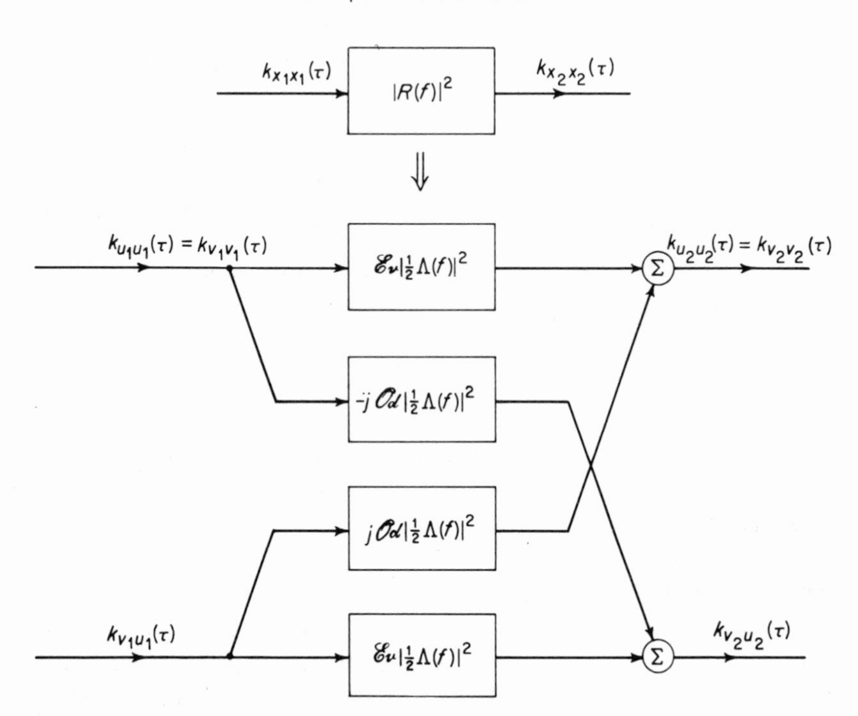

Figure 7.5. Lowpass equivalent of transformation relating input and output autocorrelation functions.

where, in terms of the lowpass transfer functions shown in Figure 4.6,

$$\begin{aligned} \text{Ev}\,|\tfrac{1}{2}\Lambda(f)|^2 &= \tfrac{1}{2}[|\tfrac{1}{2}\Lambda(f)|^2 + |\tfrac{1}{2}\Lambda(-f)|^2] \\ &= |M(f)|^2 + |N(f)|^2 \end{aligned} \tag{7.76}$$

$$\begin{aligned} \text{Od}\,|\tfrac{1}{2}\Lambda(f)|^2 &= \tfrac{1}{2}[|\tfrac{1}{2}\Lambda(f)|^2 - |\tfrac{1}{2}\Lambda(-f)|^2] \\ &= jN(f)M^*(f) - jN^*(f)M(f) \end{aligned} \tag{7.77}$$

These linear transformations on autocorrelation and crosscorrelation functions, not on the signals themselves, are indicated symbolically in Figure 7.5. These diagrams make it clear how non-symmetrical bandpass characteristics introduce correlation between in-phase and quadrature components of a bandpass process. Such considerations are important, for example, in analyzing the performance of the various coherent demodulation schemes discussed in the examples in Section 4.4.

Exercise 7.9. By analogy with (4.48), we could choose the "center" frequency of a real bandpass process as the centroid of the one-sided power spectral density; i.e.,

$$f_0 = \frac{\int_0^\infty f K_{xx}(f)\,df}{\int_0^\infty K_{xx}(f)\,df}$$

In Chapter 4, we found that f_0 defined by (4.48) could be interpreted as an envelope-weighted time average of the instantaneous frequency. Show that an analogous interpretation can be used for stationary random processes; i.e., f_0 defined above is equivalent to

$$f_0 = \frac{E[\mathbf{w}^2(t)\mathbf{f}_i(t)]}{E[\mathbf{w}^2(t)]}$$

where

$$\mathbf{f}_i(t) = \frac{1}{2\pi\mathbf{w}^2(t)}\,[\dot{\hat{\mathbf{x}}}(t)\mathbf{x}(t) - \dot{\mathbf{x}}(t)\hat{\mathbf{x}}(t)] \tag{4.41}$$

and

$$\mathbf{w}^2(t) = \mathbf{x}^2(t) + \hat{\mathbf{x}}^2(t) \tag{4.40}$$

Hint: Treat time differentiation and Hilbert transformation as time-invariant filtering operations.

REFERENCES

1. M. Loève, *Probability Theory*, Van Nostrand, 1955, Chapter X.
2. A. M. Yaglom, *An Introduction to the Theory of Stationary Random Functions*, Prentice-Hall, 1962.
3. A. Papoulis, *Probability, Random Variables, and Stochastic Processes*, McGraw-Hill, 1965.
4. Blackman and Tukey, *The Measurement of Power Spectra*, Dover, 1958.
5. K. L. Jordan, "Discrete Representations of Random Signals," MIT–RLE Report, No. 378, July, 1961.

6. J. L. Brown, Jr., "Mean Square Truncation Error in Series Expansions of Random Functions," *J. SIAM*, Vol. 8, No. 1, pp. 28–32 (March, 1960).
7. Courant and Hilbert, *Methods of Mathematical Physics*, Vol. I, Interscience, 1953.
8. H. L. Van Trees, *Detection, Estimation, and Modulation Theory*, Part I, John Wiley & Sons, 1968.
9. D. Slepian, "Estimation of Signal Parameters in the Presence of Noise," *Trans. IRE*, Vol. IT-3, pp. 68–89 (March, 1954).
10. W. B. Davenport, Jr. and W. L. Root, *An Introduction to the Theory of Random Signals and Noise*, McGraw-Hill, 1958.
11. T. Kailath, "Some Integral Equations with 'Nonrational' Kernels," *Trans. IEEE*, Vol. IT-12, No. 4, pp. 442–47 (October, 1966).

MODELS FOR RANDOM PROCESSES

8

8.1 Introduction

In Chapter 7, we emphasized the correlation function description of random signal processes. In Chapter 9, the utility of this type of description for solving a broad class of system design problems will be demonstrated. Before proceeding to these applications, however, it will be worthwhile to examine the properties of the correlation functions for various types of processes. Rather than selecting correlation functions arbitrarily or making detailed measurements on a physical process, the system designer prefers to construct a *mathematical model* to represent the physical process under consideration. The reason for this preference is that certain aspects of the physical mechanism that generates the various realizations may be well understood and these aspects may be easily characterized by numerical values (parameters). Furthermore, it may happen that these physical parameters are more readily measurable than the correlation functions for the process. Hopefully, a mathematical model, involving only a few of the most significant of these parameters, will provide an adequate representation for the process. From the model, the correlation functions can be derived analytically and the effect of changes in the physical parameters on the correlation functions will be apparent.

A complete treatment of methods for modeling processes is quite beyond the scope of this text. We shall consider models for some processes which, because of the physical mechanism of generation, are of the form of a random time sequence of signal pulses. The expressions for mean value and the autocorrelation will be derived. Several of these models will be used in examples in Chapter 9.

8.2 Signal Pulses with Random Amplitudes and Arrival Times

In some systems, notably radar and sonar, the signal pulses are sufficiently isolated in time so that a model involving only a single pulse of known shape can be effectively employed to represent the signal returned from a target. Because of unknown target cross section and range, the amplitude and arrival time of the returned pulse can be interpreted as random variables. In some cases, e.g., where multiple targets are involved, the model will involve a number of pulses with random amplitudes and arrival times. The simplest models for these cases will be considered in the following examples.

Single Pulse

Let

$$\mathbf{x}(t) = \mathbf{a}s(t - \mathbf{t}_0) \tag{8.1}$$

where $\mathbf{a}$ and $\mathbf{t}_0$ are statistically independent random variables. The mean value of this process is proportional to the convolution of the pulse with the probability density function for $\mathbf{t}_0$.

$$\overline{\mathbf{x}(t)} = E[\mathbf{x}(t)] = \bar{\mathbf{a}} \int_{-\infty}^{\infty} s(t - \sigma) p_{t_0}(\sigma)\, d\sigma \tag{8.2}$$

The autocorrelation function is given by

$$\begin{aligned} k_{xx}(t + \tau, t) &= E[\mathbf{x}(t + \tau)\mathbf{x}(t)] \\ &= \overline{\mathbf{a}^2} \int_{-\infty}^{\infty} s(t + \tau - \sigma) s(t - \sigma) p_{t_0}(\sigma)\, d\sigma \end{aligned} \tag{8.3}$$

and it is clear that the process is non-stationary. If the arrival time $\mathbf{t}_0$ is known; i.e., $p_{t_0}(\sigma) = \delta(\sigma - t_0)$, then we have

$$\overline{\mathbf{x}(t)} = \bar{\mathbf{a}} s(t - t_0)$$

and

$$k_{xx}(t + \tau, t) = \overline{\mathbf{a}^2} s(t + \tau - t_0) s(t - t_0) \tag{8.4}$$

As the arrival times become more and more indefinite ($p_{t_0}(\sigma)$ becomes broad), the process takes on some aspects of a wide-sense stationary process.

If we assume a uniform density for $\mathbf{t}_0$ over the interval $|t_0| \leqslant T$, then

$$\overline{\mathbf{x}(t)} = \frac{\bar{\mathbf{a}}}{2T}\int_{-T}^{T} s(t-\sigma)\,d\sigma$$
$$k_{xx}(t+\tau, t) = \frac{\overline{\mathbf{a}^2}}{2T}\int_{-T}^{T} s(t+\tau-\sigma)s(t-\sigma)\,d\sigma \tag{8.5}$$

For T large compared to the duration of the pulse $s(t)$, the expressions in (8.5) are essentially independent of t for $|t| \ll T$. This does not produce a non-trivial stationary process, however, since the mean-squared value approaches zero as $T \to \infty$. Another feature of interest in this case is that the autocorrelation function is approximately proportional to the time-ambiguity function (2.40), $r_s(\tau) = (\mathbf{s}, \mathbf{s}_\tau)$, for the pulse $s(t)$.

Multiple Pulses

Let

$$\mathbf{x}(t) = \sum_{k=1}^{n} \mathbf{a}_k s(t - \mathbf{t}_k) \tag{8.6}$$

where the $2n$ random variables $\{\mathbf{a}_k\}$ and $\{\mathbf{t}_k\}$ are assumed to be statistically independent.

We shall further assume that the $\{\mathbf{a}_k\}$ are identically distributed and also that the $\{\mathbf{t}_k\}$ are identically distributed with probability density function $p(\sigma)$. Then we have

$$\overline{\mathbf{x}(t)} = \sum_{k=1}^{n} E[\mathbf{a}_k]E[s(t-\mathbf{t}_k)]$$
$$= n\bar{\mathbf{a}}\int_{-\infty}^{\infty} s(t-\sigma)p(\sigma)\,d\sigma \tag{8.7}$$

and

$$k_{xx}(t+\tau, t) = \sum_{k=1}^{n}\sum_{j=1}^{n} E[\mathbf{a}_k\mathbf{a}_j]E[s(t+\tau-\mathbf{t}_k)s(t-\mathbf{t}_j)]$$

where

$$E[\mathbf{a}_k\mathbf{a}_j] = \overline{\mathbf{a}^2} \qquad \text{for } k = j$$
$$= \bar{\mathbf{a}}^2 \qquad \text{for } k \neq j \tag{8.8}$$

Considering separately the n terms with $k = j$ and the $n^2 - n$ terms with $k \neq j$, we find

$$k_{xx}(t+\tau, t) = n\overline{\mathbf{a}^2}\int_{-\infty}^{\infty} s(t+\tau-\sigma)s(t-\sigma)p(\sigma)\,d\sigma$$
$$+ (n^2 - n)\bar{\mathbf{a}}^2\left[\int_{-\infty}^{\infty} s(t-\sigma)p(\sigma)\,d\sigma\right]^2 \tag{8.9}$$

If we assume $p(\sigma)$ is uniform over the interval $|\sigma| \leqslant T$, and if we let $n = \lambda 2T$, then in the limit, as $T \to \infty$, we obtain a wide-sense stationary process. In this case

$$\begin{aligned} \bar{\mathbf{x}} &= \lambda \bar{\mathbf{a}} q \\ k_{xx}(\tau) &= \lambda \overline{\mathbf{a}^2} r(\tau) + (\lambda \bar{\mathbf{a}} q)^2 \\ m_{xx}(\tau) &= \lambda \overline{\mathbf{a}^2} r(\tau) \end{aligned} \tag{8.10}$$

where

$$\begin{aligned} q &\triangleq \int_{-\infty}^{\infty} s(t)\, dt \\ r(\tau) &\triangleq \int_{-\infty}^{\infty} s(t+\tau)s(t)\, dt \end{aligned} \tag{8.11}$$

We see from (8.10) that the autocovariance function is proportional to the time-ambiguity function for $s(t)$. We shall henceforth denote this function by $r(\tau)$. The parameter λ is interpreted as the average rate of occurrence of pulses. The power spectral density for this process is

$$K_{xx}(f) = \lambda \overline{\mathbf{a}^2} R(f) + (\lambda \bar{\mathbf{a}} q)^2\, \delta(f) \tag{8.12}$$

where

$$R(f) = |S(f)|^2$$

The δ-function at $f = 0$ in (8.12) represents the power in the *d.c. component* of the process. This model is often used to represent the *shot noise* in electronic devices. In this application, $s(t)$ is the current pulse in an external circuit caused by transit of an individual charge carrier with charge q, and we let $a_k = 1$. The expressions in (8.10) for this application are often referred to as *Campbell's theorem*.[1] In Section 8.7, we shall take a closer look at this type of process and provide another derivation of the mean and autocorrelation function.

8.3 Cyclostationary Processes

Fortunately, many of the random processes encountered in physical systems can be adequately modeled by means of stationary processes. On the other hand, it is not at all uncommon to find processes, which might otherwise be stationary, that have been subjected to some repetitive processing operation. These periodic operations are usually introduced intentionally, for the purpose of providing a time reference for making signal observations. As a result of these operations, a particular kind of non-stationary process is produced. Consider, for example, the output of a receiver connected to a narrow-beam radar antenna which circularly scans a field of stationary signal sources. If the source strengths do not have a uniform distribution, with antenna angle, then clearly the statistics of receiver output vary periodically at the rate of revolution of the antenna. Similarly, a television signal, obtained by rectangular scanning of a random video field, will exhibit periodically varying statistics. In fact, all scanning operations, except those entirely controlled by another random process, will introduce a kind of periodicity into the signal process. In view of the prevalence of these kinds of operations in signal processing systems, a more detailed examination of second-order statistics of the processes is worthwhile. This will be done in the following examples.

Sample-and-hold Operation

It is often uneconomical to continuously observe a signal process. The simplest expedient is to sample the process periodically and hold the value of the last sample until the next sampling instant. This operation can be represented symbolically by the *sample-and-hold* circuit shown in Figure 8.1.

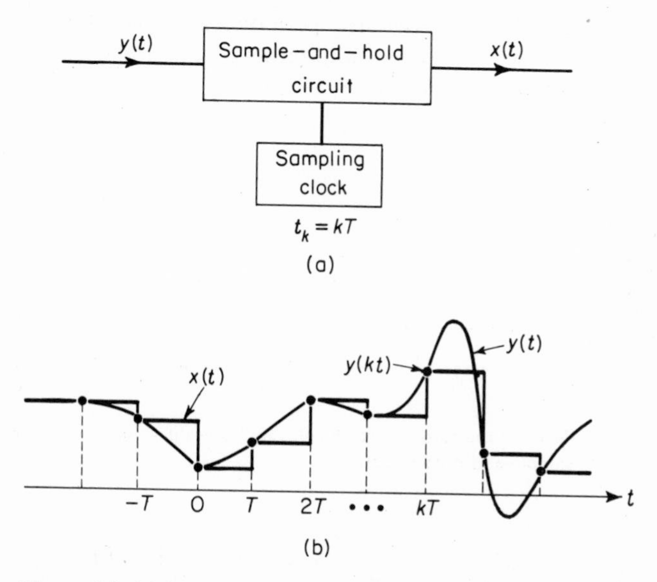

Figure 8.1. (a) Sample-and-hold circuit. (b) Typical realizations.

We assume, for the moment, that the sampling instants are given by $\{t_k = kT; k = 0, \pm 1, \pm 2, \ldots\}$ and the output process $\mathbf{x}$ is related to the sampled process $\mathbf{y}$ by

$$\mathbf{x}(t) = \sum_{k=-\infty}^{\infty} \mathbf{y}(kT)s(t - kT) \tag{8.13}$$

where

$$\begin{aligned} s(t) &= 1 \qquad \text{for } 0 \leqslant t < T \\ &= 0 \qquad \text{otherwise} \end{aligned} \tag{8.14}$$

Let us assume that $\mathbf{y}$ is a wide-sense stationary process, then

$$\overline{\mathbf{x}(t)} = \sum_{k=-\infty}^{\infty} E[\mathbf{y}(kT)]s(t - kT)$$

$$= \bar{\mathbf{y}} \sum_{k=-\infty}^{\infty} s(t - kT) = \bar{\mathbf{y}} \tag{8.15}$$

and

$$k_{xx}(t + \tau, t) = \sum_{k=-\infty}^{\infty} \sum_{j=-\infty}^{\infty} E[\mathbf{y}(kT)\mathbf{y}(jT)]s(t + \tau - kT)s(t - jT)$$

Letting $j = k + m$, we have

$$k_{xx}(t + \tau, t) = \sum_{m=-\infty}^{\infty} k_{yy}(mT) \sum_{k=-\infty}^{\infty} s(t + \tau - kT)s(t - kT - mT) \tag{8.16}$$

The sum over k in (8.16) is obviously periodic (with period T) in t. We can express the autocorrelation function in terms of a periodic *indicator function* $q(t, \tau)$ shown graphically in Figure 8.2 for $0 < \tau < T$.

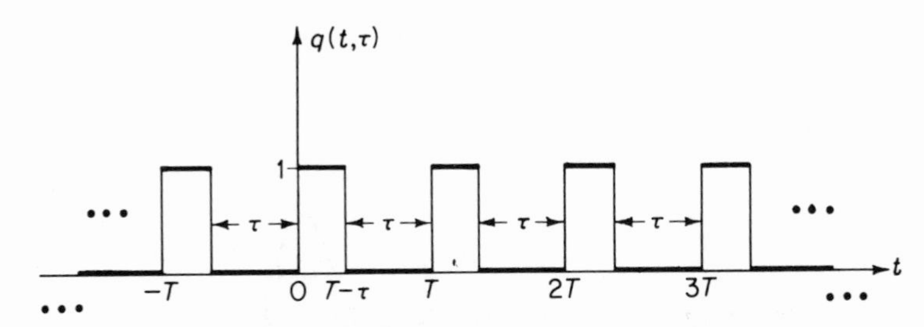

Figure 8.2. Periodic indicator function relevant to evaluation of autocorrelation at output of sample-and-hold circuit.

$$k_{xx}(t + \tau, t) = \sum_{m=-\infty}^{\infty} k_{yy}(mT)q(t, \tau + mT) \tag{8.17}$$

where

$$q(t, \tau) = \sum_{k=-\infty}^{\infty} s(t + \tau - kT)s(t - kT)$$

Note that $q(t, \tau)$ vanishes for $|\tau| \geqslant T$.

The process $\mathbf{x}$ is clearly non-stationary and belongs to the class exhibiting the following periodicity:

$$\overline{\mathbf{x}(t_1 + T)} = \overline{\mathbf{x}(t_1)}$$

$$k_{xx}(t_1 + T, t_2 + T) = k_{xx}(t_1, t_2) \tag{8.18}$$

for all t_1 and t_2. Processes satisfying (8.18) are said to be *cyclostationary*[2] (in the wide sense). These processes are also called *periodically stationary*.[3]

Phase Randomizing

Because of this particular form of non-stationarity it is tempting to find ways of treating it like a stationary process and taking advantage of such concepts as power spectral density. One thing we could do is simply average the auto-correlation function (8.17) over one period (in t). Let

$$\begin{aligned}\tilde{k}_{xx}(\tau) &= \frac{1}{T}\int_0^T k_{xx}(t+\tau, t)\,dt \\ &= \frac{1}{T}\sum_{m=-\infty}^{\infty} k_{yy}(mT)\int_0^T q(t, \tau + mT)\,dt \\ &= \frac{1}{T}\sum_{m=-\infty}^{\infty} k_{yy}(mT) r(\tau + mT)\end{aligned}$$

where

$$r(\tau) = \int_{-\infty}^{\infty} s(t+\tau)s(t)\,dt \tag{8.19}$$

This procedure for removing the dependence on t by time-averaging the autocorrelation function over one period is not as arbitrary as it may seem at first. We shall give another interpretation which is physically more appealing. We make $\mathbf{x}$ a stationary process by adding another random variable $\boldsymbol{\delta}$ such that

$$\mathbf{x}(t) = \sum_{k=-\infty}^{\infty} \mathbf{y}(kT + \boldsymbol{\delta}) s(t - kT - \boldsymbol{\delta}) \tag{8.20}$$

Thus, the sampling instants in Figure 8.1. are changed to $t_k = kT + \delta$. The physical interpretation is that the observer of the process $\mathbf{x}$, although he knows that it is derived by periodically sampling another process, may have no knowledge whatsoever of the time reference for the sampling instants. Thus, if we assume that $\boldsymbol{\delta}$ is uniformly distributed over $0 \leqslant \delta < T$, then we have

$$\begin{aligned}\overline{\mathbf{x}(t)} &= \sum_{k=-\infty}^{\infty} E[\mathbf{y}(kT + \boldsymbol{\delta})]\int_{-\infty}^{\infty} s(t - kT - \sigma) p_{\delta}(\sigma)\,d\sigma \\ &= \bar{\mathbf{y}} \sum_{k=-\infty}^{\infty} \frac{1}{T}\int_0^T s(t - kT - \sigma)\,d\sigma \\ &= \bar{\mathbf{y}}\frac{1}{T}\int_{-\infty}^{\infty} s(t)\,dt = \bar{\mathbf{y}}\end{aligned} \tag{8.21}$$

and

$$
\begin{aligned}
k_{xx}(t+\tau, t) \\
&= \sum_{m=-\infty}^{\infty} \sum_{k=-\infty}^{\infty} E[\mathbf{y}(kT+\boldsymbol{\delta})\mathbf{y}(kT+mT+\boldsymbol{\delta})] \\
&\quad \times \int_{-\infty}^{\infty} s(t+\tau-kT-\sigma)s(t-kT-mT-\sigma)p_{\delta}(\sigma)\,d\sigma \\
&= \sum_{m=-\infty}^{\infty} k_{yy}(mT)\frac{1}{T}\sum_{k=-\infty}^{\infty}\int_{0}^{T} s(t+\tau-kT-\sigma)s(t-kT-mT-\sigma)\,d\sigma \\
&= \frac{1}{T}\sum_{m=-\infty}^{\infty} k_{yy}(mT)r(\tau+mT) \qquad (8.22)
\end{aligned}
$$

where

$$r(\tau)=\int_{-\infty}^{\infty} s(t+\tau)s(t)\,dt$$

Hence, with a randomized time origin for the sampling instants (the sampling instants are still equally spaced), the process **x** is wide-sense stationary and the autocorrelation function matches that obtained by time averaging, (8.19). In some problems, this randomizing of the *phase* of the sampler is quite appropriate and we need only deal with the resulting stationary process. In other cases, it is quite important to retain the basic cyclostationary model for the process. We shall comment more on this later.

For the stationary case there is also a simple relationship between power spectral densities of the input and output processes in Figure 8.1. Taking the Fourier transform of (8.22), we have

$$K_{xx}(f)=\frac{1}{T}R(f)\sum_{m=-\infty}^{\infty} k_{yy}(mT)e^{j2\pi mTf} \qquad (8.23)$$

Now using the Poisson sum formula (6.125), (8.23) becomes

$$K_{xx}(f)=\frac{1}{T^2}R(f)\sum_{\ell=-\infty}^{\infty} K_{yy}\left(f-\frac{\ell}{T}\right) \qquad (8.24)$$

where

$$R(f)=|S(f)|^2=T^2\left(\frac{\sin \pi Tf}{\pi Tf}\right)^2$$

Thus the power spectral density for **x** is expressed in terms of a periodic function, obtained by summing frequency translates of the power spectral density for **y**, multiplied by an envelope function $R(f)$. This is shown graphically in Figure 8.3. If T is chosen small enough so that the process has little fluctuation in T second intervals, then $K_{yy}(f)$ is narrow compared to $1/T$

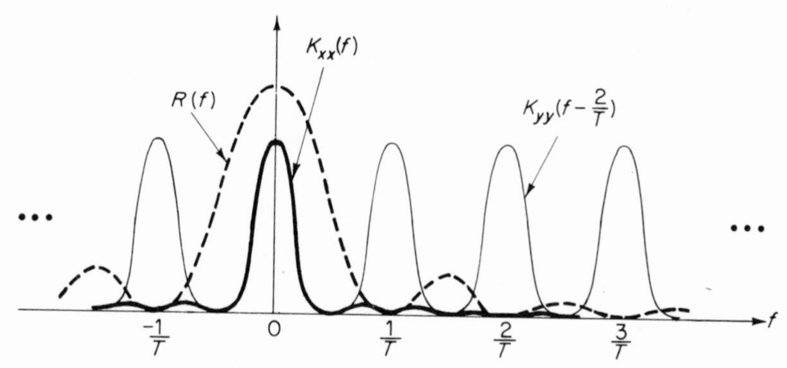

Figure 8.3. Power spectral density of process subjected to sample-and-hold operation.

and $R(f)$ tends to suppress the translates of $K_{yy}(f)$. In this case,

$$K_{xx}(f) \cong \frac{R(0)}{T^2} K_{yy}(f) = K_{yy}(f)$$

and the power spectral densities of both processes are substantially the same.

Exercise 8.1. For the random process described by $\mathbf{x}(t) = \mathbf{a} \cos 2\pi f_0 t - \mathbf{b} \sin 2\pi f_0 t$, show that the process is wide-sense stationary if, and only if, the real random variables $\mathbf{a}$ and $\mathbf{b}$ are zero mean, equal variance, and orthogonal. Show that $\mathbf{x}(t) = \mathbf{c} \cos (2\pi f_0 t + \boldsymbol{\theta})$ with $\mathbf{c}$ and $\boldsymbol{\theta}$ statistically independent is wide-sense stationary if the probability density function for $\boldsymbol{\theta}$ is the uniform density over the interval $0 \leqslant \theta < 2\pi$. Are there other probability density functions for $\boldsymbol{\theta}$ which make $\mathbf{x}$ wide-sense stationary?

Exercise 8.2. Show that the mean-squared error $E[\{\mathbf{x}(t) - \mathbf{y}(t)\}^2]$ resulting from the sample-and-hold operation is a periodic function of t. Show that the time-averaged value of this mean-squared error is equal to

$$\frac{2}{T} \int_0^T [k_{yy}(0) - k_{yy}(\tau)] \, d\tau$$

Give an example of a stationary process $\mathbf{y}$ for which the mean-squared error is zero.

Synchronous Pulse Amplitude Modulation

An important generalization of the sample-and-hold process is obtained by replacing the rectangular pulse $s(t)$ by a pulse of arbitrary shape. We also let the pulse amplitudes be given by realizations of a discrete-time process $\{\mathbf{a}_k;\, k = 0, \pm 1, \pm 2, \ldots\}$ not necessarily resulting from sampling a continuous-time process. The resulting signal

$$\mathbf{x}(t) = \sum_{k=-\infty}^{\infty} \mathbf{a}_k s(t - kT) \tag{8.25}$$

is said to carry the information in the $\{\mathbf{a}_k\}$ sequence by means of *pulse amplitude modulation* (PAM). The modulation is said to be *synchronous* because of the uniform spacing between successive pulses. The derivation of mean and autocorrelation used before is sufficiently general to include this case if we assume that the $\{\mathbf{a}_k\}$ sequence is wide-sense stationary; i.e., if

$$\left.\begin{aligned} E[\mathbf{a}_k] &= \bar{\mathbf{a}} \\ E[\mathbf{a}_k\mathbf{a}_{k+m}] &= \alpha_m = \alpha_{-m} \end{aligned}\right\} \quad \text{for all } k \tag{8.26}$$

then from (8.15) and (8.16)

$$\begin{aligned} \overline{\mathbf{x}(t)} &= \bar{\mathbf{a}} \underbrace{\sum_{k=-\infty}^{\infty} s(t-kT)} \\ k_{xx}(t+\tau, t) &= \sum_{m=-\infty}^{\infty} \alpha_m \overbrace{\sum_{k=-\infty}^{\infty} s(t+\tau-kT)s(t-kT-mT)}^{\text{periodic in } t \text{ (period } = T)} \end{aligned} \tag{8.27}$$

It follows that the PAM signal is also a cyclostationary process. The cyclostationarity property can be made intuitively more apparent in this more general case. Suppose $s(t)$ is a very narrow rectangular pulse (compared to T) as shown in Figure 8.4. Considering only the mean-squared value of the process $k_{xx}(t, t)$ in Figure 8.4, it is obvious that there is a considerable variation in this quantity over one period.

On the other hand, it is possible to select a pulse shape so that the process is wide-sense stationary. To show this, let $s(t)$ be the band-limited pulse

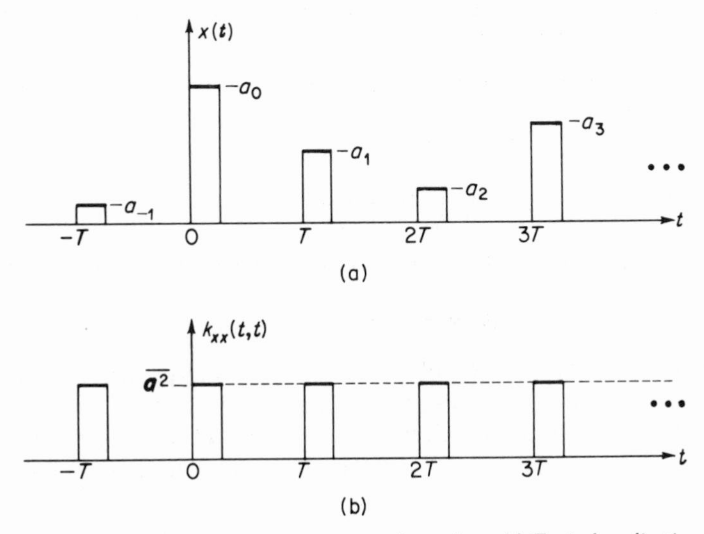

Figure 8.4. PAM signal using short rectangular pulses. (a) Typical realization of the process. (b) Mean-squared value of the process.

given by

$$s(t) = \frac{\sin(\pi t/T)}{\pi t/T} \tag{8.28}$$

so that

$$\begin{aligned} S(f) &= T \qquad \text{for } |f| \leqslant \frac{1}{2T} \\ &= 0 \qquad \text{otherwise} \end{aligned}$$

Using a slightly different version of the Poisson sum formula (6.125) which states that, for any $s(t)$,

$$\sum_{k=-\infty}^{\infty} s(t - kT) = \frac{1}{T} \sum_{\ell=-\infty}^{\infty} S\left(\frac{\ell}{T}\right) e^{j2\pi\ell\frac{t}{T}} \tag{8.29}$$

the periodic functions in (8.27) are easily shown to be given by

$$\sum_{k=-\infty}^{\infty} s(t - kT) = 1$$

$$\sum_{k=-\infty}^{\infty} s(t + \tau - kT)s(t - kT - mT) = s(\tau + mT)$$

Hence, for this special pulse shape, the PAM signal is wide-sense stationary with mean and autocorrelation given by

$$\begin{aligned} \bar{\mathbf{x}} &= \bar{\mathbf{a}} \\ k_{xx}(\tau) &= \sum_{m=-\infty}^{\infty} \alpha_m s(\tau + mT) \end{aligned} \tag{8.30}$$

Sampling Theorem

An important application of this result is in providing a sampling theorem for band-limited processes which corresponds directly to the sampling theorem (6.128) for deterministic band-limited signals. Suppose we let the $\{\mathbf{a}_k\}$ be the uniformly spaced sample values of a stationary random process $\mathbf{y}$ which is band-limited [in the sense that $K_{yy}(f) = 0$ for $|f| > 1/2T$]. We have

$$\mathbf{a}_k = \mathbf{y}(kT) \Rightarrow \alpha_m = k_{yy}(mT)$$

and from (8.30), with $S(f)$ given by (8.28),

$$\begin{aligned} K_{xx}(f) &= S(f) \sum_{m=-\infty}^{\infty} k_{yy}(mT)e^{j2\pi mTf} \\ &= \frac{S(f)}{T} \sum_{\ell=-\infty}^{\infty} K_{yy}\left(f - \frac{\ell}{T}\right) \\ &= K_{yy}(f) \qquad \text{for all } f \end{aligned} \tag{8.31}$$

The closeness of $\mathbf{x}$ to $\mathbf{y}$ is evaluated by

$$E[\{\mathbf{x}(t) - \mathbf{y}(t)\}^2] = E[\mathbf{x}^2(t)] + E[\mathbf{y}^2(t)] - 2E[\mathbf{x}(t)\mathbf{y}(t)]$$
$$= k_{xx}(0) + k_{yy}(0) - 2\sum_{i=-\infty}^{\infty} k_{yy}(t - iT)s(t - iT) \tag{8.32}$$

The crosscorrelation term in (8.32) can be evaluated with the aid of (8.29) and the band-limited property of $K_{yy}(f)$ and $S(f)$.

$$\sum_{i=-\infty}^{\infty} k_{yy}(t - iT)s(t - iT) = \frac{1}{T}\sum_{\ell=-\infty}^{\infty}\int_{-\infty}^{\infty} K_{yy}(\nu)S\left(\frac{\ell}{T} - \nu\right) d\nu e^{j2\pi\ell\frac{t}{T}}$$
$$= \frac{1}{T}\int_{-\infty}^{\infty} K_{yy}(\nu)S(-\nu)\, d\nu = \int_{-\infty}^{\infty} K_{yy}(\nu)\, d\nu = k_{yy}(0) \tag{8.33}$$

since only the $\ell = 0$ term above is non-zero. With this result and (8.31), (8.32) becomes

$$E[\{\mathbf{x}(t) - \mathbf{y}(t)\}^2] = 2[k_{yy}(0) - k_{yy}(0)] = 0 \tag{8.34}$$

Hence we can say that a band-limited process can be represented by its sample values

$$\mathbf{y}(t) = \sum_{k=-\infty}^{\infty} \mathbf{y}(kT)\frac{\sin(\pi/T)(t - kT)}{(\pi/T)(t - kT)} \tag{8.35}$$

in the sense that the mean-squared value of the error is zero.

Power Spectral Density for PAM Signal

Returning now to the case of arbitrary pulse shapes, we may choose to treat the cyclostationary process as a stationary process either by phase randomization or by averaging the mean and autocorrelation over one period as done previously. Hence, if we modify the process according to

$$\mathbf{x}(t) = \sum_{k=-\infty}^{\infty} \mathbf{a}_k s(t - kT - \boldsymbol{\delta}) \tag{8.36}$$

where $\boldsymbol{\delta}$ has a uniform density over $0 \leqslant \delta < T$, then we find

$$\bar{\mathbf{x}} = \frac{\bar{\mathbf{a}}q}{T}; \qquad q = \int_{-\infty}^{\infty} s(t)\, dt \tag{8.37}$$

and

$$k_{xx}(\tau) = \frac{1}{T}\sum_{m=-\infty}^{\infty} \alpha_m r(\tau + mT); \qquad r(\tau) = \int_{-\infty}^{\infty} s(t + \tau)s(t)\, dt \tag{8.38}$$

In the frequency domain,

$$K_{xx}(f) = \frac{1}{T} R(f) \underbrace{\sum_{m=-\infty}^{\infty} \alpha_m e^{j2\pi mTf}}_{\substack{\text{periodic in } f \\ (\text{period} = 1/T)}}; \qquad R(f) = |S(f)|^2 \tag{8.39}$$

In this more general case, $R(f)$ need not have zeros at multiples of $1/T$ as it did for the rectangular pulse in the sample-and-hold operation (8.24). This can give rise to harmonically related *discrete components* in the power spectral density for the process. For example, assume that the $\{\mathbf{a}_k\}$ are statistically independent, then $\alpha_m = \bar{\mathbf{a}}^2$ for $m \neq 0$ and $\alpha_0 = \overline{\mathbf{a}^2}$. Using the variance $\sigma_a^2 = \overline{\mathbf{a}^2} - \bar{\mathbf{a}}^2$, we can write

$$k_{xx}(\tau) = \frac{\sigma_a^2}{T} r(\tau) + \underbrace{\frac{\bar{\mathbf{a}}^2}{T} \sum_{m=-\infty}^{\infty} r(\tau + mT)}_{\substack{\text{periodic in } \tau \\ (\text{period} = T)}} \tag{8.40}$$

and

$$K_{xx}(f) = \frac{\sigma_a^2}{T} R(f) + \left(\frac{\bar{\mathbf{a}}}{T}\right)^2 \sum_{\ell=-\infty}^{\infty} R\left(\frac{\ell}{T}\right) \delta\left(f - \frac{\ell}{T}\right) \tag{8.41}$$

The δ-functions in (8.41) represent power concentrated at various multiples of $1/T$. The amount of power at these frequencies depends both on the mean value of the $\{\mathbf{a}_k\}$ sequence and the pulse shape employed. These considerations have a practical significance since the discrete components represent wasted power as far as information transmission is concerned. The system designer often modifies both the message sequence and the pulse shape in order to exercise control over the relative magnitude of the discrete components. Presence of discrete components at certain frequencies is not wholly undesirable since they can be used to extract timing information from the process,[2],[4] thus avoiding the need for a separate transmission channel to provide a clock signal at the various observation points in the system. Other types of periodic behavior of the $\{\alpha_m\}$ sequence, such as that resulting from periodic insertion of synchronizing pulses in the PAM signal, may also give rise to discrete components in the power spectral density.

Exercise 8.3. Sequences $\{a_k\}$ of real numbers in the interval $0 \leqslant a_k \leqslant 1$ can be mapped into time functions by an operation known as *pulse duration modulation.* Let $\{\mathbf{a}_k; k = 0, \pm 1, \pm 2, \ldots\}$ be a sequence of statistically independent random variables, each uniformly distributed over $0 \leqslant a_k \leqslant 1$, and let the random process $\mathbf{x}$ be described by

$$\mathbf{x}(t) = \sum_{k=-\infty}^{\infty} s\left(\frac{t - kT}{\mathbf{a}_k}\right)$$

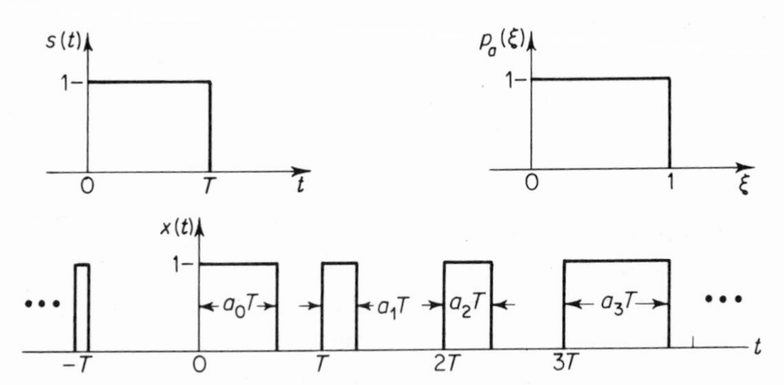

Show that the process is cyclostationary; then create a wide-sense stationary process by phase randomization. Evaluate the mean and autocorrelation function, separating out the terms which lead to the continuous and discrete parts of the power spectral density.

Exercise 8.4. Consider a binary data transmission system which transmits $s_0(t)$ to indicate "zero" and $s_1(t)$ to indicate "one." Assume that successive binary states are statistically independent and that "one" occurs with probability p. Assume that signal pulses are emitted synchronously (every T seconds) and use phase randomization to create a wide-sense stationary process. Evaluate the autocorrelation function and power spectral density of the process, indicating separately the discrete components, if any. Show how this model could be used to characterize power spectral density for binary phase-shift keying and frequency-shift keying. Outline the steps required in generalizing from binary to larger signaling alphabets. *Hint:* Express $\mathbf{x}$ as

$$\mathbf{x}(t) = \sum_{k=-\infty}^{\infty} (1 - \mathbf{a}_k)s_0(t - kT) + \mathbf{a}_k s_1(t - kT)$$

where the random variable $\mathbf{a}_k$ can take on only the values 0 and 1.

8.4 Effects of Coding on Power Spectral Density

From the previous example, it is clear that pulse shape has a predominant effect on the power spectral density of a PAM signal. It was also mentioned that operations on the $\{\mathbf{a}_k\}$ sequence could be used to modify the power spectral density. We refer to such operations as *coding* of the message sequence. One normally thinks of coding in relation to error-detecting and error-correcting schemes; however, some very simple coding operations are extensively used for the purpose of redistributing power in the frequency domain. This is usually done in order to more effectively match the signal to transmission channel characteristics.

To illustrate these ideas, we shall consider the effects of coding on a binary message sequence ($a_k = 1$ or 0). For simplicity, we shall assume that the elements of the basic uncoded sequence are statistically independent and that the probability of the event $[a_k = 1]$ is given by p. We let the pulse shape $s(t)$ be arbitrary, however, in most applications the pulse bandwidth will be at least $1/2T$ so that intersymbol interference may be small (see Section 6.7). For the uncoded binary signal, we have[1]

$$\begin{aligned} \Pr\,[a_k = 1] &= p, \qquad \Pr\,[a_k = 0] = 1 - p \\ \bar{\mathbf{a}} &= E[\mathbf{a}_k] = p \\ \alpha_m &= E[\mathbf{a}_k \mathbf{a}_{k+m}] = p \qquad \text{for } m = 0 \\ &= p^2 \qquad \text{for } m \neq 0 \end{aligned} \tag{8.42}$$

Using (8.41), the power spectral density becomes

$$K_{xx}(f) = \frac{p(1-p)}{T} R(f) + \left(\frac{p}{T}\right)^2 \sum_{\ell=-\infty} R\left(\frac{\ell}{T}\right) \delta\left(f - \frac{\ell}{T}\right) \tag{8.43}$$

The effects of the various coding schemes mentioned below will be evaluated by comparing the resulting spectral density with (8.43). Most of the results can be generalized from binary to larger signaling alphabets.[5]

Differential Binary Coding

The modified binary PAM signal

$$\mathbf{x}(t) = \sum_{k=-\infty}^{\infty} \mathbf{b}_k s(t - kT) \tag{8.44}$$

is called a *differential binary* signal if the $\{\mathbf{b}_k\}$ are related to the $\{\mathbf{a}_k\}$ in the following manner. If $a_k = 1$, then b_k is different from its previous value b_{k-1}. If $a_k = 0$, then $b_k = b_{k-1}$. In other words,

$$\begin{aligned} b_k - b_{k-1} &= \pm 1 \qquad \text{if } a_k = 1 \\ &= 0 \qquad \text{if } a_k = 0 \end{aligned} \tag{8.45}$$

In order to evaluate the power spectral density for $\mathbf{x}$, we need to know the mean and correlation for the $\{\mathbf{b}_k\}$ sequence.

$$\begin{aligned} \bar{\mathbf{b}} &= E[\mathbf{b}_k] = \Pr\,[b_k = 1] \\ &= \Pr\,[b_{k-1} = 0 \text{ and } a_k = 1] + \Pr\,[b_{k-1} = 1 \text{ and } a_k = 0] \\ &= (1 - \bar{\mathbf{b}})p + \bar{\mathbf{b}}(1 - p) \end{aligned} \tag{8.46}$$

From (8.46), we conclude that $\bar{\mathbf{b}} = \frac{1}{2}$ and that "ones" and "zeros" in the differential binary sequence are equiprobable, independently of the probabilities of "ones" and "zeros" in the original sequence.

[1] We remind the reader that a_k denotes a particular realization of the random variable $\mathbf{a}_k$.

Let

$$\begin{aligned}\beta_m &= E[\mathbf{b}_k\mathbf{b}_{k+m}] = \Pr\,[b_k = 1 \text{ and } b_{k+m} = 1] \\ &= \tfrac{1}{2}\Pr\,[b_{k+m} = 1 \mid b_k = 1]\end{aligned} \tag{8.47}$$

The conditional probability in (8.47) can be expressed as the sum of two joint probabilities, each conditioned on the event $[b_k = 1]$.

$$\begin{aligned}2\beta_m &= \Pr\,[b_{k+m-1} = 0 \text{ and } a_{k+m} = 1 \mid b_k = 1] \\ &\quad + \Pr\,[b_{k+m-1} = 1 \text{ and } a_{k+m} = 0 \mid b_k = 1]\end{aligned} \tag{8.48}$$

Comparing (8.47) and (8.48), we conclude that β_m satisfies a first-order difference equation with constant coefficients.

$$\beta_m - (1 - 2p)\beta_{m-1} = \tfrac{1}{2}p \tag{8.49}$$

The solution of this equation, using the "initial" condition $\beta_0 = \frac{1}{2}$ and the fact that $\beta_m = \beta_{-m}$, is given by[9]

$$\beta_m = \tfrac{1}{4}[(1 - 2p)^{|m|} + 1] \tag{8.50}$$

Now using (8.39), with $\{\beta_m\}$ replacing $\{\alpha_m\}$, the power spectral density is

$$\begin{aligned}K_{xx}(f) &= \frac{R(f)}{4T}\sum_{m=-\infty}^{\infty}(1 - 2p)^{|m|}e^{j2\pi mTf} + \frac{R(f)}{4T}\sum_{m=-\infty}^{\infty}e^{j2\pi mTf} \\ &= \frac{R(f)M(f)}{4T} + \frac{1}{4T^2}\sum_{\ell=-\infty}^{\infty}R\left(\frac{\ell}{T}\right)\delta\left(f - \frac{\ell}{T}\right)\end{aligned} \tag{8.51}$$

The second term in (8.51) is the discrete part of the spectral density obtained by considering only the constant term in the $\{\beta_m\}$ sequence.

We note that the discrete components are not affected by the value of the parameter p and are the same as in the uncoded signal (8.43) with $p = \frac{1}{2}$. The continuous part of the spectral density in (8.51) is modified by a periodic shaping function $M(f)$. There is a simple closed form expression for $M(f)$ obtained by splitting the summation over m into two parts, each of which is a geometric series and hence easily summable.

$$\begin{aligned}M(f) &\triangleq \sum_{m=-\infty}^{\infty}(1 - 2p)^{|m|}e^{j2\pi mTf} \\ &= \sum_{m=0}^{\infty}(1 - 2p)^m e^{j2\pi mTf} + \sum_{n=0}^{\infty}(1 - 2p)^n e^{-j2\pi nTf} - 1 \\ &= \frac{1}{1 - (1 - 2p)e^{j2\pi Tf}} + \frac{1}{1 - (1 - 2p)e^{-j2\pi Tf}} - 1 \\ &= \frac{p(1 - p)}{p^2 + (1 - 2p)(\sin \pi Tf)^2}\end{aligned} \tag{8.52}$$

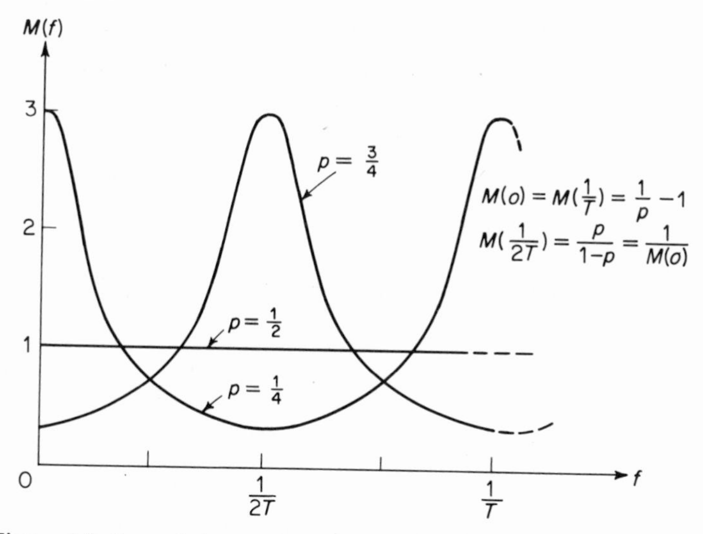

Figure 8.5. Spectral density shaping function for differential binary coding.

The shaping function is shown graphically in Figure 8.5 for various values of the parameter p. It is evident that the shaping is strongly dependent on p, causing a concentration in power about even multiples of $1/2T$ for $p < \frac{1}{2}$ and a concentration about odd multiples of $1/2T$ for $p > \frac{1}{2}$.

Bipolar Coding

For a variety of practical reasons, in the design of a PAM transmission system it is often desirable to reduce spectral density at low frequencies and also at frequencies in the neighborhood of the pulse repetition rate. This is accomplished to some extent with differential binary if it can be guaranteed that the uncoded signal has a high pulse density $p \gg \frac{1}{2}$. A much more effective scheme for this purpose is *bipolar*[4] coding. The uncoded signal, as we have described it, is *unipolar* in that a pulse is generated only when $a_k = 1$. The bipolar signal is the same except that consecutive pulses in the sequence have opposite polarity. It is intuitively evident that such an operation would reduce the low-frequency spectral density. The bipolar signal has pulse amplitudes given by the ternary sequence $\{\mathbf{c}_k\}$.

$$\mathbf{x}(t) = \sum_{k=-\infty}^{\infty} \mathbf{c}_k s(t - kT)$$

where

$$\begin{aligned} c_k &= \pm 1 \qquad \text{if } a_k = 1 \\ &= 0 \qquad \text{if } a_k = 0 \end{aligned} \tag{8.53}$$

Evaluation of the mean and correlation of the $\{\mathbf{c}_k\}$ sequence is simplified by recognizing its relation to the differential binary sequence. Since the $+1$ and -1 in (8.45) must alternate, we can write

$$c_k = b_k - b_{k-1} \tag{8.54}$$

Thus we have

$$\bar{\mathbf{c}} = E[\mathbf{c}_k] = \bar{\mathbf{b}} - \bar{\mathbf{b}} = 0 \Rightarrow \bar{\mathbf{x}} = 0 \tag{8.55}$$

and

$$\begin{aligned} \gamma_m &\triangleq E[\mathbf{c}_k\mathbf{c}_{k+m}] = E[(\mathbf{b}_k - \mathbf{b}_{k-1})(\mathbf{b}_{k+m} - \mathbf{b}_{k+m-1})] \\ &= 2\beta_m - \beta_{m+1} - \beta_{m-1} \end{aligned} \tag{8.56}$$

Now using (8.50) for β_m,

$$\begin{aligned} \gamma_m &= -p^2(1-2p)^{|m|} \qquad \text{for } |m| \geqslant 1 \\ \gamma_0 &= p \end{aligned} \tag{8.57}$$

The power spectral density for the bipolar signal is given by

$$\begin{aligned} K_{xx}(f) &= \frac{1}{T} R(f) \sum_{m=-\infty}^{\infty} \gamma_m e^{j2\pi mTf} \\ &= \frac{R(f)M(f)}{T} \end{aligned} \tag{8.58}$$

where

$$\begin{aligned} M(f) &= \sum_{m=-\infty}^{\infty} \gamma_m e^{j2\pi mTf} \\ &= p - p^2 \sum_{m=1}^{\infty} (1-2p)^{m-1} e^{j2\pi mTf} - p^2 \sum_{n=1}^{\infty} (1-2p)^{n-1} e^{-j2\pi nTf} \\ &= p - \frac{p^2 e^{j2\pi Tf}}{1-(1-2p)e^{j2\pi Tf}} - \frac{p^2 e^{-j2\pi Tf}}{1-(1-2p)e^{-j2\pi Tf}} \\ &= \frac{p(1-p)(\sin \pi Tf)^2}{p^2 + (1-2p)(\sin \pi Tf)^2} \end{aligned} \tag{8.59}$$

This periodic spectral shaping function is shown graphically in Figure 8.6 for various values of the parameter p. We note that discrete components have been eliminated by bipolar coding (for the case where the $\{\mathbf{a}_k\}$ are statistically independent). Also the spectral density has zeros at 0 and $1/T$ regardless of the value of p. For high pulse density $p \gg \frac{1}{2}$, there is a large concentration of power in the region of $1/2T$.

It is also interesting to note that bipolar coding provides some degree of error-detection capability. If consecutive pulses in a received bipolar signal should have the same polarity, then it is clear that an error has occurred in transmission. Such occurrences are called *bipolar violations*, and the rate at which violations occur can be used as a measure of system performance.

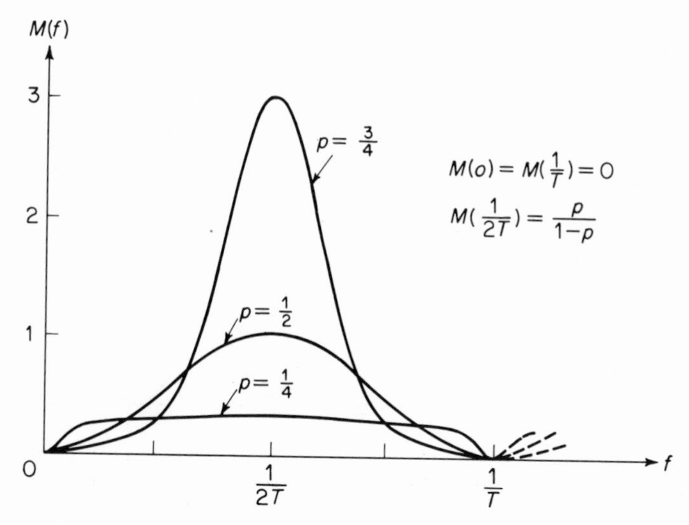

Figure 8.6. Spectral density shaping function for bipolar coding.

Partial Response Coding

In PAM systems utilizing approximately the minimum bandwidth $1/2T$, it may be advantageous to concentrate the power in the central region of $1/4T$ while producing zeros in the spectral density at 0 and $1/2T$. This can be accomplished with a coding scheme known as *partial response.*[6] The partial response signal is given by

$$\mathbf{x}(t) = \sum_{k=-\infty}^{\infty} \mathbf{d}_k s(t - kT) \tag{8.60}$$

where

$$d_k = a_k - a_{k-2}$$

Hence $\bar{\mathbf{d}} = 0 \Rightarrow \bar{\mathbf{x}} = 0$, and

$$\begin{aligned}\delta_m &\triangleq E[\mathbf{d}_k \mathbf{d}_{k+m}] = E[(\mathbf{a}_k - \mathbf{a}_{k-2})(\mathbf{a}_{k+m} - \mathbf{a}_{k+m-2})] \\ &= 2\alpha_m - \alpha_{m+2} - \alpha_{m-2}\end{aligned} \tag{8.61}$$

Hence $\delta_0 = 2p(1-p)$; $\delta_1 = \delta_{-1} = 0$; $\delta_2 = \delta_{-2} = -p(1-p)$; and $\delta_m = 0$ for $|m| \geqslant 3$. The power spectral density becomes

$$\begin{aligned}K_{xx}(f) &= \frac{R(f)}{T} \sum_{m=-\infty}^{\infty} \delta_m e^{j2\pi mTf} \\ &= \frac{R(f)M(f)}{T}\end{aligned}$$

where

$$\begin{aligned}M(f) &= 2p(1-p)[1 - \cos 2\pi(2Tf)] \\ &= 4p(1-p)(\sin 2\pi Tf)^2\end{aligned} \tag{8.62}$$

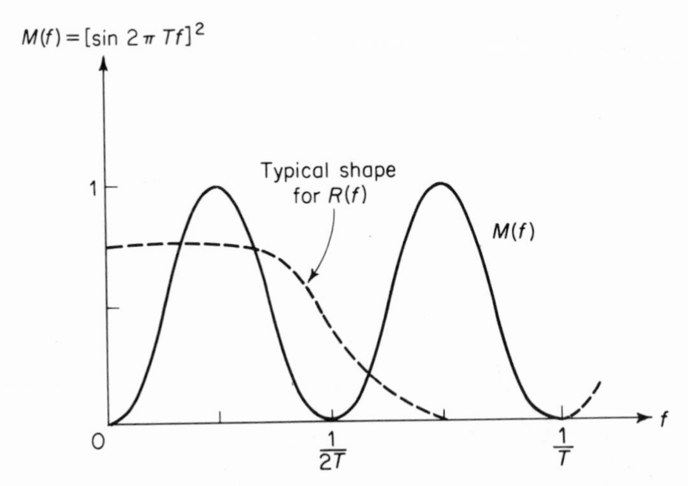

Figure 8.7. Spectral density shaping function for partial response coding.

This spectral shaping function is shown graphically in Figure 8.7 for $p = \frac{1}{2}$. The basic shape is independent of p. The discrete components in the uncoded signal (8.43) are also eliminated by this coding scheme. In this example, the same results could have been obtained by modifying the pulse shape; i.e., the partial response signal (8.60) can be expressed as

$$\mathbf{x}(t) = \sum_{k=-\infty}^{\infty} \mathbf{a}_k s_d(t - kT)$$

where

$$s_d(t) = s(t) - s(t - 2T) \tag{8.63}$$

Exercise 8.5. Another coding operation, similar in some respects to bipolar coding except that it tends to concentrate power at low frequencies, results from the conversion of binary to a ternary sequence $\{\mathbf{e}_k\}$, where

$$\mathbf{x}(t) = \sum_{k=-\infty}^{\infty} \mathbf{e}_k s(t - kT)$$

with $e_k = \pm 1$ if $a_k = 1$ and $e_k = 0$ if $a_k = 0$. Successive pulses in this process have the same polarity if they are separated by an even number of missing pulses ($e_k = 0$). If there are an odd number of missing pulses between successive pulses in $\mathbf{x}$, then these pulses have opposite polarity. For example, this coding prohibits the rapid fluctuation that would result from a $+1$ to -1 transition in adjacent pulse emission instants. This coding operation is called *duobinary coding.*[10] Evaluate the power spectral density assuming that the uncoded binary sequence $\{\mathbf{a}_k\}$ has the statistical properties of (8.42). Compare the result with that of bipolar coding. *Hint:* Show that $e_k = (-1)^k c_k$, where $\{\mathbf{c}_k\}$ is the ternary sequence resulting from bipolar coding.

Exercise 8.6. In the binary PAM process (8.42), suppose that the $\{\mathbf{a}_k\}$ are not statistically independent but instead are characterized by the fact that the probability of a_k and a_{k+1} being different is equal to $\frac{1}{4}$.

a. Show that

$$\alpha_m = E[\mathbf{a}_k \mathbf{a}_{k+m}] = \tfrac{1}{4}[(\tfrac{1}{2})^{|m|} + 1]$$

b. Using (a), write the power spectral density for the process. Indicate discrete components, if any.

Exercise 8.7. For the synchronous PAM signal (not necessarily binary), suppose that the stationary sequence $\{\mathbf{a}_k\}$ is characterized by the correlation $\alpha_m = \alpha_0 \rho^{|m|}$. Such a sequence is called a *wide-sense Markoff sequence.*[3]

a. Evaluate and sketch power spectral density for a value of ρ near unity (highly correlated sequence). Express the ratio of maximum-to-minimum power density in terms of the parameter ρ.
b. For reducing low-frequency concentration, consecutive pulses can be alternated in polarity. Let

$$\mathbf{x}(t) = \sum_{k=-\infty}^{\infty} \mathbf{b}_k s(t - kT)$$

where $b_k = (-1)^k a_k$. Evaluate power spectral density and compare with that in (a).
c. To alleviate the extreme frequency-domain concentration that may result in (a) and (b) with highly correlated sequences, a differential PAM signal can be used. Let

$$\mathbf{x}(t) = \sum_{k=-\infty}^{\infty} \mathbf{c}_k s(t - kT)$$

with $c_k = a_k - a_{k-1}$. Compare power spectral density with that obtained in (a).

8.5 PAM with Time Jitter

In constructing a more accurate model for the PAM signal, there are situations where it is important to consider a random deviation from perfectly synchronous pulse occurrences. Such deviations are called *timing jitter*, and the jittered PAM signal can be characterized by

$$\mathbf{x}(t) = \sum_{k=-\infty}^{\infty} \mathbf{a}_k s(t - kT - \boldsymbol{\delta}_k) \tag{8.64}$$

where we assume that the $\{\boldsymbol{\delta}_k\}$ are zero-mean, identically distributed random variables. A typical realization is shown in Figure 8.8.

The mean value of the process is

$$\begin{aligned}\overline{\mathbf{x}(t)} &= \bar{\mathbf{a}} \sum_{k=-\infty}^{\infty} \int_{-\infty}^{\infty} s(t - kT - \sigma) p(\sigma)\, d\sigma \\ &= \bar{\mathbf{a}} \sum_{k=-\infty}^{\infty} \tilde{s}(t - kT) \end{aligned} \tag{8.65}$$

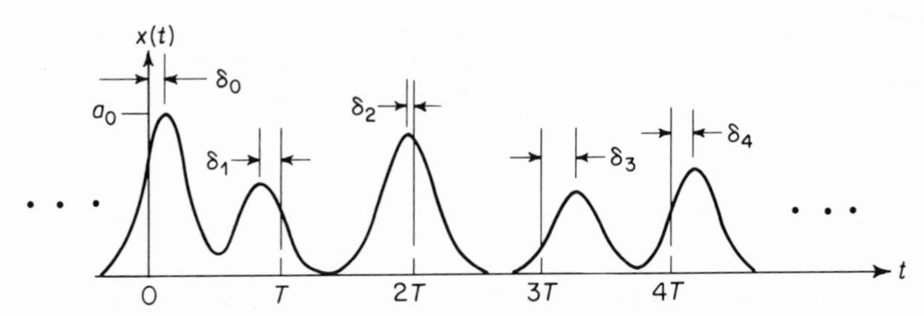

Figure 8.8. Typical realization of PAM signal with timing jitter.

where $\tilde{s}(t)$ is a *jitter averaged* version of the pulse, broadened somewhat by convolution with the probability density function $p(\sigma)$ for the jitter variable.

To evaluate the autocorrelation function for $\mathbf{x}$, we need the joint probability density function for $\boldsymbol{\delta}_k$ and $\boldsymbol{\delta}_{k+m}$. We shall assume that the $\{\boldsymbol{\delta}_k\}$ is a stationary sequence so that this joint density depends only on m. Then we have

$$k_{xx}(t+\tau, t) = \sum_{m=-\infty}^{\infty} \alpha_m \sum_{k=-\infty}^{\infty} \iint_{-\infty}^{\infty} s(t+\tau-kT-\sigma_1) \times s(t-kT-mT-\sigma_2)p_m(\sigma_1, \sigma_2)\, d\sigma_1\, d\sigma_2 \quad (8.66)$$

Examination of (8.65) and (8.66) shows that the jittered PAM signal is also a cyclostationary process. If we choose to treat this as a stationary process, we can introduce an additional random variable uniformly distributed over one pulse period as before. Thus we let

$$\mathbf{x}(t) = \sum_{k=-\infty}^{\infty} \mathbf{a}_k s(t-kT-\boldsymbol{\delta}_k-\boldsymbol{\delta}) \quad (8.67)$$

and we find

$$\overline{\mathbf{x}(t)} = \frac{\bar{\mathbf{a}}}{T}\int_{-\infty}^{\infty} \tilde{s}(t)\, dt \quad (8.68)$$

To obtain a simplified expression for the autocorrelation function, we shall assume that the $\{\boldsymbol{\delta}_k\}$ are statistically independent. Then

$$p_m(\sigma_1, \sigma_2) = p(\sigma_1)p(\sigma_2) \qquad \text{for } m \neq 0$$

and

$$k_{xx}(\tau) = \frac{\alpha_0}{T}\tilde{r}(\tau) + \frac{1}{T}\sum_{m\neq 0}\alpha_m r_1(\tau+mT)$$

where

$$\tilde{r}(\tau) = \int_{-\infty}^{\infty} r(\tau-\sigma)p(\sigma)\, d\sigma$$

$$r(\tau) = \int_{-\infty}^{\infty} s(t+\tau)s(t)\, dt \quad (8.69)$$

$$r_1(\tau) = \int_{-\infty}^{\infty} \tilde{s}(t+\tau)\tilde{s}(t)\, dt$$

In (8.69), $r_1(\tau)$ is the time-ambiguity function for the jitter averaged pulse $\tilde{s}(t)$; while $\tilde{r}(\tau)$ is the jitter averaged time-ambiguity function for $s(t)$. The power spectral density for this process is

$$K_{xx}(f) = \frac{R(f)}{T}\left[\alpha_0 P(f) + \sum_{m\neq 0} \alpha_m \,|P(f)|^2\, e^{j2\pi mTf}\right] \tag{8.70}$$

8.6 Time-multiplexed Signals

As a final example of practical signal processing operations which give rise to cyclostationary processes, we consider the time-multiplexing operation which forms a *composite* signal by repetitively interleaving segments from two or more signal processes. This is illustrated in Figure 8.9 for multiplexing two stationary processes **y** and **z**, which we shall assume to be statistically independent. The switching operation can be characterized by a periodic indicator function $q(t)$. The composite signal **x** is then given by

$$\mathbf{x}(t) = q(t)\mathbf{y}(t) + [1 - q(t)]\mathbf{z}(t)$$

where

$$\begin{aligned} q(t) &= 1 \qquad \text{for } kT \leqslant t < kT + T_1;\ k = 0, \pm 1, \pm 2, \ldots \\ &= 0 \qquad \text{otherwise} \end{aligned} \tag{8.71}$$

The mean value of the composite signal is

$$\overline{\mathbf{x}(t)} = \bar{\mathbf{y}}q(t) + \bar{\mathbf{z}}[1 - q(t)] \tag{8.72}$$

Note that $\overline{\mathbf{x}(t)}$ is constant if the processes **y** and **z** have the same mean value.

The autocorrelation function for the composite signal is

$$\begin{aligned} k_{xx}(t+\tau, t) &= E[\mathbf{x}(t+\tau)\mathbf{x}(t)] \\ &= q(t)q(t+\tau)k_{yy}(\tau) + [1 - q(t+\tau)][1 - q(t)]k_{zz}(\tau) \\ &\quad + \bar{\mathbf{y}}\bar{\mathbf{z}}\{q(t+\tau)[1 - q(t)] + q(t)[1 - q(t+\tau)]\} \end{aligned} \tag{8.73}$$

Because of the periodicity of $q(t)$, it is clear that **x** is a cyclostationary process. Generalization to a larger number of multiplexed processes is straightforward. Also correlation between the processes can be accommodated. Cases where the switching function is a random process have been treated.[7] In some cases of practical interest, one of the processes may not be random. For example, in modeling a television signal, the signal $z(t)$ might be periodic in order to represent the periodic insertion of synchronizing and video blanking pulses.[8]

Finally, as in the previous examples, if the clock phase is completely indeterminate, we can consider the composite signal as a stationary process,

$$\mathbf{x}(t) = q(t - \boldsymbol{\delta})\mathbf{y}(t - \boldsymbol{\delta}) + [1 - q(t - \boldsymbol{\delta})]\mathbf{z}(t - \boldsymbol{\delta}) \tag{8.74}$$

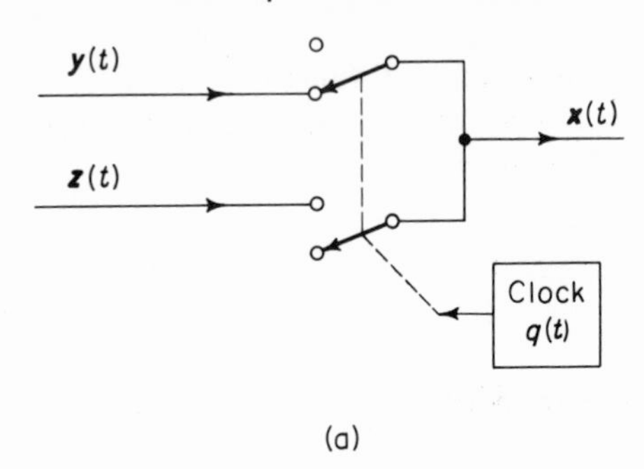

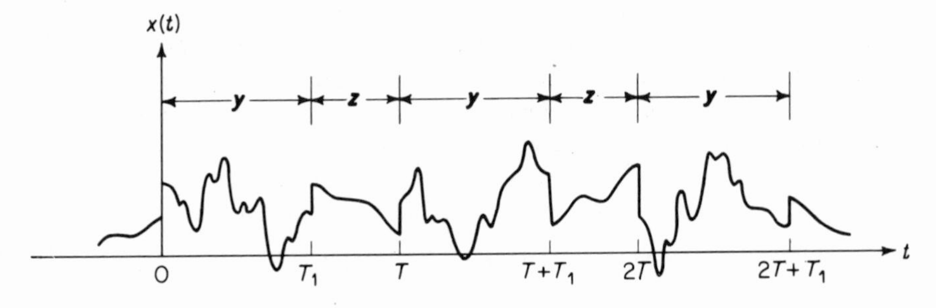

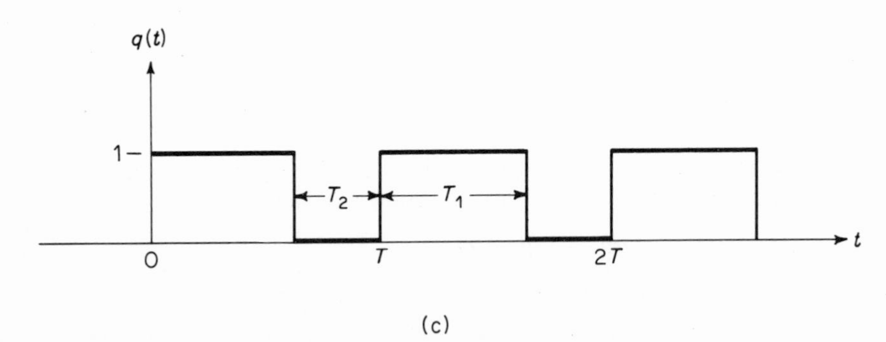

Figure 8.9. (a) Time multiplexer. (b) Typical realization of composite signal. (c) Periodic indicator function.

where $\boldsymbol{\delta}$ has a uniform density over $0 \leqslant \delta < T$. In this case, we have

$$\bar{\mathbf{x}} = \bar{\mathbf{y}} \frac{1}{T} \int_0^T q(t - \sigma)\, d\sigma + \bar{\mathbf{z}} \frac{1}{T} \int_0^T [1 - q(t - \sigma)]\, d\sigma$$

$$= \frac{T_1}{T} \bar{\mathbf{y}} + \frac{T_2}{T} \bar{\mathbf{z}} \tag{8.75}$$

and

$$k_{xx}(\tau) = \frac{T_1}{T} w(\tau) k_{yy}(\tau) + \left\{ \frac{T_2}{T} - \frac{T_1}{T} [1 - w(\tau)] \right\} k_{zz}(\tau) + 2\bar{\mathbf{y}}\bar{\mathbf{z}} \frac{T_1}{T} [1 - w(\tau)] \tag{8.76}$$

where

$$w(\tau) \triangleq \frac{1}{T_1}\int_0^T q(t)q(t+\tau)\,dt$$

The periodic shaping function $w(\tau)$ has the form shown in Figure 8.10.

As a final comment on cyclostationary processes, it should be emphasized that the phase randomizing introduced in these examples is appropriate only for certain considerations. An example of where the phase randomized process can be used is in calculating interference power coupled into another transmission channel. In any situation where the signal is not observed in

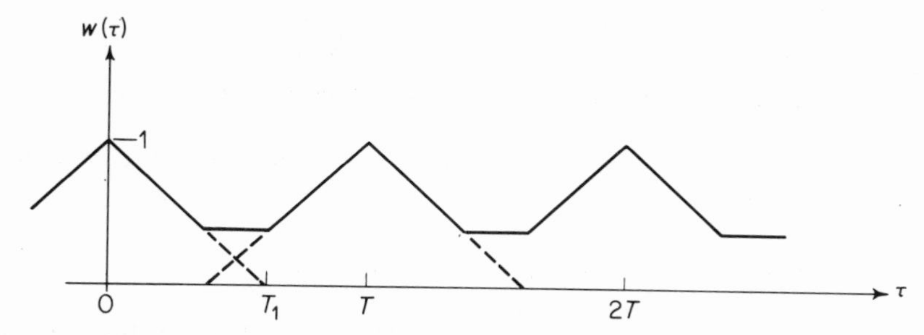

Figure 8.10. Periodic shaping function relevant to evaluation of the autocorrelation function of the composite signal.

synchronism with the generating clock, the phase randomizing approach will usually be valid. On the other hand, the periodic variation in statistical averages must be taken into account when evaluating the effects of subsequent periodic processing operations on the signal process. For example, a received PAM signal is demodulated by sampling the signal every T seconds. The phase of the receiver clock is adjusted to minimize intersymbol interference. The quantities of interest are the various expectations at the sampling instants, not a time-averaged version of these quantities.

8.7 Signal Processes Related to the Poisson Process

In contrast to the more or less synchronous signal processes described in previous sections, we often encounter processes where only the average rate of pulse occurrences is known. In such situations, where pulse occurrences are completely random, the *Poisson* process can be very useful in constructing a model. The Poisson process is a particular form of a *counting process*,[1] i.e., an integer-valued process with unit increments. A typical realization of a counting process is shown in Figure 8.11.

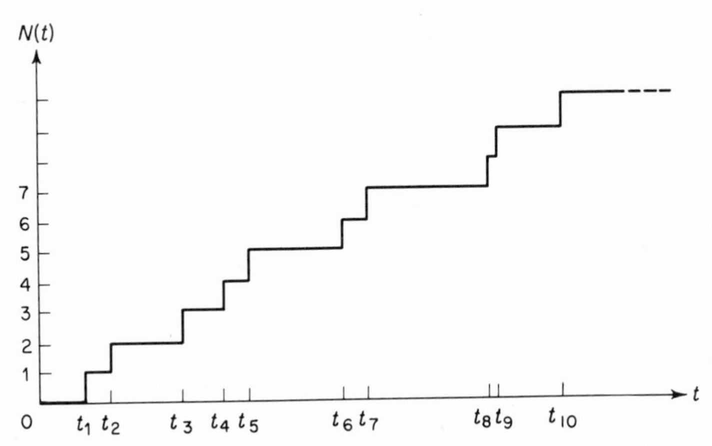

Figure 8.11. Typical realization of Poisson counting process, $\mathbf{N}(t)$.

This type of process can be described by the random sequence of points $\{\mathbf{t}_k\}$ along the time axis.

$$\mathbf{N}(t) = \sum_{k=1}^{\infty} w(t - \mathbf{t}_k); \qquad 0 \leqslant t_1 \leqslant t_2 \leqslant \cdots \tag{8.77}$$

where $w(t)$ is the unit step function, $w(t) = \frac{1}{2}(1 + \operatorname{sgn} t)$. Thus, the random variable $\mathbf{N}(t)$ is equal to the number of point occurrences between 0 and t. The process can be characterized by the probabilities associated with the events $[N(t) = n]$ for all $t \geqslant 0$ and $n = 0, 1, 2, \ldots$. Thus we define

$$P_n(t) = \Pr\,[N(t) = n] \tag{8.78}$$

Although there are more general forms for the Poisson process,[1] we shall consider only the particular case which satisfies the following conditions.

a. $\Pr\,[N(t) - N(s) = n \quad \text{and} \quad N(u) - N(v) = m]$
$$= P_n(t - s)P_m(u - v); \qquad t \geqslant s \geqslant u \geqslant v \geqslant 0 \tag{8.79}$$

b. $P_1(\Delta t) = \lambda\,\Delta t$

c. $P_0(\Delta t) = 1 - \lambda\,\Delta t$

(b and c) as $\Delta t \to 0$

Condition (a) in (8.79) imposes stationarity on the point occurrences; i.e., the probability of a particular number of occurrences depends only on the length of the interval, regardless of its position on the time axis. This condition also states that the number of occurrences in non-overlapping time intervals are statistically independent random variables. Conditions (b) and (c) state that for an interval Δt sufficiently small, the probability of one occurrence is proportional to the length of the interval, and the occurrence of two or more points in this interval can be considered impossible. From these conditions, we can derive expressions for $P_n(t)$ which depend only on

n, t, and the parameter λ. To derive these expressions, we note that

$$P_n(t + s) = \sum_{m=0}^{n} P_m(t)P_{n-m}(s) \tag{8.80}$$

In (8.80) we can replace s by Δt; then for Δt small we have $P_k(\Delta t) = 0$ for $k \geqslant 2$, and (8.80) becomes

$$\begin{aligned} P_n(t + \Delta t) &= P_n(t)P_0(\Delta t) + P_{n-1}(t)P_1(\Delta t) \\ &= P_n(t)(1 - \lambda\, \Delta t) + P_{n-1}(t)\lambda\, \Delta t \end{aligned}$$

$$\frac{P_n(t + \Delta t) - P_n(t)}{\Delta t} = -\lambda P_n(t) + \lambda P_{n-1}(t) \tag{8.81}$$

Letting $\Delta t \to 0$ in (8.81), we obtain a first-order differential equation for $P_n(t)$.

$$\frac{d}{dt} P_n(t) = -\lambda P_n(t) + \lambda P_{n-1}(t) \tag{8.82}$$

Using the initial condition $P_n(0) = 0$, we find

$$P_n(t) = \lambda \int_0^t e^{-\lambda(t-\tau)} P_{n-1}(\tau)\, d\tau \qquad \text{for } n \geqslant 1 \tag{8.83}$$

For $n = 0$, (8.82) is a homogeneous equation with the initial condition $P_0(0) = 1$; hence

$$P_0(t) = e^{-\lambda t}; \qquad t \geqslant 0 \tag{8.84}$$

Using (8.83) and (8.84), we can establish by induction that

$$P_n(t) = \frac{(\lambda t)^n}{n!} e^{-\lambda t}; \qquad t \geqslant 0 \tag{8.85}$$

The parameter λ is called the *rate parameter* for the process since it is equal to the average number of point occurrences in a unit interval. To show this, let T be an arbitrary interval; then

$$\begin{aligned} E[\mathbf{N}(t + T) - \mathbf{N}(t)] &= \sum_{n=0}^{\infty} nP_n(T) \\ &= \sum_{n=0}^{\infty} n \frac{(\lambda T)^n}{n!} e^{-\lambda T} = \lambda T e^{-\lambda T} \sum_{m=0}^{\infty} \frac{(\lambda T)^m}{m!} \\ &= \lambda T \end{aligned} \tag{8.86}$$

In constructing models for signal processes, we shall use the sequence of random variables $\{\mathbf{t}_k\}$ in various ways. We shall say that the $\{\mathbf{t}_k\}$ are distributed according to a Poisson process with rate parameter λ. In evaluating mean and autocorrelation for the process, the probability density functions for the random variables will be needed. The probability density function for $\mathbf{t}_k$ is related to the probability that the kth point occurs in a small interval at the arbitrary point t; i.e., for small Δt,

$$\begin{aligned} p_{t_k}(t)\,\Delta t &= \Pr\,[t < t_k \leqslant t + \Delta t] \\ &= \Pr\,[N(t) = k - 1]\,\Pr\,[N(t + \Delta t) - N(t) = 1] \\ &= P_{k-1}(t)\lambda\,\Delta t \end{aligned}$$

Thus we have

$$p_{t_k}(t) = \lambda P_{k-1}(t); \qquad t \geqslant 0 \tag{8.87}$$

We shall also need the joint probability density function for pairs of points $\mathbf{t}_k$ and $\mathbf{t}_j$. By similar reasoning we write (assuming $j < k$)

$$\begin{aligned} p_{t_k t_j}(t, s)\,\Delta t\,\Delta s &= \Pr\,[t < t_k \leqslant t + \Delta t \text{ and } s < t_j \leqslant s + \Delta s] \\ &= \Pr\,[N(s) = j - 1]\,\Pr\,[N(s + \Delta s) - N(s) = 1] \\ &\quad \times \Pr\,[N(t) - N(s) = k - j - 1]\,\Pr\,[N(t + \Delta t) - N(t) = 1] \end{aligned}$$

Thus we have

$$p_{t_k t_j}(t, s) = \lambda^2 P_{j-1}(s) P_{k-j-1}(t - s); \qquad k > j, \quad t > s \geqslant 0 \tag{8.88}$$

Exercise 8.8. For the counting process described by (8.77), evaluate the mean $\overline{\mathbf{N}(t)}$ and autocorrelation $k_{NN}(t_1, t_2)$. Assume that the step occurrences are distributed according to a Poisson process with rate parameter λ.

Random Facsimile Signal

A facsimile signal is obtained by optically scanning a black and white picture with a constant velocity scanner. For a random picture, the output of the scanner $\mathbf{x}(t)$ will be a two-valued random process, say $x(t) = 1$ for black portions and $x(t) = 0$ for white portions. The length of time between successive black-to-white and white-to-black transitions will be a random variable. Typically, the process may have a strong bias toward mostly black or mostly white. We might model this process as follows. Let $\{\mathbf{t}_k; k = 0, \pm 1, \pm 2, \ldots\}$ be an ordered sequence of random variables distributed over the entire real line according to a Poisson process with rate parameter λ. Here we have removed the restriction in (8.77) that the points lie on the positive part of the real line. This causes no difficulty because of the stationarity of point occurrences. A time origin can be chosen arbitrarily. In the intervals defined by consecutive points, we say that $\mathbf{x}$ is constant over the interval, assuming the value 1 or 0 with probability p or $1 - p$, respectively. Furthermore, the values of $\mathbf{x}$ in different intervals are statistically independent. This gives us a model for the random facsimile signal which is characterized by only two parameters, p and λ. A typical realization is shown in Figure 8.12.

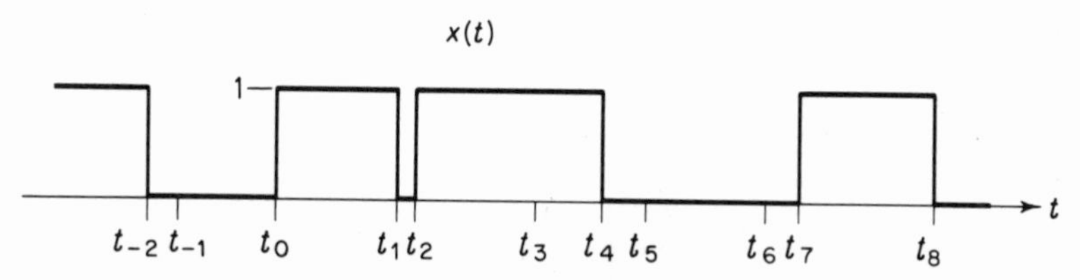

Figure 8.12. Typical realization of random facsimile signal. The $\{\mathbf{t}_k\}$ are distributed according to a Poisson process with rate parameter λ.

The mean value of the process is obtained by choosing an arbitrary t and evaluating

$$E[\mathbf{x}(t)] = \Pr\,[x(t) = 1] = p \tag{8.89}$$

The mean value is constant, $\bar{\mathbf{x}} = p$, since it does not matter which interval t falls in. The autocorrelation function is

$$\begin{aligned} k_{xx}(t+\tau, t) &= E[\mathbf{x}(t+\tau)\mathbf{x}(t)] \\ &= \Pr\,[x(t+\tau) = 1 \text{ and } x(t) = 1] \end{aligned} \tag{8.90}$$

The joint probability in (8.90) depends only on whether t and $t + \tau$ are in the same interval or not.

$$\begin{aligned} \Pr\,[x(t+\tau) = 1 \text{ and } x(t) = 1] &= p \qquad \text{if } t \text{ and } t+\tau \text{ fall in the same interval} \\ &= p^2 \qquad \text{if } t \text{ and } t+\tau \text{ fall in different intervals} \end{aligned} \tag{8.91}$$

The probability that t and $t + \tau$ are in the same interval is simply $P_0(|\tau|)$, independent of t. Using (8.84) and (8.91), (8.90) becomes

$$\begin{aligned} k_{xx}(\tau) &= pe^{-\lambda|\tau|} + p^2[1 - e^{-\lambda|\tau|}] \\ &= p(1-p)e^{-\lambda|\tau|} + p^2 \end{aligned} \tag{8.92}$$

The process is clearly wide-sense stationary and the power spectral density is

$$K_{xx}(f) = \frac{2\lambda p(1-p)}{(2\pi f)^2 + \lambda^2} + p^2\,\delta(f) \tag{8.93}$$

shown graphically in Figure 8.13. It is apparent that λ can be also interpreted as a bandwidth parameter for the process.

Instead of a two-valued process, we can let $\mathbf{x}$ take on any real value between 0 and 1 (gray levels) over an interval according to an arbitrary probability density function. The autocorrelation function has the same form as (8.92) in this case (see Exercise 8.10). This model has been found useful for representing television signals.[8] Another process frequently mentioned is the *random telegraph signal*,[1],[3] where transitions between 0 and 1 take

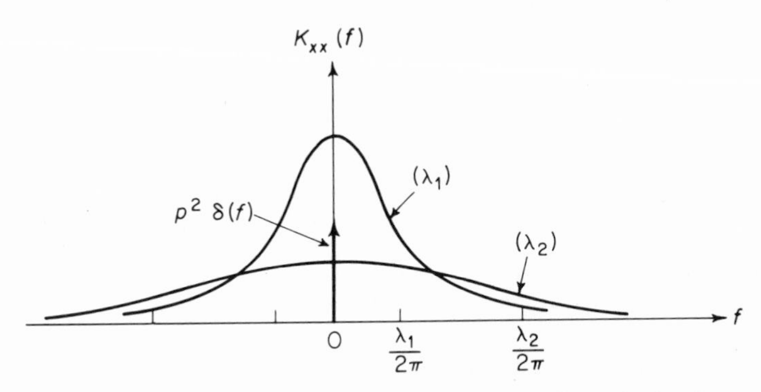

Figure 8.13. Power spectral density for random facsimile signal.

place at each of the random points. This process is equivalent to the random facsimile signal with $p = \frac{1}{2}$ and λ increased by a factor of 2.

Exercise 8.9. For the random facsimile signal, evaluate the average time interval between black-to-white and white-to-black transitions. Also evaluate the average transition rate.

Exercise 8.10. In generalizing the random facsimile signal from a two-valued process, we let $x(t) = a_k;\ t_k \leqslant t < t_{k+1}$, where $\{\mathbf{a}_k\}$ is a sequence of statistically independent random variables, each with probability density function $p_a(\xi)$. This process has been called the *random step function*.[11]. Evaluate the mean and autocorrelation for $\mathbf{x}$.

Exercise 8.11. In the random step function described in Exercise 8.10, we can le $\{\mathbf{a}_k\}$ be a stationary wide-sense Markoff sequence as in Exercise 8.7 rather than a statistically independent sequence. Show that, as far as mean and autocorrelation are concerned, this modification is equivalent to a change in the rate parameter. Let $\tilde{\lambda}$ be the new rate parameter in a random step function with independent levels and show the relation between $\tilde{\lambda}$, λ, and ρ.

Random Pulse Sequence

For a more general utilization of the Poisson process, we consider the model for a pulse sequence with an arbitrary pulse shape and random amplitudes and occurrence times.

$$\mathbf{x}(t) = \sum_{k=-\infty}^{\infty} \mathbf{a}_k s(t - \mathbf{t}_k); \qquad k \neq 0 \tag{8.94}$$

where $\{\mathbf{a}_k\}$ is a stationary sequence statistically independent of $\{\mathbf{t}_k\}$ which is an ordered sequence distributed according to a Poisson process with rate parameter λ. The reason for omitting the $k = 0$ term in (8.94) will be explained in the following discussion. In order to evaluate the mean and autocorrelation for $\mathbf{x}$, we choose an arbitrary value for t, then index the

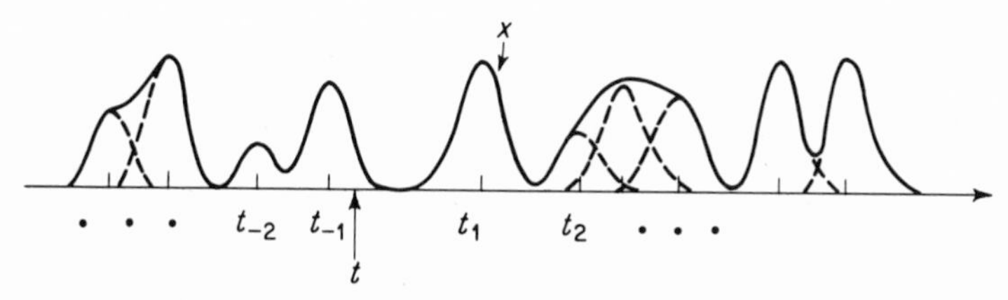

Figure 8.14. Typical realization of random pulse sequence showing indexing scheme for pulse occurrences.

$\{\mathbf{t}_k\}$ in a typical realization so that t_1 is the first point to the right of t and t_{-1} is the first point to the left of t as shown in Figure 8.14.

It may seem strange at first, but the length of the interval between t_{-1} and t_1 is statistically different from the lengths of the intervals between all other consecutive pulse occurrences. The reason for this is that the interval t_{-1} to t_1 is constrained to include a given point t, while there is no such constraint imposed on the other intervals. The probability density function for the length of time $\boldsymbol{\tau}$ from *any* point on the time axis to the next point occurrence is given by $p_\tau(\sigma) = \lambda e^{-\lambda\sigma}$. Thus if we let $t_0 = t$ and define the random variable $\boldsymbol{\tau}_k = \mathbf{t}_k - \mathbf{t}_{k-1}$, the $\boldsymbol{\tau}_k$ are identically distributed for each k (including zero). In view of this, we can more correctly write (8.94) as

$$\mathbf{x}(t) = \sum_{k=-\infty}^{\infty} \mathbf{a}_k s(t - \mathbf{t}_k); \qquad \begin{cases} t_{-1} < t_0 = t < t_1 \\ a_0 = 0 \end{cases} \tag{8.95}$$

Another way of looking at this is to let the arbitrary point t be the time origin; then the random variables $\mathbf{t}_k - t$ for $k > 0$ and $t - \mathbf{t}_k$ for $k < 0$ are described by the probability density functions given in (8.87) and (8.88). Now we can write

$$\begin{aligned} E[\mathbf{x}(t)] &= \sum_{k=-\infty}^{\infty} E[\mathbf{a}_k]E[s(t - \mathbf{t}_k)] \\ &= \sum_{k=1}^{\infty} \bar{\mathbf{a}} \int_0^{\infty} s(\sigma)\lambda P_{k-1}(\sigma)\, d\sigma + \sum_{k=-\infty}^{-1} \bar{\mathbf{a}} \int_0^{\infty} s(-\xi)P_{-k-1}(\xi)\, d\xi \\ &= \lambda\bar{\mathbf{a}} \int_0^{\infty} s(t)e^{-\lambda t} \sum_{k=1}^{\infty} \frac{(\lambda t)^{k-1}}{(k-1)!}\, dt + \lambda\bar{\mathbf{a}} \int_{-\infty}^{0} s(t)e^{\lambda t} \sum_{\ell=1}^{\infty} \frac{(-\lambda t)^{\ell-1}}{(\ell-1)!}\, dt \\ &= \lambda\bar{\mathbf{a}} \int_{-\infty}^{\infty} s(t)\, dt = \lambda\bar{\mathbf{a}}q \end{aligned} \tag{8.96}$$

In evaluating the autocorrelation function, we shall assume that the $\{\mathbf{a}_k\}$ are statistically independent so that

$$\begin{aligned} E[\mathbf{a}_k\mathbf{a}_j] &= \overline{\mathbf{a}^2} && \text{for } k = j \neq 0 \\ &= \bar{\mathbf{a}}^2 && \text{for } k \neq j \\ &= 0 && \text{for } k \text{ or } j = 0 \end{aligned} \tag{8.97}$$

Then we have

$$k_{xx}(t+\tau, t) = E[\mathbf{x}(t+\tau)\mathbf{x}(t)]$$
$$= \sum_{k=-\infty}^{\infty} \sum_{j=-\infty}^{\infty} E[\mathbf{a}_k \mathbf{a}_j] E[s(t - \mathbf{t}_k + \tau) s(t - \mathbf{t}_j)] \tag{8.98}$$

The sum in (8.98) is evaluated by considering separately the terms for which $k = j > 1; k = j < -1; k > j > 1; j > k > 1; j < k < -1; k < j < -1;$ $k > 1, j < -1; j > 1, k < -1$. After some manipulation, we obtain the simple result

$$k_{xx}(\tau) = \lambda \overline{\mathbf{a}^2} r(\tau) + (\lambda \bar{\mathbf{a}} q)^2 \tag{8.99}$$

where

$$r(\tau) = \int_{-\infty}^{\infty} s(t+\tau) s(t)\, dt$$

$$q = \int_{-\infty}^{\infty} s(t)\, dt$$

Finally, we note the equivalence of this process to the one described in Section 8.2 where pulse occurrences were uniformly and independently distributed over the entire time axis. See equations (8.10) and (8.11).

Exercise 8.12. Provide the missing steps in the derivation of (8.99).

REFERENCES

1. E. Parzen, *Stochastic Processes*, Holden-Day, 1962.
2. W. R. Bennett, "Statistics of Regenerative Digital Transmission," *Bell Sys. Tech. Jour.*, Vol. 37, pp. 1501–42 (November, 1958).
3. A. Papoulis, *Probability, Random Variables, and Stochastic Processes*, McGraw-Hill, 1965.
4. M. R. Aaron, "PCM Transmission in the Exchange Plant," *Bell Sys. Tech. Jour.*, Vol. 41, pp. 99–141 (January, 1962).
5. A. Lender, "Correlative Digital Communication Techniques," *Trans. IEEE*, Vol. COM-12, pp. 128–35 (December, 1964).
6. E. R. Kretzmer, "Binary Data Communication by Partial Response Transmission," Paper No. CP65-419 at the 1965 IEEE Communications Convention, Boulder, Colorado.
7. M. V. Johns, Jr., "Spectral Analysis of a Process of Randomly Delayed Pulses," *Trans. IEEE*, Vol. IT-6, No. 4, pp. 440–44 (September, 1960).
8. L. E. Franks, "A Model for the Random Video Process," *Bell Sys. Tech. Jour.*, Vol. 45, pp. 609–30 (April, 1966).

9. F. B. Hildebrand, *Methods of Applied Mathematics*, Prentice-Hall, 1952.
10. A. Lender, "The Duobinary Technique for High-Speed Data Transmission," *IEEE Trans. on Communication and Electronics*, Vol. 82, pp. 214–18 (May, 1963).
11. J. H. Laning, Jr. and R. H. Battin, *Random Processes in Automatic Control*, McGraw-Hill, 1956.

OPTIMUM FILTERING OF SIGNAL PROCESSES

9

9.1 Introduction

Using the techniques described in the preceding two chapters for characterizing random processes, we are in a position to apply a much more realistic approach to the design of optimal signal processing systems. This is because we are able to characterize the properties of both deterministic and random elements of the system in terms of linear and quadratic functionals. With various aspects of system performance described by these functionals, we can apply the variational methods of Chapter 6 to the problem of finding the deterministic elements which yield best performance. Examples of deterministic elements to be optimized might be the pulse shapes used in a pulse amplitude modulation system or the impulse response of a signal processing filter designed to extract wanted signals out of a background of noise. In this chapter, we confine our attention to the signal filtering problem. We are able to treat a broad class of problems which involve finding the optimal linear filter under a mean-squared error performance criterion. The physical interpretation of these results is greatly aided by the signal space concepts introduced in earlier chapters.

In the following section, a generalized formulation of the linear filtering problem in signal parameter estimation is presented. A general condition on the impulse response of the filter in order that mean-squared error in parameter estimation be minimized is derived. The remainder of the chapter is devoted to finding functions which satisfy the condition for specific filtering problems. Several of the examples are well-known and have received extensive treatment in the communication and control system literature. Other examples have received much less attention but they can be solved by essentially the same techniques. The examples are somewhat idealized for the sake of simplicity and for giving a basic understanding of how specific circumstances affect the form of the optimum filter. For practical applications, a problem combining features of several of the examples might be formulated.

In many of the problems, it might be argued that the optimum signal processing operation is not linear. Regardless of this fact, treatment of the linear filtering problem has a considerable practical significance because realization techniques for linear filters are much more highly developed than for non-linear filters. In other words, we may regard linearity, like physical realizability (non-anticipatory response), as a practical constraint imposed by the designer. Concerning the question of physical realizability, the examples are initially formulated without regard to this constraint. It is felt that this approach, besides its simplicity, is valid for two reasons: (1) In most cases, the response of the optimum filter can be closely approximated by that of a physically realizable filter, usually by allowing a sufficiently long time delay in the filter operation. (2) The performance of the system using the unconstrained optimum filter provides the designer with a useful guide in evaluating the suboptimum performance resulting from filters which are constrained to be physically realizable. In other cases, however, physical realizability is a severe constraint and it is important that the optimization be carried out subject to this constraint. This problem is treated in Section 9.6.

9.2 Minimum Mean-squared Error Estimation

It is convenient to put parameter estimation problems into a signal transmission context as shown in Figure 9.1. The transmitted signal process $\mathbf{x}(t)$ is subjected to a mapping T_1 which characterizes the transmission channel. The image of $\mathbf{x}(t)$ under this mapping is the received signal $\mathbf{z}(t)$ which is to be processed by a filtering operation with impulse response $h(t, s)$. The impulse response is chosen so that the output (at time t) is a minimum mean-squared error estimate of the signal parameter $\boldsymbol{\omega}(t)$. The signal parameter to be estimated can be characterized by the mapping T_0. It is sometimes helpful to think of T_0 as representing an ideal channel and the problem is to find the linear filter which, placed in cascade with the actual channel T_1, provides a mapping which closely matches that of T_0.

Assuming that the signal processes are real, the mean-squared error

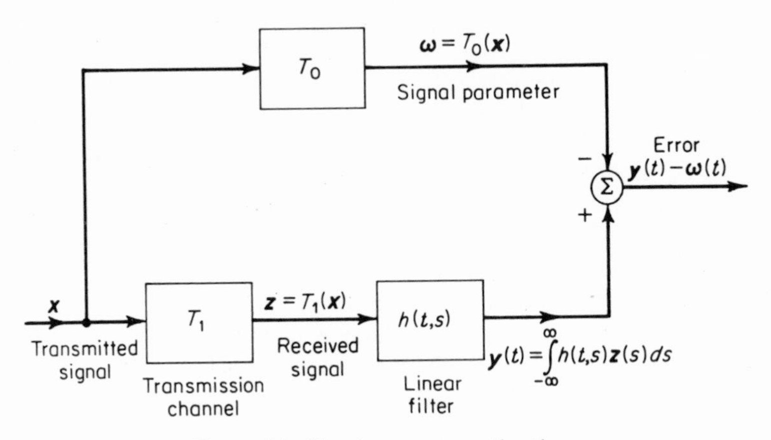

Figure 9.1. Signal parameter estimation.

functional I at time t is given by

$$\begin{aligned} I &= E[|\mathbf{y}(t) - \boldsymbol{\omega}(t)|^2] \\ &= E[|\mathbf{y}(t)|^2] - 2E[\mathbf{y}(t)\boldsymbol{\omega}(t)] + E[|\boldsymbol{\omega}(t)|^2] \end{aligned} \tag{9.1}$$

Using the input-output relation

$$\mathbf{y}(t) = \int_{-\infty}^{\infty} h(t, s)\mathbf{z}(s)\, ds \tag{9.2}$$

in (9.1), the functional I can be expressed as the sum of a quadratic, a linear, and a constant functional on $h(t, s)$, where $h(t, s)$ is considered as a function of s with t regarded as a fixed parameter.

$$I = \iint_{-\infty}^{\infty} k_{zz}(s, \sigma)h(t, \sigma)h(t, s)\, d\sigma\, ds - 2\int_{-\infty}^{\infty} h(t, s)k_{z\omega}(s, t)\, ds + k_{\omega\omega}(t,t) \tag{9.3}$$

These functionals are completely characterized by the autocorrelation and crosscorrelation of the processes $\mathbf{z}(t)$ and $\boldsymbol{\omega}(t)$. Noting that the quadratic functional corresponds to a self-adjoint operator, the stationary points of I with respect to variations in $h(t, s)$ are given by the solutions of

$$\int_{-\infty}^{\infty} k_{zz}(s, \sigma)h(t, \sigma)\, d\sigma = k_{z\omega}(s, t) \tag{9.4}$$

obtained by setting the gradient of I equal to zero. In most cases of interest, $h(t, s)$ will be an $L^2(-\infty, \infty)$ function of s, although occasionally we shall have to broaden the function space in order to include solutions of (9.4).

The Orthogonality Principle[1]

An alternative viewpoint on optimum filtering is gained by considering the error functional as a functional on the space of random variables. Let $\mathscr{L}_0$ be the linear operator for the filter that satisfies condition (9.4). Then (9.4) can be written in the equivalent form

$$E[\{\mathscr{L}_0 \cdot \mathbf{z}(t) - \boldsymbol{\omega}(t)\}\mathbf{z}(s)] = 0 \tag{9.5}$$

This says that $\mathscr{L}_0$ should be chosen so that the random error variable $\mathbf{y}(t) - \boldsymbol{\omega}(t)$ is orthogonal to all the random variables $\{\mathbf{z}(s); -\infty < s < \infty\}$ in the received signal process $\mathbf{z}$. This result is intuitively satisfying since it seems reasonable that if there were correlation present between error and received data, then further processing should be able to extract a better estimate. Using (9.5), we can independently establish that (9.4) is the required condition for optimality. Let $\mathscr{L}$ be any other linear operator, then

$$\begin{aligned} I &= E[|\mathscr{L} \cdot \mathbf{z}(t) - \boldsymbol{\omega}(t)|^2] \\ &= E[|\{\mathscr{L} \cdot \mathbf{z}(t) - \mathscr{L}_0 \cdot \mathbf{z}(t)\} + \{\mathscr{L}_0 \cdot \mathbf{z}(t) - \boldsymbol{\omega}(t)\}|^2] \\ &= E[|\mathscr{L}_0 \cdot \mathbf{z}(t) - \boldsymbol{\omega}(t)|^2] + 2E[(\mathscr{L} - \mathscr{L}_0) \cdot \mathbf{z}(t)\{\mathscr{L}_0 \cdot \mathbf{z}(t) - \boldsymbol{\omega}(t)\}] \\ &\quad + E[|(\mathscr{L} - \mathscr{L}_0) \cdot \mathbf{z}(t)|^2] \end{aligned} \tag{9.6}$$

Because of (9.5), the middle term on the last line of (9.6) vanishes. The last term is non-negative so $\mathscr{L}$ cannot produce a smaller mean-squared error than $\mathscr{L}_0$. This, incidently, proves sufficiency, as well as necessity, of condition (9.4).

The minimum mean-squared error is evaluated by noting that

$$I = E[\{\boldsymbol{\omega}(t) - \mathbf{y}(t)\}\boldsymbol{\omega}(t)] + E[\{\mathbf{y}(t) - \boldsymbol{\omega}(t)\}\mathbf{y}(t)] \tag{9.7}$$

For the optimum filter, the last term vanishes and we have

$$\begin{aligned} I_{\min} &= E[\{\boldsymbol{\omega}(t) - \mathbf{y}(t)\}\boldsymbol{\omega}(t)] \\ &= k_{\omega\omega}(t, t) - \int_{-\infty}^{\infty} k_{z\omega}(s, t)h(t, s)\, ds \end{aligned} \tag{9.8}$$

In the following sections, we formulate problems by specifying mappings T_0 and T_1 which correspond to situations of practical interest. In solving for the optimum filter characteristics, it is usually more convenient to use a frequency-domain version of the condition (9.4).

9.3 Continuous Waveform Estimation

In what might be considered the classical filtering problem, it is desired to reproduce the transmitted waveform with good fidelity. If $\mathbf{x}$ is a random process, we want the random variable $\mathbf{y}(t)$ to be close to the random variable $\mathbf{x}(t)$ for all values of t. Often it is desired to reproduce a time-translated version of the transmitted signal. In this case, we let T_0: $\boldsymbol{\omega}(t) = \mathbf{x}(t - T)$.

The delay parameter T (which may be positive or negative) has only a minor effect on the solution when physical realizability is not considered.

For this class of problems we shall assume that $\mathbf{x}$ is a wide-sense stationary random process. We also assume that T_1 is a stationary mapping, in the sense that the image of any wide-sense stationary process at the input is also a wide-sense stationary process. With these assumptions, $\mathbf{z}$ and $\boldsymbol{\omega}$ are jointly wide-sense stationary. Furthermore, because of the stationarity, the optimum weighting of past and future received data $\mathbf{z}(s)$ by $h(t, s)$ should be independent of t; i.e., the optimum filter is time-invariant and the impulse response can be expressed as $h(t - s)$. Using (7.25) and (7.28) for the stationary processes, condition (9.4) can be written as

$$\int_{-\infty}^{\infty} k_{zz}(s - \sigma)h(t - \sigma)\, d\sigma = k_{z\omega}(s - t) \tag{9.9}$$

Changing variables, $\tau = t - s$ and $\eta = \sigma - s$, and using (7.27), we find

$$\int_{-\infty}^{\infty} k_{zz}(\eta)h(\tau - \eta)\, d\eta = k_{\omega z}(\tau) \tag{9.10}$$

The solution for the transfer function of the optimum filter is obtained immediately by taking the Fourier transform of (9.10).

$$H(f) = \frac{K_{\omega z}(f)}{K_{zz}(f)} \tag{9.11}$$

The frequency-domain expression for minimum mean-squared error corresponding to (9.8) is

$$I_{\min} = \int_{-\infty}^{\infty} \frac{K_{\omega\omega}(f)K_{zz}(f) - |K_{\omega z}(f)|^2}{K_{zz}(f)}\, df \tag{9.12}$$

Additive Noise Disturbance

The most familiar filtering problem concerns a channel mapping which characterizes a linear, time-invariant channel dispersion $G(f)$ and additive, zero-mean, wide-sense stationary noise expressed in terms of an equivalent noise generator at the input to the receiver as shown in Figure 9.2. Hence with

$$\begin{aligned} T_0&: \boldsymbol{\omega}(t) = \mathbf{x}(t - T) \\ T_1&: \mathbf{z}(t) = \int_{-\infty}^{\infty} g(t - \sigma)\mathbf{x}(\sigma)\, d\sigma + \mathbf{u}(t) \end{aligned} \tag{9.13}$$

and assuming that signal and noise are statistically independent, we obtain

$$\begin{aligned} k_{zz}(\tau) &= \iint_{-\infty}^{\infty} g(t + \tau - \sigma)g(t - \eta)k_{xx}(\sigma, \eta)\, d\sigma\, d\eta + k_{uu}(\tau) \\ k_{\omega z}(\tau) &= E[\mathbf{x}(t + \tau - T)\mathbf{z}(t)] = k_{xz}(\tau - T) \end{aligned} \tag{9.14}$$

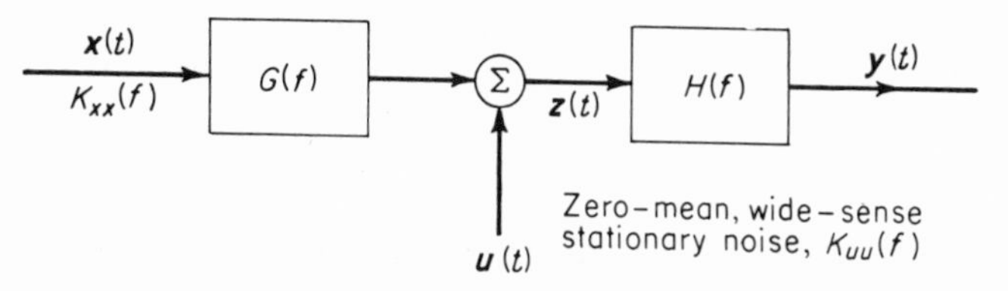

Figure 9.2. Continuous waveform estimation in the presence of additive noise and channel dispersion.

where

$$k_{xz}(\tau) = \int_{-\infty}^{\infty} k_{xx}(\tau + \sigma) g(\sigma)\, d\sigma$$

Taking Fourier transforms,

$$\begin{aligned} K_{zz}(f) &= |G(f)|^2 K_{xx}(f) + K_{uu}(f) \\ K_{xz}(f) &= \int_{-\infty}^{\infty} K_{xx}(f) g(\sigma) e^{j2\pi f\sigma}\, d\sigma = K_{xx}(f) G^*(f) \end{aligned} \tag{9.15}$$

The optimum filter, often called the *Wiener filter*, from (9.11) has a transfer function given by

$$H(f) = \frac{K_{xx}(f) G^*(f) e^{-j2\pi Tf}}{|G(f)|^2 K_{xx}(f) + K_{uu}(f)} \tag{9.16}$$

For a physical interpretation of this result, we can examine the frequency-domain expression for the functional I (assuming $T = 0$).

$$I = \int_{-\infty}^{\infty} K_{xx}(f)\, |1 - G(f)H(f)|^2\, df + \int_{-\infty}^{\infty} K_{uu}(f)\, |H(f)|^2\, df \tag{9.17}$$

The first term in (9.17) represents error due to imperfect compensation for channel dispersion, while the second term represents error due to the filtered noise remaining at the output. In frequency regions where the received signal density $K_{xx}(f)\,|G(f)|^2$ is large compared to the noise density $K_{uu}(f)$, the optimum filter concentrates on compensating for the channel dispersion: $H(f) \simeq G^{-1}(f)$. In regions where noise density is large, the optimum filter introduces additional attenuation. If noise density tends to be small compared to signal density for all frequencies, then the optimum filter is essentially the inverse of the dispersion characteristic $G(f)$ and the filter is commonly referred to as an *equalizer*. With any noise present, the equalizer is, of course, suboptimal. Using the equalizer, we have

$$I = \int_{-\infty}^{\infty} \frac{K_{uu}(f)}{|G(f)|^2}\, df \tag{9.18}$$

whereas, substituting (9.16) into (9.17), we find

$$I_{\min} = \int_{-\infty}^{\infty} \frac{K_{uu}(f)\, df}{|G(f)|^2 + [K_{uu}(f)/K_{xx}(f)]} \tag{9.19}$$

Example 9.1. *Equalizer with gain-bandwidth constraint:* Even in the case where noise is very small or absent, there is an interesting optimization problem in connection with equalizer design. In the realization of a transfer function inverse to $G(f)$, it may be required to provide extremely large gain over sizable intervals in the frequency domain. This may be prohibitively expensive and the designer may want to impose some kind of *gain-bandwidth product* constraint on the optimum filter. One version of this constraint might be expressed as

$$I_1 = \int_{-\infty}^{\infty} |H(f)|^2\, df = \int_{-\infty}^{\infty} h^2(t)\, dt = \text{constant} \tag{9.20}$$

The stationary points of $I + \lambda I_1$, where λ is a Lagrange multiplier, are given by

$$\int_{-\infty}^{\infty} k_{zz}(\sigma) h(t - \sigma)\, d\sigma - k_{\omega z}(t) + \lambda h(t) = 0 \tag{9.21}$$

for a time-invariant filter. In the frequency domain, (9.21) becomes

$$H(f) = \frac{K_{\omega z}(f)}{K_{zz}(f) + \lambda} = \frac{K_{xx}(f) G^*(f) e^{-j2\pi Tf}}{K_{xx}(f)\, |G(f)|^2 + \lambda} \tag{9.22}$$

We see that for λ sufficiently large any constraint on gain-bandwidth product can be met. If the constraint is severe, then the equalizer may be quite substantially different from the unconstrained equalizer. Note that the optimum equalizer has the same form as the optimum filter for additive white noise. Also, in contrast to the unconstrained situation, the optimal equalizer will depend on the power spectral density of the transmitted signal. ▮

Example 9.2. *Joint optimization of transmitter and receiver filters:* It is apparent that the performance of the optimum filter for the problem shown in Figure 9.2 is strongly dependent upon the dispersion characteristic $G(f)$. This leads us to consider the possibility of improving performance by modifying $G(f)$ with "pre-equalization" at the transmitter as shown in Figure 9.3. We now consider the problem of finding the best pair of transfer functions, $H(f)$ and $G(f)$, to minimize mean-squared error in continuous waveform estimation. For convenience we assume $T = 0$. An obvious solution is to make $G(f)$ very large; then with $H(f) = G^{-1}(f)$, the dispersion error is zero and the filtered noise has a very small mean-squared value. This amounts merely to "swamping" the noise with a large amount of signal power. More

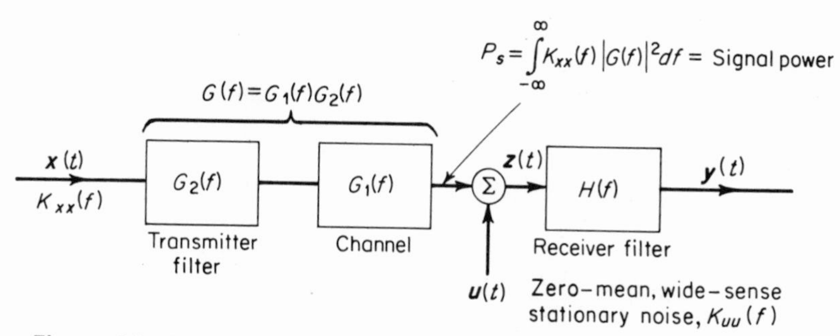

Figure 9.3. Optimum filtering with pre-transmission and post-transmission filters.

often, as in the case of a channel with a limited power-handling capability, there is a constraint on allowable signal power. With this constraint, we obtain a meaningful optimization problem. We want to find the stationary points of $I + \lambda P_S$ with respect to variations in $G(f)$ and $H(f)$ simultaneously, where

$$P_S = \int_{-\infty}^{\infty} K_{xx}(f)\,|G(f)|^2\,df \tag{9.23}$$

specifies the constraint on received signal power.

Using (9.17) and (9.23), it is clear which terms are quadratic, linear, and constant in $G(f)$ and $H(f)$.

$$\begin{aligned} I + \lambda P_S = &\int_{-\infty}^{\infty} K_{xx}(f)\,df - 2\int_{-\infty}^{\infty} K_{xx}(f)G^*(f)H^*(f)\,df \\ &+ \int_{-\infty}^{\infty} K_{xx}(f)\,|G(f)|^2\,|H(f)|^2\,df \\ &+ \int_{-\infty}^{\infty} K_{uu}(f)\,|H(f)|^2\,df + \lambda\int_{-\infty}^{\infty} K_{xx}(f)\,|G(f)|^2\,df \end{aligned} \tag{9.24}$$

Setting the gradient of $I + \lambda P_S$ with respect to $H(f)$ equal to zero, we have

$$2K_{xx}(f)\,|G(f)|^2\,H(f) - 2K_{xx}(f)G^*(f) + 2K_{uu}(f)H(f) = 0 \tag{9.25}$$

whereas setting the gradient with respect to $G(f)$ equal to zero gives

$$2K_{xx}(f)\,|H(f)|^2\,G(f) - 2K_{xx}(f)H^*(f) + 2\lambda K_{xx}(f)G(f) = 0 \tag{9.26}$$

In order to find the simultaneous solutions to (9.25) and (9.26), we first multiply (9.25) by $H^*(f)$ and (9.26) by $G^*(f)$, then subtract to find the relation

$$K_{uu}(f)\,|H(f)|^2 = \lambda K_{xx}(f)\,|G(f)|^2 \tag{9.27}$$

From (9.27) it is apparent that λ is positive. It follows from (9.26) that $G^*(f)H^*(f)$ is real and non-negative. This means that $G(f)H(f) = |G(f)H(f)| = |G(f)|\,|H(f)|$, hence (9.27) can be combined with (9.25) or (9.26) to obtain the solutions

$$\left.\begin{aligned} |H(f)|^2 &= \left[\frac{\lambda K_{xx}(f)}{K_{uu}(f)}\right]^{\frac{1}{2}} - \lambda \\ |G(f)|^2 &= \left[\frac{K_{uu}(f)}{\lambda K_{xx}(f)}\right]^{\frac{1}{2}} - \frac{K_{uu}(f)}{K_{xx}(f)} \end{aligned}\right\} \quad \text{for } f \in B$$

$$|H(f)| = |G(f)| = 0 \qquad \text{for } f \in B' \tag{9.28}$$

where B is the set of frequencies for which the ratio of signal and noise power densities exceeds a certain threshold. B' is the complement of B; i.e., $B \cup B' = R$ and $B \cap B' = \varnothing$.

$$B = \left\{f;\ \frac{K_{xx}(f)}{K_{uu}(f)} > \lambda\right\} \tag{9.29}$$

The physical interpretation of this result is that it is best not to expend any of the allowable signal power in frequency bands where the signal-to-noise ratio is too low. The reason that the equations involve only the magnitudes of the transfer functions of the filters is also easy to interpret physically. The portion of mean-squared error due to signal distortion

$$\int_{-\infty}^{\infty} K_{xx}(f)\,|1 - G(f)H(f)|^2\,df$$

is minimized by zero phase shift in $G(f)H(f)$; whereas the error due to noise is independent of the phase shift in $H(f)$. We conclude that $H(f)$ can have arbitrary phase provided that $G(f)$ has the complementary phase.

Now choosing λ to satisfy the signal power constraint (9.23), we find

$$\lambda^{\frac{1}{2}} = \frac{\int_B [K_{xx}\ f)K_{uu}(f)]^{\frac{1}{2}}\,df}{P_S + P_N} \tag{9.30}$$

where

$$P_N = \int_B K_{uu}(f)\,df$$

is the channel noise power contained in the passband of the filters. Combining (9.30) and (9.28) with (9.17), the minimum mean-squared error is given by

$$I_{\min} = \frac{\{\int_B [K_{xx}(f)K_{uu}(f)]^{\frac{1}{2}}\,df\}^2}{P_S + P_N} + \int_{B'} K_{xx}(f)\,df \tag{9.31}$$

If the signal-to-noise ratio on the channel is moderately large, then λ is small and we see from (9.28) that the transmitting and receiving filters are essentially inverse to each other. If we had initially imposed the constraint $G(f) = H^{-1}(f)$ and optimized with respect to $H(f)$ alone, we would obtain

$$|H(f)|^2 = \left[\frac{\lambda K_{xx}(f)}{K_{uu}(f)}\right]^{\frac{1}{2}}$$

and

$$I_{\min} = \frac{\{\int_{-\infty}^{\infty} [K_{xx}(f)K_{uu}(f)]^{\frac{1}{2}}\, df\}^2}{P_S} \tag{9.32}$$

These results are close to optimum, (9.28) and (9.31), when signal-to-noise ratio is moderately large. ∎

Example 9.3. We can illustrate the advantage to be gained by joint optimization of $G(f)$ and $H(f)$ as described in Example 9.2 by considering specific signal and noise processes. Let the additive interference be white noise with spectral density N_0 watts/Hz. Let the signal be a unit variance random facsimile process (Section 8.7) with rate parameter $2\pi f_0$ and with the mean value subtracted out. Then we have

$$\begin{aligned} K_{xx}(f) &= \frac{1}{\pi f_0} \frac{1}{1 + (f/f_0)^2} \\ K_{uu}(f) &= N_0 \end{aligned} \tag{9.33}$$

In order to make comparisons, we shall assume that the channel has no dispersion so that when no transmitting filter is used we have $G(f) = 1$ and the optimum receiving filter is given by (9.16). When both filters are used, the optimum transfer functions are given by (9.28). In this latter case, the minimum mean-squared error, call it I_2, from (9.31) becomes

$$\begin{aligned} I_2 &= \frac{\dfrac{f_0 N_0}{\pi}\left[\displaystyle\int_{-f_B}^{f_B} \frac{df}{f_0\sqrt{1 + (f/f_0)^2}}\right]^2}{1 + \int_{-f_B}^{f_B} N_0\, df} + \frac{1}{\pi}\int_{B'} \frac{df}{f_0[1 + (f/f_0)^2]} \\ &= \frac{\dfrac{f_0 N_0}{\pi}\left[\log\left(\dfrac{\sqrt{1 + \xi_B^2} + \xi_B}{\sqrt{1 + \xi_B^2} - \xi_B}\right)\right]^2}{1 + 2f_0 N_0 \xi_B} + 1 - \frac{2}{\pi}\tan^{-1}\xi_B \end{aligned} \tag{9.34}$$

where the normalized passband $\xi_B = f_B/f_0$ of the filters, according to (9.29), must satisfy the relation

$$\frac{1}{\sqrt{1 + \xi_B^2}} = \frac{f_0 N_0 \log\left(\dfrac{\sqrt{1 + \xi_B^2} + \xi_B}{\sqrt{1 + \xi_B^2} - \xi_B}\right)}{1 + 2f_0 N_0 \xi_B} \tag{9.35}$$

On the other hand, using a filter at the receiver only, (9.19) is applicable and the minimum mean-squared error is

$$I_1 = \frac{1}{\pi f_0} \int_{-\infty}^{\infty} \frac{df}{1 + (1/\pi N_0 f_0) + (f/f_0)^2} = \left[\frac{1}{1 + (1/\pi N_0 f_0)}\right]^{\frac{1}{2}} \quad (9.36)$$

The relative values of I_1 and I_2 for various values of the parameter $f_0 N_0$ are shown on Figure 9.4. These results indicate that combined pre-transmission

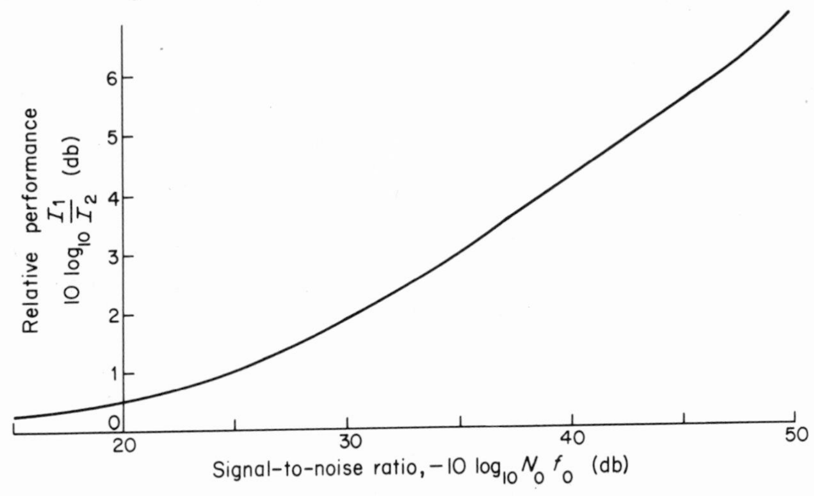

Figure 9.4. Comparison of best performance of combined transmitting and receiving filters (I_2) with receiving filter only (I_1).

and post-transmission filtering display an appreciable advantage only when the signal-to-noise ratio on the channel is large (when $N_0 f_0$ is small). The results pertaining to this example appear to have been first presented in Reference [2]. ∎

Exercise 9.1. Suppose that it is desired to continuously estimate a "smoothed" version of the transmitted signal; i.e., find the optimum filter for the problem

$$T_0: \boldsymbol{\omega}(t) = \int_{-\infty}^{\infty} h_0(t - \tau)\mathbf{x}(\tau - T)\, d\tau$$

$$T_1: \mathbf{z}(t) = \int_{-\infty}^{\infty} g(t - \tau)\mathbf{x}(\tau)\, d\tau + \mathbf{u}(t)$$

where $\mathbf{x}$ and $\mathbf{u}$ are statistically independent, zero-mean, wide-sense stationary random processes.

Exercise 9.2. In the situation shown in Figure 9.2, suppose that the transmitted signal $x(t)$ is a known, finite-energy signal and that it is desired that the output $\mathbf{y}(t)$ of the receiving filter be a faithful reproduction of this waveform. One approach to this problem is to require that the $L^2(-\infty, \infty)$ norm of the mean error

$$E[x(t) - \mathbf{y}(t)]$$

should not exceed a specified bound. Find the optimum filter function for a time-invariant filter which meets this constraint and also minimizes the variance of the error $x(t) - \mathbf{y}(t)$.

Exercise 9.3. For the problem involving design of transmitting and receiving filters described in Example 9.2, verify the results, (9.32), under the constraint $G(f) = H^{-1}(f)$.

Exercise 9.4. In the joint optimization of transmitting and receiving filters, suppose that the signal power constraint is imposed on power at the output of the transmitter filter; i.e., in Figure 9.3, $G_1(f)$ is prescribed and the constraint is

$$P_{S_0} = \int_{-\infty}^{\infty} K_{xx}(f)\,|G_2(f)|^2\,df$$

Find the optimum functions $G_2(f)$ and $H(f)$.

Multiplicative Noise Disturbance

In some signal transmission systems, some of the disturbing influences are more accurately characterized by a multiplicative process (gain fluctuations)

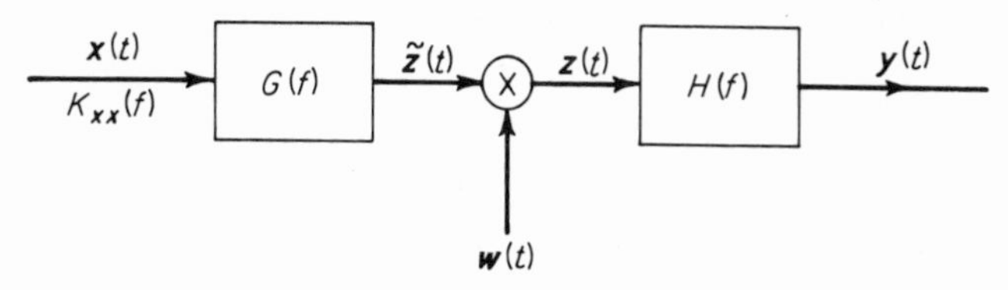

Multiplicative noise, $K_{ww}(f)$

Figure 9.5. Filtering of signal with stationary multiplicative noise.

than by an additive process. We shall consider the simplified problem shown in Figure 9.5 where $\mathbf{w}$ is a wide-sense stationary process statistically independent of the signal $\mathbf{x}$. Thus with

$$\begin{aligned} &T_0: \boldsymbol{\omega}(t) = \mathbf{x}(t) \\ &T_1: \mathbf{z}(t) = \mathbf{w}(t)\int_{-\infty}^{\infty} g(t-\sigma)\mathbf{x}(\sigma)\,d\sigma \triangleq \mathbf{w}(t)\tilde{\mathbf{z}}(t) \end{aligned} \tag{9.37}$$

then $\mathbf{z}$ and $\boldsymbol{\omega}$ are jointly wide-sense stationary, and the correlation functions are given by

$$\begin{aligned} k_{zz}(s, \sigma) &= E[\mathbf{w}(s)\tilde{\mathbf{z}}(s)\mathbf{w}(\sigma)\tilde{\mathbf{z}}(\sigma)] \\ &= E[\mathbf{w}(s)\mathbf{w}(\sigma)]E[\tilde{\mathbf{z}}(s)\tilde{\mathbf{z}}(\sigma)] \\ &= k_{ww}(s-\sigma)k_{\tilde{z}\tilde{z}}(s-\sigma) \end{aligned} \tag{9.38}$$

$$\begin{aligned} k_{\omega z}(t, s) &= E[\mathbf{x}(t)\mathbf{w}(s)\tilde{\mathbf{z}}(s)] \\ &= \overline{\mathbf{w}}k_{x\tilde{z}}(t-s) \end{aligned} \tag{9.39}$$

Taking Fourier transforms,

$$K_{zz}(f) = K_{ww}(f) \otimes K_{\tilde{z}\tilde{z}}(f)$$
$$= K_{ww}(f) \otimes K_{xx}(f)\,|G(f)|^2 \tag{9.40}$$
$$K_{\omega z}(f) = \overline{\mathbf{w}}K_{x\tilde{z}}(f) = \overline{\mathbf{w}}K_{xx}(f)G^*(f) \tag{9.41}$$

The transfer function of the optimum filter, from (9.11), is

$$H(f) = \frac{\overline{\mathbf{w}}K_{xx}(f)G^*(f)}{K_{ww}(f) \otimes K_{xx}(f)\,|G(f)|^2} \tag{9.42}$$

If the multiplicative noise is zero mean, then the best estimate is simply zero. In cases of practical interest, $\mathbf{w}$ will have a substantial mean value relative to the size of the fluctuations about the mean. Writing the autocorrelation for $\mathbf{w}$ as the sum of the autocovariance $m_{ww}(\tau)$ plus the square of the mean, the expression for the optimum filter resembles that for the case of additive noise.

$$k_{ww}(\tau) = m_{ww}(\tau) + (\overline{\mathbf{w}})^2$$
$$\Rightarrow K_{ww}(f) = M_{ww}(f) + (\overline{\mathbf{w}})^2\,\delta(f)$$
$$\Rightarrow K_{zz}(f) = M_{ww}(f) \otimes K_{xx}(f)\,|G(f)|^2 + (\overline{\mathbf{w}})^2 K_{xx}(f)\,|G(f)|^2 \tag{9.43}$$

Now (9.42) can be rewritten as

$$H(f) = \frac{(1/\overline{\mathbf{w}})K_{xx}(f)G^*(f)}{K_{xx}(f)\,|G(f)|^2 + \dfrac{1}{(\overline{\mathbf{w}})^2}M_{ww}(f) \otimes K_{xx}(f)\,|G(f)|^2} \tag{9.44}$$

As expected, if the covariance of $\mathbf{w}$ is very small compared to the square of its mean, then $H(f)$ is approximately the inverse of the channel characteristic $G(f)$ with a constant gain factor of $1/\overline{\mathbf{w}}$. To obtain some idea of the effect of the covariance $m_{ww}(\tau)$ on the characteristics of the optimum filter, we let $G(f) = 1$ and consider the extreme cases of multiplicative noise which fluctuates very rapidly and very slowly compared to the signal process. If $\mathbf{w}$ fluctuates much more rapidly than $\mathbf{x}$, then $M_{ww}(f)$ is broad compared to $K_{xx}(f)$ and the convolution of these functions is approximately proportional to $M_{ww}(f)$.

$$M_{ww}(f) \otimes K_{xx}(f) \simeq M_{ww}(f)\int_{-\infty}^{\infty} K_{xx}(f)\,df = \overline{\mathbf{x}^2}\,M_{ww}(f) \tag{9.45}$$

In this case the optimum filter closely resembles the one for additive noise (9.16), having a power spectral density proportional to $M_{ww}(f)$.

$$H(f) \cong \frac{(1/\overline{\mathbf{w}})K_{xx}(f)}{K_{xx}(f) + (\overline{\mathbf{x}^2}/(\overline{\mathbf{w}})^2)M_{ww}(f)} \tag{9.46}$$

On the other hand, if the fluctuations of $\mathbf{w}$ are very slow compared to those of $\mathbf{x}$, then the convolution of $M_{ww}(f)$ and $K_{xx}(f)$ is approximately proportional to $K_{xx}(f)$.

$$M_{ww}(f) \otimes K_{xx}(f) \simeq \sigma_w^2 K_{xx}(f) \tag{9.47}$$

and the optimum filter has approximately constant gain given by

$$H(f) \simeq \frac{\overline{\mathbf{w}}}{(\overline{\mathbf{w}})^2 + \sigma_w^2} = \frac{\overline{\mathbf{w}}}{\overline{\mathbf{w}^2}} \tag{9.48}$$

Example 9.4. *Channel with sinusoidal gain fluctuations:* Very often random gain fluctuations have a periodic behavior. In this example we let

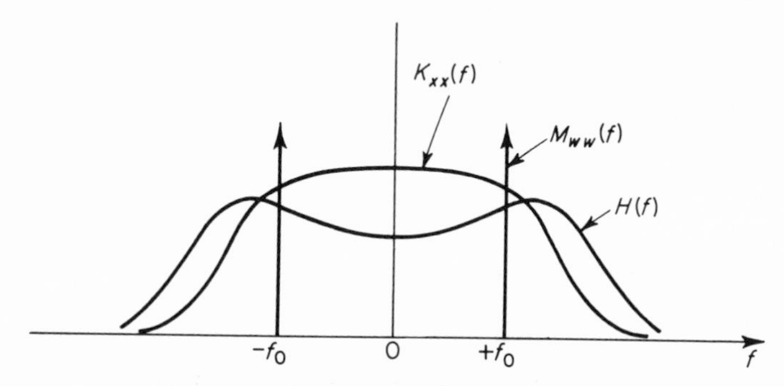

Figure 9.6. Optimum filter for sinusoidal multiplicative noise.

$G(f) = 1$ and let $\mathbf{w}$ be a sinusoidal multiplicative interference having a frequency f_0.

$$\mathbf{w}(t) = w_0 + \mathbf{a} \cos(2\pi f_0 t + \boldsymbol{\theta}) \tag{9.49}$$

where $\mathbf{a}$ and $\boldsymbol{\theta}$ are independent random variables with $\boldsymbol{\theta}$ distributed uniformly over $0 \leqslant \theta < 2\pi$. In this case, $\mathbf{w}$ is wide-sense stationary and

$$k_{ww}(\tau) = w_0^2 + \tfrac{1}{2}\overline{\mathbf{a}^2} \cos 2\pi f_0 \tau$$

$$\Rightarrow K_{ww}(f) = w_0^2\, \delta(f) + \tfrac{1}{4}\overline{\mathbf{a}^2}[\delta(f - f_0) + \delta(f + f_0)] \tag{9.50}$$

The optimum filter, from (9.44), is given by

$$H(f) = \frac{(1/w_0)K_{xx}(f)}{K_{xx}(f) + \frac{1}{4}(\overline{\mathbf{a}^2}/w_0^2)[K_{xx}(f - f_0) + K_{xx}(f + f_0)]} \tag{9.51}$$

A typical transfer function for this case is illustrated in Figure 9.6. ▮

Random Channel Dispersion

In the solution for the filtering problem with additive noise illustrated in Figure 9.2, we note that the optimum filter is strongly dependent on the channel dispersion characteristic $G(f)$. In many situations, $G(f)$ is not exactly known. It might happen that $G(f)$ could be any one of a large set of transfer functions. The optimal filter for one element of the set would, of course, be suboptimal for the remaining elements of the set. One approach to this problem would be to consider $G(f)$ as a realization of a random process and we could find the filter function $H(f)$ which yields the minimum average value [over the set of $G(f)$'s] of mean-squared error. Since each realization of the process is a time-invariant transformation, the received signal $\mathbf{z}(t)$, shown in Figure 9.7, is still a wide-sense stationary process.

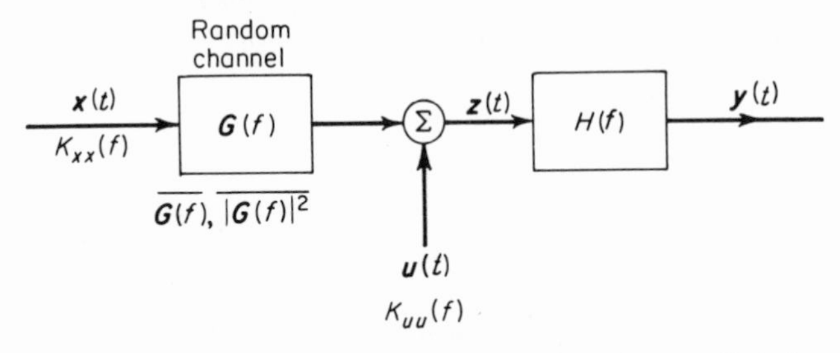

Figure 9.7. Optimum filtering for random channel and statistically independent additive noise.

The channel dispersion is characterized by the set of random variables $\{\mathbf{G}(f); -\infty < f < \infty\}$ or, equivalently by the random variables $\{\mathbf{g}(t); -\infty < t < \infty\}$, where $g(t)$ is the impulse response of a typical realization of the channel. Thus for the problem

$$T_0\colon \boldsymbol{\omega}(t) = \mathbf{x}(t)$$

$$T_1\colon \mathbf{z}(t) = \int_{-\infty}^{\infty} \mathbf{x}(t-\sigma)\mathbf{g}(\sigma)\,d\sigma + \mathbf{u}(t) \tag{9.52}$$

the required correlation functions are

$$\begin{aligned} k_{zz}(\tau) &= E[\mathbf{z}(t+\tau)\mathbf{z}(t)] \\ &= \iint_{-\infty}^{\infty} E[\mathbf{x}(t+\tau-\sigma)\mathbf{x}(t-\xi)]E[\mathbf{g}(\sigma)\mathbf{g}(\xi)]\,d\sigma\,d\xi + k_{uu}(\tau) \\ &= \iint_{-\infty}^{\infty} E[\mathbf{g}(\sigma)\mathbf{g}(\xi)]k_{xx}(\tau-\sigma+\xi)\,d\sigma\,d\xi + k_{uu}(\tau) \end{aligned} \tag{9.53}$$

$$\begin{aligned} k_{\omega z}(\tau) &= E[\mathbf{x}(t+\tau)\mathbf{z}(t)] \\ &= \int_{-\infty}^{\infty} k_{xx}(\tau+\sigma)E[\mathbf{g}(\sigma)]\,d\sigma \end{aligned} \tag{9.54}$$

Taking the Fourier transform of (9.53),

$$K_{zz}(f) = \iint_{-\infty}^{\infty} E[\mathbf{g}(\sigma)\mathbf{g}(\xi)]K_{xx}(f)e^{-j2\pi f(\sigma-\xi)}\, d\sigma\, d\xi + K_{uu}(f) \tag{9.55}$$

Interchanging expectation and integration operations and then performing the integrations, (9.55) becomes

$$K_{zz}(f) = E[|\mathbf{G}(f)|^2]K_{xx}(f) + K_{uu}(f) \tag{9.56}$$

Similarly, from (9.54) we find

$$K_{\omega z}(f) = E[\mathbf{G}^*(f)]K_{xx}(f) \tag{9.57}$$

Using the overbar notation for expectations on the channel variables, the transfer function of the optimum filter, from (9.11), is given by

$$H(f) = \frac{K_{xx}(f)[\overline{\mathbf{G}(f)}]^*}{K_{xx}(f)\,\overline{|\mathbf{G}(f)|^2} + K_{uu}(f)} \tag{9.58}$$

Physical interpretation of this result is aided by expressing the mean-squared channel gain $\overline{|\mathbf{G}(f)|^2}$ as the sum of *channel variance* $\sum^2(f)$ and the squared magnitude of the mean channel characteristic $|\overline{\mathbf{G}(f)}|^2$. Then (9.58) can be written as

$$H(f) = \frac{K_{xx}(f)[\overline{\mathbf{G}(f)}]^*}{K_{xx}(f)\,|\overline{\mathbf{G}(f)}|^2 + K_{uu}(f) + K_{xx}(f)\sum^2(f)} \tag{9.59}$$

In this form, it is apparent that we can consider the optimum filter to be the one designed for the mean channel using (9.16), where we add a noise process having a power spectral density of $K_{xx}(f)\sum^2(f)$. In this way we see that channel variance has very much the same effect as additive noise. We comment here that the usual approach in dealing with unknown channel transfer functions has been to estimate the mean, or sometimes the median, transfer function of the channel and then construct the Wiener filter corresponding to this transfer function. The error in using this approach can be significant if the channel variance is large. A typical application of these results is illustrated in the following example. Several other problems of this nature are treated in Reference [3].

Example 9.5. *Equalization of a random width scanner:* In Section 5.4, we noted that the operation of scanning a spatially distributed signal with a "window" function $k(t)$ moving at constant velocity was equivalent to a time-invariant filtering operation with impulse response $k(-t)$. In many signal processing devices, the window has a rectangular shape. In the manufacture of such devices as magnetic readout heads and optical slits there is bound to be some variability in the gap width. Frequently, these devices are followed by equalizers to compensate to some extent for the loss of high-frequency components due to a non-zero scanner width. It is of interest to determine the filter which minimizes mean-squared error at the output when the width of the scanner is subject to random variations. This corresponds to the random channel problem if we let $\mathbf{g}(t)$ be a rectangular function with random width

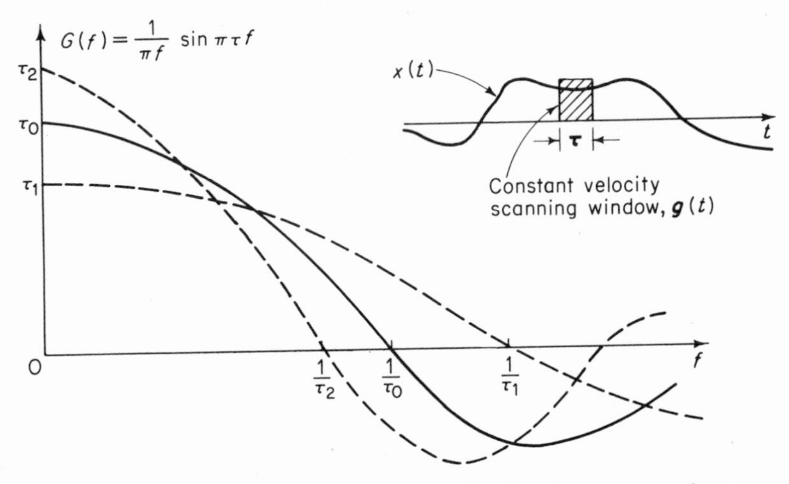

Figure 9.8. Typical realizations of the transfer function corresponding to a random width scanner.

$\boldsymbol{\tau}$, as shown in Figure 9.8. The corresponding transfer function is

$$\mathbf{G}(f) = \frac{\sin \pi \boldsymbol{\tau} f}{\pi f} \tag{9.60}$$

Assigning a probability density function $p_\tau(\xi)$ for the random variable $\boldsymbol{\tau}$, we find

$$\begin{aligned}\overline{\mathbf{G}(f)} &= \int_{-\infty}^{\infty} \frac{\sin \pi \xi f}{\pi f}\, p_\tau(\xi)\, d\xi \\ &= \frac{1}{2j\pi f}\left[P_\tau\left(-\frac{f}{2}\right) - P_\tau\left(\frac{f}{2}\right)\right]\end{aligned} \tag{9.61}$$

and

$$\begin{aligned}\overline{|\mathbf{G}(f)|^2} &= \int_{-\infty}^{\infty} \frac{\sin^2 \pi \xi f}{(\pi f)^2}\, p_\tau(\xi)\, d\xi \\ &= \frac{1}{2(\pi f)^2}\left[1 - \tfrac{1}{2}P_\tau(f) - \tfrac{1}{2}P_\tau(-f)\right]\end{aligned} \tag{9.62}$$

In (9.62) we have used the fact that $P_\tau(0) = 1$ [$P_\tau(f)$ is the Fourier transform of $p_\tau(\xi)$].

In the high signal-to-noise ratio situation, we can neglect $K_{uu}(f)$ in

(9.58) and the expression for the optimum filter becomes

$$H(f) = \frac{[\overline{\mathbf{G}(f)}]^*}{\overline{|\mathbf{G}(f)|^2}} = \frac{j\pi f[P_\tau(f/2) - P_\tau(-f/2)]}{1 - \frac{1}{2}[P_\tau(f) + P_\tau(-f)]} \tag{9.63}$$

This result is shown in Figure 9.9 for the case of widths distributed uniformly over $0.5\tau_0 < \tau < 1.5\tau_0$. For comparison, the equalizers (inverse

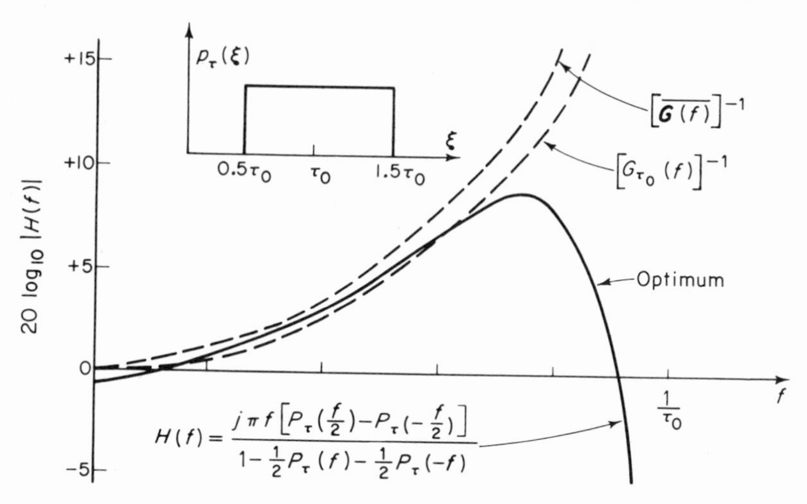

Figure 9.9. Equalizer characteristics for the random width scanner.

filters) for the mean channel $\overline{\mathbf{G}(f)}$ and the channel function $G_{\tau_0}(f)$ corresponding to the mean scanner width are also shown. The gain of the optimum equalizer is substantially different from these at high frequencies, as a result of taking into account the channel variance. ▌

Exercise 9.5. In the random, time-invariant channel problem illustrated in Figure 9.7, suppose that the channel has an ideal lowpass characteristic with random cutoff frequency **W**; i.e.,

$$\mathbf{G}(f) = \tfrac{1}{2}[1 + \text{sgn}\,(\mathbf{W} - |f|)]$$

where **W** is uniformly distributed over the interval $0 < f_a \leqslant W \leqslant f_b$. Assuming that $K_{uu}(f)/K_{xx}(f) = 0.01$, find and sketch the optimum $H(f)$ for continuous waveform estimation of a wide-sense stationary process **x**.

Exercise 9.6. Find the optimum filter for continuous waveform estimation for a channel with multiplicative noise followed by a random time-invariant transformation followed by additive noise; i.e.,

$$T_0\colon \boldsymbol{\omega}(t) = \mathbf{x}(t - T)$$

$$T_1\colon \mathbf{z}(t) = \int_{-\infty}^{\infty} \mathbf{g}(t - \tau)\mathbf{w}(\tau)\mathbf{x}(\tau)\, d\tau + \mathbf{u}(t)$$

where $\mathbf{w}(t)$, $\mathbf{g}(t)$, $\mathbf{u}(t)$, and $\mathbf{x}(t)$ are statistically independent random variables.

9.4 Pulse Amplitude Estimation

In situations such as radar and PAM data transmission, where the desired information is pulse amplitude, it is rather beside the point to perform a continuous waveform estimation. In fact, doing so may lead to considerably suboptimum results as far as mean-squared estimation of pulse amplitude is concerned. In this section, we investigate the possibility of finding the optimum filter such that when its output is sampled at $t = t_0$, we obtain a number which is a minimum mean-squared error estimate of the transmitted pulse amplitude $\mathbf{a}$. The problem is illustrated in Figure 9.10, where we have

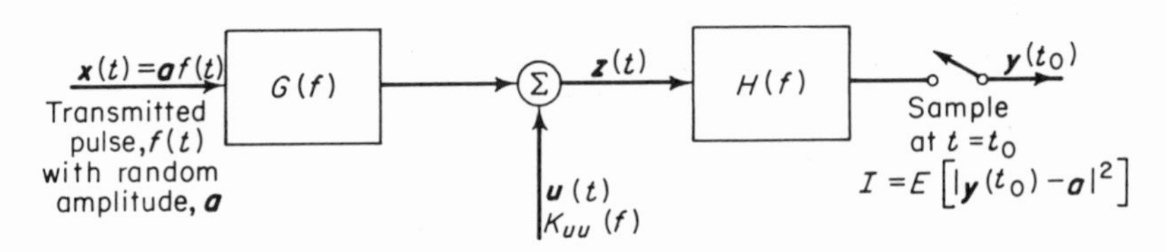

Figure 9.10. Estimation of pulse amplitude.

assumed that the optimum filter is time-invariant. This imposes no restriction because we are only concerned with how the filter responds, at $t = t_0$, to the received data. The response at other times is irrelevant so there is no need to consider time-variable filters. Thus with

$$T_0: \boldsymbol{\omega}(t_0) = \mathbf{a} \tag{9.64}$$

the condition on $h(t)$ for optimality, (9.4) becomes

$$\int_{-\infty}^{\infty} k_{zz}(s, \sigma)h(t_0 - \sigma)\, d\sigma = k_{z\omega}(s, t_0) \tag{9.65}$$

Although the signal process will not be wide-sense stationary, as it was in the preceding problems, it is often convenient to use a frequency-domain approach to the solution of (9.65). The equivalent frequency-domain condition is obtained by first substituting

$$h(t_0 - \sigma) = \int_{-\infty}^{\infty} H^*(\nu)e^{-j2\pi\nu(t_0-\sigma)}\, d\nu \tag{9.66}$$

into (9.65) and then taking the Fourier transform with respect to the s variable. This gives

$$\int_{-\infty}^{\infty} H^*(\nu)e^{-j2\pi\nu t_0}B_{zz}(f, \nu)\, d\nu = C_{z\omega}(f, t_0) \tag{9.67}$$

where we have defined

$$B_{zz}(f, \nu) \triangleq \iint_{-\infty}^{\infty} k_{zz}(s, \sigma)e^{-j2\pi(fs-\nu\sigma)}\, ds\, d\sigma \tag{9.68}$$

and

$$C_{z\omega}(f, t_0) \triangleq \int_{-\infty}^{\infty} k_{z\omega}(s, t_0)e^{-j2\pi fs}\, ds \tag{9.69}$$

The Matched Filter

The classical problem in estimation of the amplitude of an isolated pulse involves a pulse of known shape and arrival time which is corrupted with additive, zero-mean, wide-sense stationary noise. From Figure 9.10, we see that

$$\begin{aligned} T_1\colon \mathbf{z}(t) &= \mathbf{a}\int_{-\infty}^{\infty} f(t-\sigma)g(\sigma)\, d\sigma + \mathbf{u}(t) \\ &\triangleq \mathbf{a}r(t) + \mathbf{u}(t) \end{aligned} \tag{9.70}$$

Evaluating the correlation functions,

$$\begin{aligned} k_{zz}(s, \sigma) &= E[\mathbf{a}r(s)\mathbf{a}r(\sigma)] + k_{uu}(s-\sigma) \\ &= \overline{\mathbf{a}^2}r(s)r(\sigma) + k_{uu}(s-\sigma) \\ k_{z\omega}(s, t_0) &= E[\mathbf{a}\cdot \mathbf{a}r(s)] = \overline{\mathbf{a}^2}r(s) \end{aligned} \tag{9.71}$$

Now using (9.68) and (9.69),

$$\begin{aligned} B_{zz}(f, \nu) &= \overline{\mathbf{a}^2}R(f)R^*(\nu) + K_{uu}(f)\,\delta(f-\nu) \\ C_{z\omega}(f, t_0) &= \overline{\mathbf{a}^2}R(f) \end{aligned}$$

The condition (9.67) on the transfer function of the optimum filter becomes

$$\overline{\mathbf{a}^2}R(f)\int_{-\infty}^{\infty} R^*(\nu)H^*(\nu)e^{-j2\pi\nu t_0}\, d\nu + H^*(f)K_{uu}(f)e^{-j2\pi f t_0} = \overline{\mathbf{a}^2}R(f) \tag{9.72}$$

Using a proportionality constant α, the solution has the form

$$H(f) = \frac{\alpha R^*(f)e^{-j2\pi f t_0}}{K_{uu}(f)} \tag{9.73}$$

Substituting (9.73) back into (9.72), we find that

$$\alpha = \frac{\overline{\mathbf{a}^2}}{1 + \overline{\mathbf{a}^2}\displaystyle\int_{-\infty}^{\infty} \frac{|R(f)|^2}{K_{uu}(f)}\, df} \tag{9.74}$$

This filter is known as the *matched filter*[4]–[6] for the received pulse $r(t)$ in noise with power spectral density $K_{uu}(f)$. We see that this is a generalized version of the matched filter obtained in Example 6.3 where we found the filter that maximized output at $t = t_0$ for a unit-energy input signal. It might have been suspected that such a filter would be optimal for pulse amplitude estimation. It is, for the case of white noise, but (9.73) shows that there is some advantage to be gained by emphasizing those frequencies where the noise density is smaller if it is not a constant function of frequency. The amount of this advantage for a particular case is illustrated in the following example.

Example 9.6. *Triangular pulse in exponentially correlated noise:* Let $r(t)$ be the triangular pulse shown in Figure 9.11 and let the additive noise have a power spectral density given by

$$K_{uu}(f) = \frac{N_0}{1 + (f/f_0)^2} \tag{9.75}$$

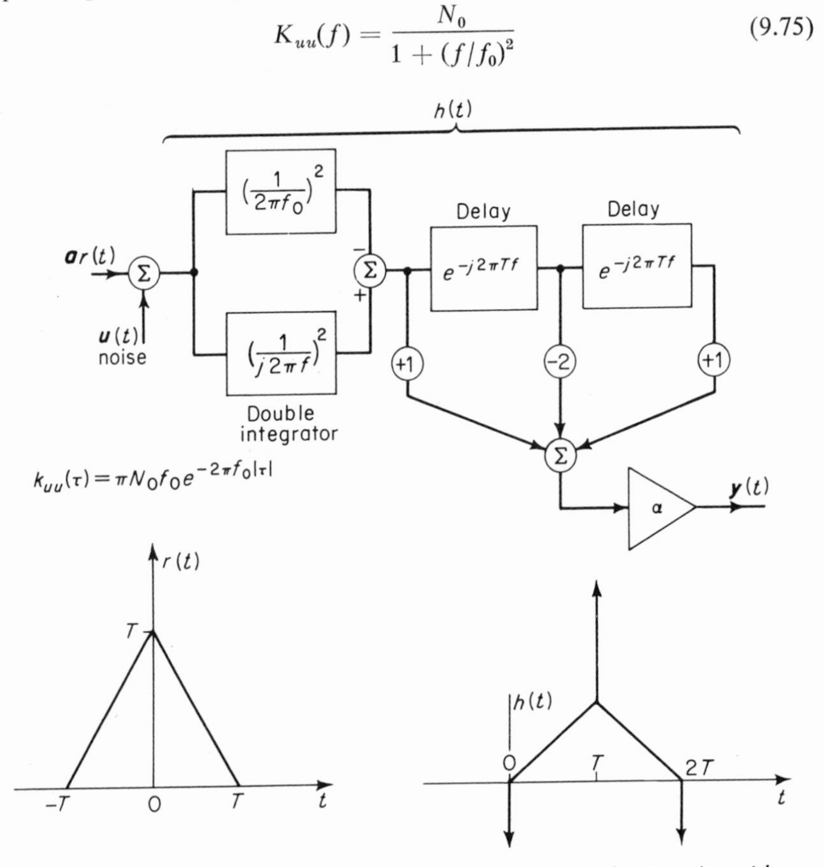

Figure 9.11. Realization of matched filter for triangular pulse in noise with exponential autocorrelation.

The transfer function for the matched filter, from (9.73), is

$$H(f) = \frac{\alpha}{N_0}\left[1 + \left(\frac{f}{f_0}\right)^2\right] R^*(f) e^{-j2\pi f t_0}$$
$$= \frac{\alpha}{N_0}\left[1 + \left(\frac{f}{f_0}\right)^2\right] \frac{\sin^2 \pi T f}{(\pi f)^2} e^{-j2\pi f t_0} \tag{9.76}$$

Letting $T = t_0$ and expanding $\sin \pi T f$ into exponential functions, the transfer function can be expressed as

$$H(f) = \frac{\alpha}{N_0}\left[\left(\frac{1}{j2\pi f}\right)^2 - \left(\frac{1}{2\pi f_0}\right)^2\right](1 - 2e^{-j2\pi T f} + e^{-j2\pi(2T)f}) \tag{9.77}$$

Inspection of (9.77) leads directly to the realization of the filter shown in Figure 9.11.

From (9.8), the resulting mean-squared error in pulse amplitude estimation is given by

$$I_{\min} = k_{\omega\omega}(t_0, t_0) - \int_{-\infty}^{\infty} k_{z\omega}(s, t_0) h(t_0 - s)\, ds$$
$$= \overline{\mathbf{a}^2} - \overline{\mathbf{a}^2}\alpha \int_{-\infty}^{\infty} \frac{|R(f)|^2}{K_{uu}(f)}\, df$$
$$= \frac{\overline{\mathbf{a}^2}}{1 + \overline{\mathbf{a}^2}\displaystyle\int_{-\infty}^{\infty} \frac{|R(f)|^2}{K_{uu}(f)}\, df} \tag{9.78}$$

For comparison, we consider the performance that would result from using the white noise matched filter for this pulse. This filter is obtained by removing the $(1/2\pi f_0)^2$ element in Figure 9.11 and readjusting the gain α to its optimal value. When this is done, the mean-squared error (see Exercise 9.9) is

$$I_W = \frac{\overline{\mathbf{a}^2}}{1 + \dfrac{\overline{\mathbf{a}^2}[\int |R(f)|^2\, df]^2}{\int |R(f)|^2 K_{uu}(f)\, df}} \tag{9.79}$$

For moderately large signal-to-noise ratio ($\overline{\mathbf{a}^2}/N_0$ large), the ratio of suboptimal to optimal performance can be expressed approximately as

$$\frac{I_W}{I_{\min}} \cong \frac{\left[\displaystyle\int \frac{|R(f)|^2}{K_{uu}(f)}\, df\right][\int |R(f)|^2 K_{uu}(f)\, df]}{[\int |R(f)|^2\, df]^2} \tag{9.80}$$

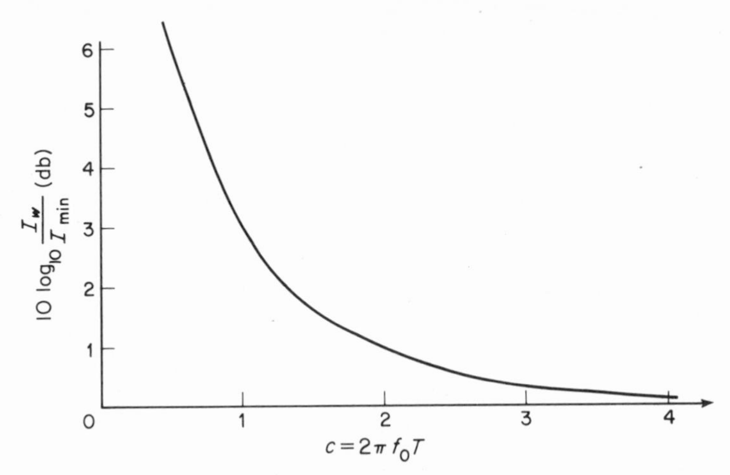

Figure 9.12. Comparison of optimum and suboptimum matched filter performance at high signal-to-noise ratios.

This ratio is shown graphically in Figure 9.12 as a function of the parameter $c = 2\pi f_0 T$. It is evident that the advantage of the optimal filter is significant only when the noise bandwidth is equal to or less than the pulse bandwidth. ▌

Exercise 9.7. Solve the matched filter problem (Figure 9.10) subject to a gain-bandwidth product constraint expressed by

$$I_1 = \int_{-\infty}^{\infty} |H(f)|^2 \, df$$

Exercise 9.8. Show that the filter (in Figure 9.10) which maximizes peak pulse response

$$I_1 = E[\mathbf{y}(t_0)]$$

subject to a constraint on noise power at the output of the filter

$$I_2 = \int_{-\infty}^{\infty} K_{uu}(f) \, |H(f)|^2 \, df$$

has a transfer function proportional to that of the matched filter for this case. The matched filter is often interpreted in this sense of maximizing "peak signal to r.m.s. noise" ratio rather than in the sense of pulse amplitude estimation.

Exercise 9.9. If the matched filter designed for a pulse $r(t)$ in white noise

$$H(f) = \alpha R^*(f) e^{-j2\pi f t_0}$$

is used in a different noise density situation, show that the minimum mean-squared error in estimating the amplitude of $r(t)$ is given by

$$I_W = \cfrac{\overline{\mathbf{a}^2}}{1 + \cfrac{\overline{\mathbf{a}^2}[\int |R(f)|^2\, df]^2}{\int K_{uu}(f)\,|R(f)|^2\, df}}$$

where the gain constant α is adjusted to its optimum value.

Matched Filter for Randomly Distorted Pulses

If the received pulse shape $r(t)$ is not accurately known, then the performance of the matched filter will be suboptimal. We might choose to consider $r(t)$ as a realization of a random process and then find the filter that minimizes mean-squared error in pulse amplitude estimation on the average. The problem can be formulated in terms of a known transmitted pulse shape $f(t)$ and the random channel dispersion function $G(f)$ used previously in connection with continuous waveform estimation. Thus the problem is the same as that shown in Figure 9.10 with $G(f)$ replaced by the random variable $\mathbf{G}(f)$.

$$\begin{aligned} T_0&: \boldsymbol{\omega}(t_0) = \mathbf{a} \\ T_1&: \mathbf{z}(t) = \mathbf{a}\int_{-\infty}^{\infty} f(t-\xi)\mathbf{g}(\xi)\, d\xi + \mathbf{u}(t) \end{aligned} \tag{9.81}$$

The required correlation functions are

$$\begin{aligned} k_{zz}(s, \sigma) &= \overline{\mathbf{a}^2}\int_{-\infty}^{\infty} f(s-\xi)f(\sigma-\eta)E[\mathbf{g}(\xi)\mathbf{g}(\eta)]\, d\xi\, d\eta + k_{uu}(s-\sigma) \\ k_{z\omega}(s, t_0) &= \overline{\mathbf{a}^2}\int_{-\infty}^{\infty} f(s-\xi)E[\mathbf{g}(\xi)]\, d\xi \end{aligned} \tag{9.82}$$

Taking Fourier transforms according to (9.68) and (9.69),

$$\begin{aligned} B_{zz}(f, \nu) &= \overline{\mathbf{a}^2}E[\mathbf{G}(f)\mathbf{G}^*(\nu)]F(f)F^*(\nu) + K_{uu}(f)\,\delta(f-\nu) \\ C_{z\omega}(f, t_0) &= \overline{\mathbf{a}^2}E[\mathbf{G}(f)]F(f) \end{aligned} \tag{9.83}$$

Hence condition (9.67) becomes

$$\overline{\mathbf{a}^2}\int_{-\infty}^{\infty} E[\mathbf{G}(f)\mathbf{G}^*(\nu)]F(f)F^*(\nu)H^*(\nu)e^{-j2\pi\nu t_0}\, d\nu + K_{uu}(f)H^*(f)e^{-j2\pi f t_0} = \overline{\mathbf{a}^2}F(f)E[\mathbf{G}(f)] \tag{9.84}$$

This equation can be put in a more convenient form by changing the unknown argument function to

$$W(f) \triangleq F^*(f)H^*(f)e^{-j2\pi f t_0} \tag{9.85}$$

Substituting (9.85) into (9.84) and dividing by $\overline{\mathbf{a}^2}F(f)$, we obtain the standard form for a non-homogeneous Fredholm integral equation.

$$\int_{-\infty}^{\infty} E[\mathbf{G}(f)\mathbf{G}^*(\nu)]W(\nu)\,d\nu + \frac{K_{uu}(f)}{\overline{\mathbf{a}^2}\,|F(f)|^2}\,W(f) = \overline{\mathbf{G}(f)} \tag{9.86}$$

The kernel in this integral equation is simply the *autocorrelation function* $E[\mathbf{G}(f)\mathbf{G}^*(\nu)]$ for the random channel. As might be expected, more detailed statistical information about the channel is needed in this case than in the case of continuous waveform estimation where only the channel mean and variance were required. Exact solutions to (9.86) are difficult to obtain, in general, but there are some specific problems of practical interest where the equation is greatly simplified.[3] The following example illustrates a typical application of this problem.

Example 9.7. *Matched filter for pulse with random arrival time:* There are often circumstances where the pulse shape is known except for its position on the time axis. To treat the problem of random pulse arrival time, we can let $\mathbf{G}(f)$ correspond to a delay operator where the delay $\boldsymbol{\tau}$ is a random variable. Suppose that $\boldsymbol{\tau}$ is assigned a probability density function $p_\tau(\xi)$, then we have

$$\mathbf{G}(f) = e^{-j2\pi\boldsymbol{\tau}f}$$

$$E[\mathbf{G}(f)] = \int_{-\infty}^{\infty} p_\tau(\xi)e^{-j2\pi\xi f}\,d\xi = P_\tau(f) \tag{9.87}$$

$$E[\mathbf{G}(f)\mathbf{G}^*(\nu)] = \int_{-\infty}^{\infty} p_\tau(\xi)e^{-j2\pi\xi(f-\nu)}\,d\xi = P_\tau(f-\nu) \tag{9.88}$$

where $P_\tau(f)$ is the Fourier transform of $p_\tau(\xi)$. Using these expressions in (9.86), we obtain

$$\int_{-\infty}^{\infty} P_\tau(f-\nu)W(\nu)\,d\nu + \frac{K_{uu}(f)}{\overline{\mathbf{a}^2}\,|F(f)|^2}\,W(f) = P_\tau(f) \tag{9.89}$$

An exact solution to (9.89) can be demonstrated by making the (not too unrealistic) assumption that $|F(f)|^2$ is proportional to $K_{uu}(f)$. It makes sense to speak of

$$\rho = \frac{\overline{\mathbf{a}^2}\,|F(f)|^2}{K_{uu}(f)} \tag{9.90}$$

as a signal-to-noise ratio parameter. Using this relation and taking the inverse Fourier transform of (9.89), we obtain a simple time-domain solution to the problem.

$$p_\tau(t)w(t) + \frac{1}{\rho}\,w(t) = p_\tau(t)$$

so that

$$w(t) = \frac{\rho p_\tau(t)}{1 + \rho p_\tau(t)} \tag{9.91}$$

The physical interpretation of this result is made clear by examining the pulse response of the filter, not the impulse response $h(t)$. Let $Y_0(f) = F(f)H(f)$, then from (9.85) we have

$$w(t) = y_0(t_0 - t) \tag{9.92}$$

where $y_0(t)$ is the noise-free response of the filter to a unit-amplitude, zero-delay pulse $f(t)$. These pulse responses are shown in Figure 9.13 for a particular probability density function for the delay variable. As signal-to-noise

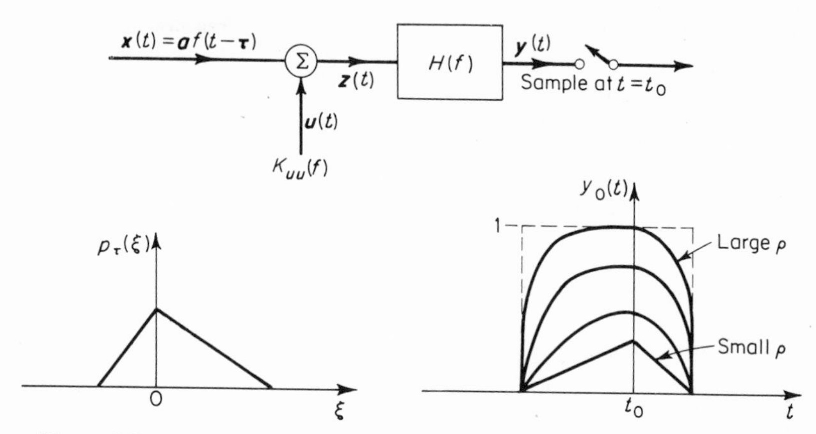

Figure 9.13. Matched filter for a pulse of random arrival time for various values of signal-to-noise ratio, ρ. $y_0(t)$ is the response of the filter to a unit-amplitude pulse with $\tau = 0$.

ratio becomes small, the optimal output pulse approaches a time-reversed version of the probability density function. In other words, the filter only "looks" at the time regions where the pulse is most likely to be. At high signal-to-noise ratio, the noise is less of a problem and the optimal output pulse approaches a flat-topped, unit-amplitude pulse giving a strong response no matter when the input pulse arrives. ▮

Exercise 9.10. For a pulse received over a random time-invariant channel $\mathbf{G}(f)$ with additive noise $\mathbf{u}(t)$ at the channel output, find the filter which maximizes peak pulse response subject to a constraint on output noise power; i.e., find $H(f)$ to maximize

$$I_1 = E[\mathbf{y}(t_0)]$$

subject to

$$I_2 = \int_{-\infty}^{\infty} K_{uu}(f)\,|H(f)|^2\,df$$

Interpret the result in terms of a matched filtering operation. Note that this is one of the few problems where channel variance does not have to be considered.

Adjacent Pulse Interference

A somewhat different problem arises when the received pulse is not isolated but instead occurs in a time sequence of similarly shaped pulses. In this case the optimum filter must not only combat additive noise but also the interference from the "tails" of pulses adjacent to the one whose amplitude is being estimated. The optimum filter will represent the best trade-off between large bandwidth to produce narrow output pulses for good resolution and narrow bandwidth to reduce the effects of noise interference. To make the problem specific, suppose that we want to estimate (at $t = t_0$) the amplitude of a pulse occurring at $t = t_0$. Suppose also that the remaining pulses in the sequence have the same shape but have statistically independent random amplitudes and occurrence times distributed according to a Poisson process with rate parameter λ on each side of t_0. The problem can be stated as

$$\begin{aligned} T_0&: \boldsymbol{\omega}(t_0) = \mathbf{a}_0 \\ T_1&: \mathbf{z}(t) = \mathbf{a}_0 f(t - t_0) + \mathbf{v}(t) + \mathbf{u}(t) \end{aligned} \tag{9.93}$$

where

$$\mathbf{v}(t) = \sum_{k=-\infty}^{\infty} \mathbf{a}_k f(t - \mathbf{t}_k); \qquad k \neq 0 \tag{9.94}$$

is statistically independent of $\mathbf{a}_0$ and $\mathbf{u}(t)$. The pulse interference process $\mathbf{v}(t)$ is the same process as described in Section 8.7. It was shown that the process is wide-sense stationary and expressions for the mean (8.96) and autocorrelation (8.99) are given.

Evaluating the correlation functions,

$$\begin{aligned} k_{zz}(s, \sigma) = \overline{\mathbf{a}_0^2} f(s - t_0) f(\sigma - t_0) + \bar{\mathbf{a}}_0 \bar{\mathbf{v}} f(s - t_0) + \bar{\mathbf{a}}_0 \bar{\mathbf{v}} f(\sigma - t_0) \\ + k_{vv}(s - \sigma) + k_{uu}(s - \sigma) \end{aligned} \tag{9.95}$$

$$k_{z\omega}(s, t_0) = \overline{\mathbf{a}_0^2} f(s - t_0) + \bar{\mathbf{a}}_0 \bar{\mathbf{v}} \tag{9.96}$$

Taking Fourier transforms, and using (8.99),

$$\begin{aligned} B_{zz}(f, \nu) = \overline{\mathbf{a}_0^2} F(f) F^*(\nu) e^{-j2\pi t_0 (f-\nu)} \\ + \bar{\mathbf{a}}_0 \bar{\mathbf{v}} F(f) e^{-j2\pi f t_0} \delta(\nu) + \bar{\mathbf{a}}_0 \bar{\mathbf{v}} F^*(\nu) e^{+j2\pi \nu t_0} \delta(f) \\ + [\lambda \overline{\mathbf{a}^2} |F(f)|^2 + (\bar{\mathbf{v}})^2 \delta(f) + K_{uu}(f)] \, \delta(f - \nu) \end{aligned} \tag{9.97}$$

$$C_{z\omega}(f, t_0) = \overline{\mathbf{a}_0^2} F(f) e^{-j2\pi f t_0} + \bar{\mathbf{a}}_0 \bar{\mathbf{v}} \, \delta(f) \tag{9.98}$$

The terms in (9.97) and (9.98) involving $\delta(f)$ and $\delta(\nu)$ can be dropped since they pertain to the mean value of pulse interference. In practice, this component of interference would be eliminated by d.c. blocking. Mathematically, this is equivalent to making $H(0) = 0$ without substantially affecting its values at other frequencies and this does not affect the performance in estimation of $\mathbf{a}_0$. Using (9.97) and (9.98) in condition (9.67), the optimum filter has a transfer function given by

$$H(f) = \frac{\alpha \overline{\mathbf{a}_0^2} F^*(f)}{\lambda \overline{\mathbf{a}^2}\, |F(f)|^2 + K_{uu}(f)}\,; \qquad H(0) = 0 \tag{9.99}$$

where the proportionality constant α can be evaluated by substituting (9.99) back into (9.67). The form of the optimum filter is intuitively satisfying since if the pulses tend to be isolated due to small λ (average rate of occurrence), or if the noise is relatively large, then the $K_{uu}(f)$ term in the denominator predominates and the optimum filter is approximately the matched filter for the isolated pulse. On the other hand, if the noise is relatively unimportant, due to large λ or large $\overline{\mathbf{a}^2}$, then the $K_{uu}(f)$ term can be neglected and the optimum filter approaches $F^{-1}(f)$, thus producing very narrow output pulses.

9.5 Periodic Waveform Estimation

PAM data transmission is a typical situation where we are neither interested in continuous waveform estimation nor in estimation at a single instant. What is desired is periodic estimation of the received signal in order to extract the information carried in the amplitudes of the synchronous pulse sequence. As shown in Figure 9.14, the optimum filter provides an output

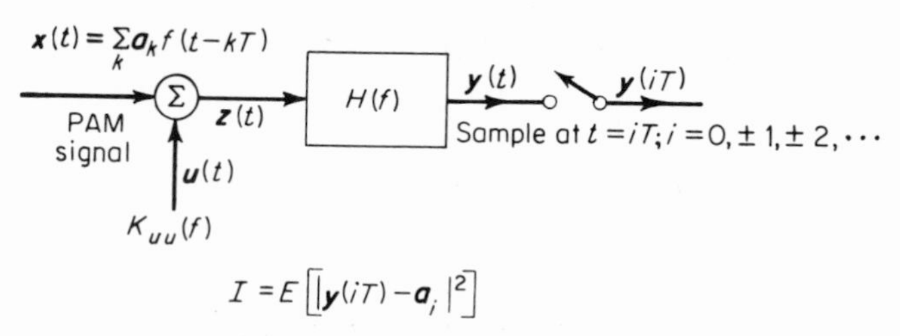

Figure 9.14. Periodic estimation of PAM data signal.

signal which, sampled synchronously every T seconds, gives minimum mean-squared error estimates of the data sequence $\{\mathbf{a}_k\}$. In Section 8.3, we showed that the PAM signal was a cyclostationary process provided that the data sequence was wide-sense stationary, (8.26). Since the received signal is statistically the same at each sampling instant, it follows that the optimum filter can be time-invariant.

For this problem,

$$T_0: \boldsymbol{\omega}(iT) = \mathbf{a}_i; \qquad i = 0, \pm 1, \pm 2, \ldots \tag{9.100}$$

$$T_1: \mathbf{z}(t) = \sum_{k=-\infty}^{\infty} \mathbf{a}_k f(t - kT) + \mathbf{u}(t) \tag{9.101}$$

Using (8.27), the correlation functions are given by

$$k_{zz}(s, \sigma) = \sum_m \alpha_m \sum_k f(s - kT) f(\sigma - kT - mT) + k_{uu}(s - \sigma) \tag{9.102}$$

$$k_{z\omega}(s, iT) = \sum_k \alpha_{i-k} f(s - kT) \tag{9.103}$$

where $\alpha_m = E[\mathbf{a}_k \mathbf{a}_{k+m}]$ characterizes the correlation in the data sequence. Now taking Fourier transforms according to (9.68) and (9.69),

$$B_{zz}(f, \nu) = \sum_m \alpha_m \sum_k F(f) e^{-j2\pi kTf} F^*(\nu) e^{j2\pi(k+m)T\nu} + K_{uu}(f)\,\delta(f - \nu) \tag{9.104}$$

$$C_{z\omega}(f, iT) = F(f) \sum_k \alpha_{i-k} e^{-j2\pi kTf} \tag{9.105}$$

As in Section 8.4, we represent the correlation in the data sequence in the frequency domain by a periodic function $M(f)$.

$$M(f) = \sum_m \alpha_m e^{j2\pi mTf} \tag{9.106}$$

Now (9.104) and (9.105) can be expressed as

$$B_{zz}(f, \nu) = F(f) F^*(\nu) M(\nu) \frac{1}{T} \sum_\ell \delta\left(f - \nu - \frac{\ell}{T}\right) + K_{uu}(f)\,\delta(f - \nu) \tag{9.107}$$

$$C_{z\omega}(f, iT) = F(f) M(f) e^{-j2\pi iTf} \tag{9.108}$$

where we have used the fact that

$$\sum_k e^{-j2\pi kTf} = \frac{1}{T} \sum_\ell \delta\left(f - \frac{\ell}{T}\right)$$

(see Exercise 4.4).

With (9.107) and (9.108) substituted into (9.67), the condition on the transfer function for the optimum filter becomes

$$\frac{1}{T} F(f) \sum_\ell M\left(f - \frac{\ell}{T}\right) F^*\left(f - \frac{\ell}{T}\right) H^*\left(f - \frac{\ell}{T}\right) + K_{uu}(f) H^*(f) = F(f) M(f) \tag{9.109}$$

Noting that $M(f)$ is periodic, and making the substitution $W(f) = F^*(f)H^*(f)$, (9.109) can be rewritten as

$$\frac{M(f)}{T}\sum_{\ell} W\left(f-\frac{\ell}{T}\right)+\frac{W(f)}{R(f)}=M(f) \tag{9.110}$$

where we have defined

$$R(f) \triangleq \frac{|F(f)|^2}{K_{uu}(f)} \tag{9.111}$$

Fortunately, this infinite-order difference equation (9.110) has a simple solution. This is because the right-hand side and the first term on the left-hand side are both periodic with period $1/T$. It follows that $W(f)/R(f)$ must also be periodic. Let $W(f) = R(f)Q(f)$, where $Q(f)$ is a periodic function. Then

$$\frac{1}{T}\sum_{\ell} W\left(f-\frac{\ell}{T}\right) = L(f)Q(f)$$

where we have defined

$$L(f) \triangleq \frac{1}{T}\sum_{\ell} R\left(f-\frac{\ell}{T}\right) \tag{9.112}$$

Substituting back into (9.110), we find that

$$W(f) = \frac{R(f)M(f)}{1+L(f)M(f)} \tag{9.113}$$

Finally, the filter transfer function is given by

$$H(f) = \frac{F^*(f)}{K_{uu}(f)}\left[\frac{M(f)}{1+L(f)M(f)}\right] \tag{9.114}$$

From this expression, we see that the optimum filter is the cascade combination of the matched filter for $f(t)$ in noise with power spectral density $K_{uu}(f)$ and a filter which has a periodic transfer function [$L(f)$ and $M(f)$ are periodic]. If the second factor in (9.114) is approximated by a finite Fourier series, then the filter (with an additional delay of nT seconds) can be realized as a tapped delay-line transversal filter as shown in Figure 9.15.[7]–[10] The tap gain factors are given by the Fourier coefficients of $D(f)$, where

$$D(f) \triangleq \frac{M(f)}{1+L(f)M(f)} = \sum_{k} d_k e^{j2\pi kTf}$$

$$\simeq \sum_{k=-n}^{n} d_k e^{j2\pi kT}; \qquad d_k = d_{-k} \tag{9.115}$$

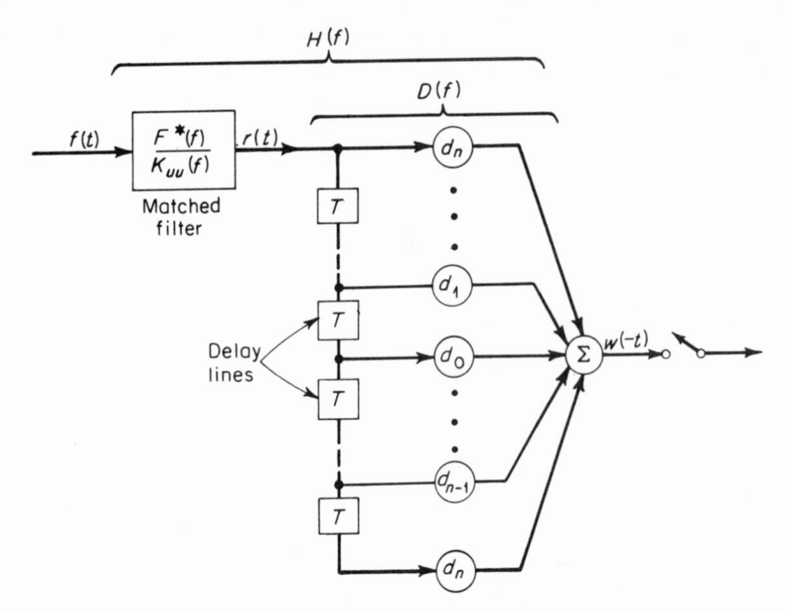

Figure 9.15. Optimum filter for synchronous PAM signal. The signals shown are the responses to a single, unit-amplitude data pulse.

The performance of this filter has a simple expression in terms of $D(f)$. When (9.114) is used in (9.8), we find

$$I_{\min} = \alpha_0 - \int_{-\infty}^{\infty} R(f)M(f)D(f)\,df \tag{9.116}$$

Some insight into the operation of the optimum filter can be gained by considering the special case where $r(t)$, which is the response of the matched filter portion to an individual data pulse, happens to satisfy the Nyquist criterion (6.123) for no intersymbol interference. From (9.112) we have $L(f) = r(0)$ in this case, Since there is no intersymbol interference at the output of the matched filter, we might expect that the tapped delay-line portion of the filter should be eliminated since it would introduce intersymbol interference. This is true when the pulse amplitudes are uncorrelated, for then $M(f) = \alpha_0$ and $D(f)$ is a constant. If there is correlation in the data sequence, however, the optimum filter takes advantage of this by reducing the noise interference at the expense of a small amount of intersymbol interference. This effect is appreciable only when the noise is relatively large. When the signal is large, we approach the condition

$$D(f) = \frac{M(f)}{1 + r(0)M(f)} \simeq \frac{1}{r(0)} \tag{9.117}$$

in which the tapped delay-line filter would be eliminated in this special case where $r(t)$ produces no intersymbol interference.

Exercise 9.11. In periodic estimation of a PAM signal, suppose that the signal has a timing jitter as described in Section 8.5. In this case,

$$\mathbf{x}(t) = \sum_k \mathbf{a}_k f(t - \mathbf{t}_k)$$

where the random pulse arrival times are given by

$$\mathbf{t}_k = kT + \boldsymbol{\delta}_k; \qquad k = 0, \pm 1, \pm 2, \ldots$$

Assume that the $\boldsymbol{\delta}_k$ are statistically independent random variables with probability density function $p(\xi)$. Derive the condition for the optimum filter $H(f)$ in a form similar to that of (9.110).

9.6 The Constraint of Physical Realizability

When it comes to implementation of the various optimum filters discussed in the previous sections, the designer is faced with numerous limitations on the realizability of the filter functions which he may not have incorporated as constraints in the optimization problem. These limitations cause the performance of the system to fall short of the predicted performance. Many limitations, such as restricted number of allowable network elements and the presence of unavoidable parasitic elements, are not really fundamental to the filtering problem and can be alleviated by increased expenditure of material and design effort. In contrast, the feature of non-anticipatory response is a fundamental restriction on all physically realizable filters. In this section, we shall consider physical realizability in this latter sense, which can be stated as a restriction on the impulse response $h(t, s)$ of the filter. It is necessary and sufficient that $h(t, s)$ vanish for all $t < s$. As in the case of duration-limiting constraints encountered in the signal optimization problems in Chapter 6, it is usually easier to incorporate the constraint of physical realizability by the direct method rather than by the method of Lagrange multipliers. Accordingly, we introduce the auxiliary condition

$$h(t, s) = w(t - s)h(t, s) \tag{9.118}$$

where

$$\begin{aligned} w(t) &= 1 \qquad \text{for } t \geqslant 0 \\ &= 0 \qquad \text{for } t < 0 \end{aligned}$$

In the minimum mean-squared error filtering problem formulated in Section 9.2, the functional I, (9.3), is modified to

$$I = \iint_{-\infty}^{\infty} w(t - s)w(t - \sigma)h(t, s)h(t, \sigma)k_{zz}(s, \sigma)\, ds\, d\sigma - 2\int_{-\infty}^{\infty} w(t - s)h(t, s)k_{z\omega}(s, t)\, ds + k_{\omega\omega}(t, t) \tag{9.119}$$

Setting the gradient of (9.119) with respect to $h(t, s)$ equal to zero gives the necessary condition for a minimum.

$$\int_{-\infty}^{\infty} w(t-s)w(t-\sigma)h(t,\sigma)k_{zz}(s,\sigma)\,d\sigma = w(t-s)k_{z\omega}(s,t) \tag{9.120}$$

Noting the form of $w(t)$ from (9.118), the equation (9.120) can be written as

$$\int_{-\infty}^{t} h(t,\sigma)k_{zz}(s,\sigma)\,d\sigma = k_{z\omega}(s,t) \qquad \text{for } t \geqslant s \tag{9.121}$$

This equation has been solved only for very specialized circumstances. On the other hand, when $\mathbf{z}$ and $\boldsymbol{\omega}$ are jointly stationary and $k_{zz}(s, \sigma)$ and $k_{z\omega}(s, t)$ are replaced by $k_{zz}(s - \sigma)$ and $k_{z\omega}(s - t)$, respectively, then (9.121) reduces to the *Wiener-Hopf* equation which has received extensive general treatment. In this case,

$$\int_{-\infty}^{t} h(t-\sigma)k_{zz}(s-\sigma)\,d\sigma = k_{z\omega}(s-t) \qquad \text{for } t \geqslant s$$

or, by a change of variables,

$$\int_{0}^{\infty} h(\tau)k_{zz}(t-\tau)\,d\tau = k_{\omega z}(t) \qquad \text{for } t \geqslant 0 \tag{9.122}$$

Methods for solving the Wiener-Hopf equation, especially in the case that $K_{zz}(f)$ and $K_{\omega z}(f)$ are rational functions of f, are presented in many texts on communication and control systems.[11]–[15] We shall consider an approach presented by Bode and Shannon [16],[17] which, besides not being restricted to rational functions, has the advantage of providing a physical insight to the optimum filtering operations. We start by noticing that if the received signal $z(t)$ corresponded to unit-variance white noise $k_{zz}(s - \sigma) = \delta(s - \sigma)$, then (9.122) has an especially simple solution.

$$\begin{aligned} h(t) &= k_{\omega z}(t) \qquad \text{for } t \geqslant 0 \\ &= w(t)k_{\omega z}(t) \end{aligned} \tag{9.123}$$

Hence the optimum realizable filter, in this special case, has simply a truncated (at $t = 0$) version of the impulse response of the optimum filter without the constraint of physical realizability. The simplicity of the Bode-Shannon approach results from the fact that, in most cases of interest, the signal $\mathbf{z}$ can be passed through a physically realizable time-invariant filter such that the output autocorrelation is equivalent to that of unit-variance white noise. It now remains to place in cascade with this *pre-whitening* filter, a physically realizable filter corresponding to the special case above, (9.123), operating on the signal $\tilde{\mathbf{z}}$ as shown in Figure 9.16.

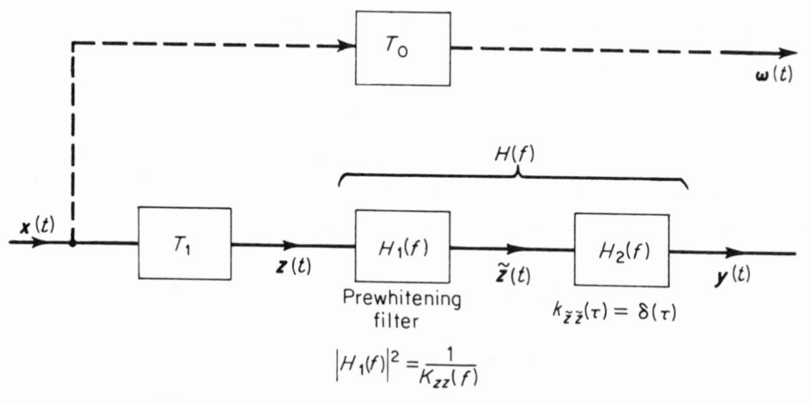

Figure 9.16. Physically realizable optimum filtering.

The optimum realizable filter operating on $\tilde{\mathbf{z}}$ has an impulse response, from (9.123), given by

$$\begin{aligned} h_2(t) &= w(t)k_{\omega\tilde{z}}(t) \\ &= w(t)\int_0^\infty h_1(\tau)k_{\omega z}(t+\tau)\,d\tau \end{aligned} \tag{9.124}$$

For an arbitrary function $x(t)$, it is conventional to let $[X(f)]^+$ denote the Fourier transform of $w(t)x(t)$, a truncated version of $x(t)$. Then we have

$$H_2(f) = [K_{\omega\tilde{z}}(f)]^+ = [H_1^*(f)K_{\omega z}(f)]^+ \tag{9.125}$$

The transfer function of the pre-whitening part of the filter is usually found by a process called *spectrum factorization.*[11] If $K_{zz}(f)$ can be factored into the form

$$K_{zz}(f) = K(f)K^*(f) \tag{9.126}$$

where $K(f)$ has an associated phase such that $K^{-1}(f)$ is physically realizable, then the overall optimum filter has the form

$$H(f) = \frac{1}{K(f)}\left[\frac{K_{\omega z}(f)}{K^*(f)}\right]^+ \tag{9.127}$$

This factorization is always possible if $K_{zz}(f)$ is rational. More generally, if $k_{zz}(\tau)$ can be split into a sum of singularity functions at the origin (δ-functions and their even-ordered derivatives) and a remainder $k'_{zz}(\tau)$ which is square-integrable and satisfies the *Paley-Wiener condition,*[18]

$$\int_{-\infty}^{\infty} \frac{|\log K'_{zz}(f)|}{1+f^2}\,df < \infty \tag{9.128}$$

then the factorization indicated is still possible.

The minimum mean-squared error using a realizable filter can be evaluated by expressing $h_2(t)$ as

$$h_2(t) = k_{\omega\tilde{z}}(t) - g(t) \tag{9.129}$$

where $g(t) = [1 - w(t)]k_{\omega\tilde{z}}(t)$ corresponds to the part of the optimum filter response eliminated by the constraint of physical realizability. Expressing mean-squared error (9.3) in terms of frequency-domain functions,

$$I = \int_{-\infty}^{\infty} |H_1(f)|^2 \, |H_2(f)|^2 \, K_{zz}(f)\, df - 2\int_{-\infty}^{\infty} H_1(f)H_2(f)K_{\omega z}^*(f)\, df + \int_{-\infty}^{\infty} K_{\omega\omega}(f)\, df \tag{9.130}$$

and using

$$|H_1(f)|^2 K_{zz}(f) = 1$$

$$H_2(f) = K_{\omega\tilde{z}}(f) - G(f) = H_1^*(f)K_{\omega z}(f) - G(f)$$

the minimum mean-squared error becomes

$$I_{\text{min}} = \int_{-\infty}^{\infty} \frac{K_{zz}(f)K_{\omega\omega}(f) - |K_{\omega z}(f)|^2}{K_{zz}(f)}\, df + \int_{-\infty}^{\infty} |G(f)|^2\, df \tag{9.131}$$

Comparing this result with (9.12) for the unconstrained filter, we see that physical realizability has introduced an additional amount of mean-squared error given by

$$\int_{-\infty}^{\infty} |G(f)|^2\, df$$

Thus the degradation in performance is easily evaluated in terms of the amount of the optimum $h_2(t)$ that is removed by truncation.

Example 9.8. *Random facsimile signal in white noise:* As an illustrative example, we consider the problem of continuous estimation of the signal, with lag T; i.e., $\boldsymbol{\omega}(t) = \mathbf{x}(t - T)$ and $K_{\omega z}(f) = K_{xx}(f)e^{-j2\pi fT}$. Suppose that the signal is a unit-variance random facsimile process (with the mean subtracted out) which is corrupted with additive, zero-mean white noise with spectral density of N_0 watts/Hz.

$$\begin{aligned} K_{xx}(f) &= \frac{1}{\pi f_0}\frac{1}{1 + (f/f_0)^2} \\ K_{uu}(f) &= N_0 \end{aligned} \tag{9.132}$$

then

$$\begin{aligned} K_{zz}(f) &= K_{xx}(f) + K_{uu}(f) \\ &= N_0 \frac{(f/f_0)^2 + \alpha^2}{(f/f_0)^2 + 1} \end{aligned} \tag{9.133}$$

where

$$\alpha^2 \triangleq \left(1 + \frac{1}{\pi f_0 N_0}\right)$$

The realizable pre-whitening filter is given by

$$H_1(f) = \frac{1}{N_0^{\frac{1}{2}}}\left[\frac{j(f/f_0) + 1}{j(f/f_0) + \alpha}\right] \tag{9.134}$$

whereas $H_2(f)$, from (9.125), is given by

$$\begin{aligned} H_2(f) &= [H_1^*(f)K_{xx}(f)e^{-j2\pi fT}]^+ \\ &= \frac{1}{\pi f_0 N_0^{\frac{1}{2}}}\left\{\left[\frac{-j(f/f_0) + 1}{-j(f/f_0) + \alpha}\right]\left[\frac{e^{-j2\pi fT}}{1 + (f/f_0)^2}\right]\right\}^+ \\ &= \frac{1}{\pi f_0 N_0^{\frac{1}{2}}}\left\{\frac{e^{-j2\pi fT}}{[-j(f/f_0) + \alpha][j(f/f_0) + 1]}\right\}^+ \end{aligned} \tag{9.135}$$

The transfer function $H_2(f)$ can be determined by making a partial fraction expansion of (9.135).

$$H_2(f) = \frac{1}{\pi f_0 N_0^{\frac{1}{2}}(1 + \alpha)}\left[\frac{e^{-j2\pi fT}}{-j(f/f_0) + \alpha} + \frac{e^{-j2\pi fT}}{j(f/f_0) + 1}\right]^+ \tag{9.136}$$

Since $[1 + j(f/f_0)]^{-1}$ is the Fourier transform of $2\pi f_0 w(t)e^{-2\pi f_0 t}$ and $[\alpha - j(f/f_0)]^{-1}$ is the Fourier transform of $2\pi f_0 w(-t)e^{2\pi\alpha f_0 t}$, the impulse response $h_2(t)$ is given by

$$\begin{aligned} h_2(t) &= 0 \qquad \text{for } t < 0 \\ &= \frac{2}{N_0^{\frac{1}{2}}(1 + \alpha)}\, e^{2\pi\alpha f_0(t-T)} \qquad \text{for } 0 \leqslant t < T \\ &= \frac{2}{N_0^{\frac{1}{2}}(1 + \alpha)}\, e^{-2\pi f_0(t-T)} \qquad \text{for } t \geqslant T \end{aligned} \tag{9.137}$$

shown graphically on Figure 9.17.

For the zero-lag filter, only the second term in (9.136) is used and the overall filter is

$$H(f) = H_1(f)H_2(f) = \frac{1}{\pi f_0 N_0(1 + \alpha)}\left[\frac{1}{j(f/f_0) + \alpha}\right] \tag{9.138}$$

which is simply a single-section *RC* lowpass filter with -3 db point at $f = \alpha f_0$. With non-zero lag, however, the optimum filter is not nearly so easy to realize and an approximation involving many network elements may be required.

To evaluate the mean-squared error with the optimum realizable filter, we evaluate the error for the unconstrained filter, sometimes called the

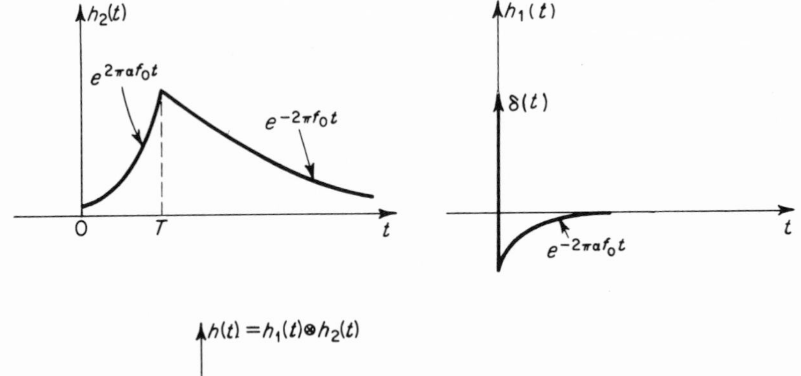

Figure 9.17. Impulse response of optimum realizable filter.

irremovable error,[2] and add on the square of the $L^2(-\infty, \infty)$ norm of the portion of $h_2(t)$ removed by truncation as indicated in (9.131). From (9.137) we see that

$$g(t) = \frac{2}{N_0^{\frac{1}{2}}(1 + \alpha)} e^{2\pi\alpha f_0(t-T)} \qquad \text{for } t < 0$$

Hence

$$\int_{-\infty}^{\infty} |G(f)|^2\, df = \int_{-\infty}^{0} g^2(t)\, dt$$

$$= \frac{4}{N_0(1 + \alpha)^2} \int_{-\infty}^{0} e^{4\pi\alpha f_0(t-T)}\, dt$$

$$= \frac{e^{-4\pi\alpha f_0 T}}{\pi f_0 N_0 \alpha (1 + \alpha)^2} = \frac{1}{\alpha}\left(\frac{\alpha - 1}{\alpha + 1}\right) e^{-4\pi\alpha f_0 T} \tag{9.139}$$

Finally,

$$I_{\min} = \frac{1}{\pi f_0} \int_{-\infty}^{\infty} \frac{df}{(f/f_0)^2 + \alpha^2} + \frac{1}{\alpha}\left(\frac{\alpha - 1}{\alpha + 1}\right) e^{-4\pi\alpha f_0 T}$$

$$= \frac{1}{\alpha}\left[1 + \left(\frac{\alpha - 1}{\alpha + 1}\right) e^{-4\pi\alpha f_0 T}\right] \tag{9.140}$$

where

$$\alpha \triangleq \left(1 + \frac{1}{\pi f_0 N_0}\right)^{\frac{1}{2}}$$

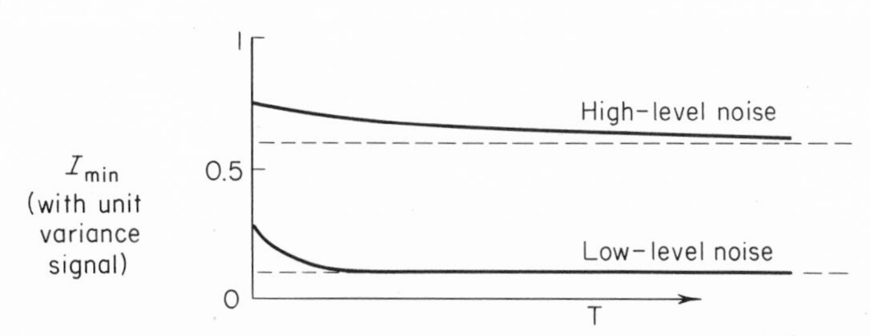

Figure 9.18. Performance of realizable filters with various lags.

It is clear that the infinite-lag realizable filter matches the performance of the unconstrained optimum filter. For low-level additive noise (large α), the zero-lag filter does about 3 db poorer than the infinite-lag filter; however, only relatively small lags are required to approach optimum performance. With a high noise level (α close to unity) the difference between zero-lag and infinite-lag performance is small and relatively long lags are required to approach optimum performance. This behavior is indicated on Figure 9.18. ∎

The feature of near-optimum performance with large lags is common to most filtering problems since then the truncation of $h_2(t)$ is generally a minor modification in filter response. This justifies the study of the simpler filtering problems without the constraint of physical realizability since the inclusion of a moderate time delay often has no bearing on system performance. For example, in many communication channels, the time lag needed for close approximation to the optimum filter may be quite small compared with the transit time of signals over the channel. A commonly used approach to the approximation of optimum filters in this case is to independently design a physically realizable filter whose magnitude approximates that of the optimum characteristic. Then this filter is cascaded with a series of all-pass filters with a phase shift designed to approximate the difference between the desired phase and the phase of the magnitude shaping filter to within some linear phase (constant delay) component. In systems involving some form of signal feedback, however, it is often critically important to limit the time delay introduced by filtering operations. In these situations it is essential that the constraint of physical realizability be originally incorporated in the optimization problem.

Physically Realizable Filter with Gain-bandwidth Constraint

The pre-whitening approach is also useful in problems where additional realizability constraints are imposed. Consider, for example, the gain-bandwidth product constraint discussed in Example 9.1. Assuming $\mathbf{z}$ and $\boldsymbol{\omega}$

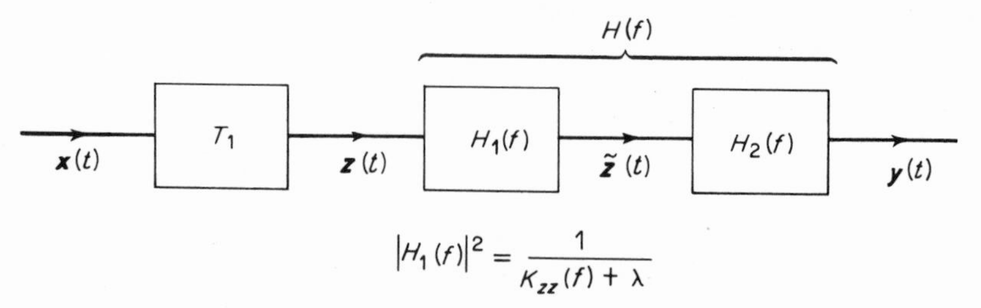

Figure 9.19. Optimum filtering with physical realizability and gain-bandwidth constraints.

jointly stationary, we minimize I, (9.3), subject to the realizability constraints

$$h(t) = w(t)h(t) \tag{9.141a}$$

$$I_1 = \int_{-\infty}^{\infty} |H(f)|^2 \, df = \int_{-\infty}^{\infty} h^2(t) \, dt = \text{constant} \tag{9.141b}$$

Incorporating (9.141a) directly and (9.141b) by the method of Lagrange multipliers, for a stationary point of $I + \lambda I_1$, we require

$$\int_0^{\infty} h(\tau)k_{zz}(t - \tau) \, d\tau + \lambda h(t) = k_{\omega z}(t) \qquad \text{for } t \geqslant 0 \tag{9.142}$$

This equation is solved by employing a *pseudo-pre-whitener* $H_1(f)$, as shown in Figure 9.19, where

$$|H_1(f)|^2 = \frac{1}{K_{zz}(f) + \lambda} \triangleq L(f) \tag{9.143}$$

Now the problem is to find $h_2(t)$ which minimizes I subject to

$$h_2(t) = w(t)h_2(t) \tag{9.144a}$$

$$I_1 = \int_{-\infty}^{\infty} L(f) \, |H_2(f)|^2 \, df$$

$$= \iint_{\infty}^{-\infty} \ell(t - \tau)h_2(\tau)h_2(t) \, dt \, d\tau = \text{constant} \tag{9.144b}$$

The condition for a stationary point of $I + \lambda I_1$ with respect to $h_2(t)$ becomes

$$\int_0^{\infty} h_2(\tau)k_{\tilde{z}\tilde{z}}(t - \tau) \, d\tau + \lambda \int_0^{\infty} \ell(t - \tau)h_2(\tau) \, d\tau = k_{\omega\tilde{z}}(t) \qquad \text{for } t \geqslant 0 \tag{9.145}$$

Since

$$K_{\tilde{z}\tilde{z}}(f) + \lambda L(f) = |H_1(f)|^2 K_{zz}(f) + \lambda \, |H_1(f)|^2 = 1$$

then $k_{\tilde{z}\tilde{z}}(t-\tau) + \lambda\ell(t-\tau) = \delta(t-\tau)$ and the solution to (9.145) is simply

$$h_2(t) = w(t)k_{\omega\tilde{z}}(t) \tag{9.146}$$

and the overall optimum filter has a transfer function given by

$$H(f) = H_1(f)H_2(f) = \frac{1}{K_\lambda(f)}\left[\frac{K_{\omega z}(f)}{K_\lambda^*(f)}\right]^+ \tag{9.147}$$

where $|K_\lambda(f)|^2 = K_{zz}(f) + \lambda$ indicates the spectrum factorization process described by (9.126). By choosing λ sufficiently large, the gain-bandwidth constraint, (9.141b), can be met for an arbitrarily prescribed value of I_1. To show this, we note that

$$\begin{aligned}
I_1 &= \int_{-\infty}^{\infty} |H(f)|^2\,df = \int_{-\infty}^{\infty} \left|\frac{1}{K_\lambda(f)}\right|^2 \left|\left[\frac{K_{\omega z}(f)}{K_\lambda^*(f)}\right]^+\right|^2 df \\
&\leqslant \frac{1}{\lambda}\int_{-\infty}^{\infty} \left|\left[\frac{K_{\omega z}(f)}{K_\lambda^*(f)}\right]^+\right|^2 df \leqslant \frac{1}{\lambda}\int_{-\infty}^{\infty} \frac{|K_{\omega z}(f)|^2}{|K_\lambda(f)|^2}\,df \\
&\leqslant \frac{1}{\lambda}\int_{-\infty}^{\infty} \frac{|K_{\omega z}(f)|^2}{K_{zz}(f)}\,df \leqslant \frac{1}{\lambda}\int_{-\infty}^{\infty} K_{\omega\omega}(f)\,df \\
&= \frac{1}{\lambda}k_{\omega\omega}(0) = \frac{1}{\lambda}E[\boldsymbol{\omega}^2]
\end{aligned} \tag{9.148}$$

Hence if $E[\boldsymbol{\omega}^2]$ is finite, then I_1 is bounded by a number which can be made arbitrarily small by choosing λ large enough.

Physically Realizable Matched Filter

To further illustrate the versatility of the pre-whitening approach, we reconsider the matched filter problem under the constraint of physical realizability. In this situation, the received signal is not a stationary random process since the signal has a deterministic shape $r(t)$ with a random scaling factor $\mathbf{a}$ to be estimated at the instant $t = t_0$ by the output of a time-invariant filter. Incorporating the constraint of physical realizability, the necessary condition on $h(t)$ is modified to

$$\int_{-\infty}^{t_0} h(t_0-\sigma)[\overline{\mathbf{a}^2}r(s)r(\sigma) + k_{uu}(s-\sigma)]\,d\sigma = \overline{\mathbf{a}^2}r(s) \qquad \text{for } t_0 \geqslant s \tag{9.149}$$

First consider the case where the additive noise $\mathbf{u}$ is white, $k_{uu}(s-\sigma) = N_0\,\delta(s-\sigma)$, then the solution to (9.149) is

$$N_0h(t_0-s) = \overline{\mathbf{a}^2}r(s)\left[1 - \int_{-\infty}^{t_0} h(t_0-\sigma)r(\sigma)\,d\sigma\right] \qquad \text{for } t_0 \geqslant s$$

or, by a change of variables,

$$h(t) = \alpha w(t)r(t_0-t) \tag{9.150}$$

Comparing this with (9.73), for the white noise case, we see that $h(t)$ is simply a truncated version of the impulse response of the optimum unconstrained filter. Furthermore, if t_0 is large enough so that $r(t) = 0$ for $t > t_0$, then the truncation has no effect. In other words, if we allow the filter enough time to "see" all the signal pulse, then the unconstrained optimum filter is physically realizable.

Now for the case where the additive noise has arbitrary spectral density, we employ a physically realizable, noise-whitening filter $H_1(f)$, as shown in

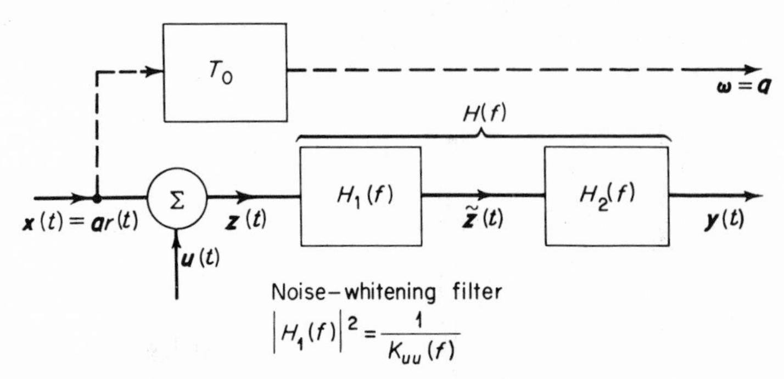

Figure 9.20. Physically realizable matched filter.

Fig. 9.20, where

$$|H_1(f)|^2 = \frac{1}{K_{uu}(f)} \tag{9.151}$$

We now proceed to find the physically realizable $h_2(t)$ which minimizes $I = E[\{\mathbf{y}(t_0) - \mathbf{a}\}^2]$.

$$\begin{aligned} \tilde{\mathbf{z}}(t) &= \mathbf{a}\int_0^\infty h_1(\tau)r(t-\tau)\,d\tau + \tilde{\mathbf{u}}(t) \\ &= \mathbf{a}q(t) + \tilde{\mathbf{u}}(t) \end{aligned} \tag{9.152}$$

Hence

$$\begin{aligned} k_{\tilde{z}\tilde{z}}(s, \sigma) &= E[\mathbf{a}^2 q(s)q(\sigma)] + k_{\tilde{u}\tilde{u}}(s-\sigma) \\ &= \overline{\mathbf{a}^2}q(s)q(\sigma) + \delta(s-\sigma) \end{aligned} \tag{9.153}$$

and

$$\begin{aligned} k_{\tilde{z}w}(s, t_0) &= E[\{\mathbf{a}q(s) + \tilde{u}(s)\}\mathbf{a}] \\ &= \overline{\mathbf{a}^2}q(s) \end{aligned}$$

Using (9.149) for the equivalent condition on $h_2(t)$, we find

$$\int_{-\infty}^{t_0} h_2(t_0 - \sigma)[\overline{\mathbf{a}^2}q(s)q(\sigma) + \delta(s-\sigma)]\,d\sigma = \overline{\mathbf{a}^2}q(s) \qquad \text{for } t_0 \geqslant s$$

Hence

$$h(t_0 - s) = \overline{\mathbf{a}^2} q(s)\left[1 - \int_{-\infty}^{t_0} h_2(t_0 - \sigma)q(\sigma)\,d\sigma\right] \qquad \text{for } t_0 \geqslant s$$

or

$$h_2(t) = \alpha w(t) q(t_0 - t) \tag{9.154}$$

Using a spectrum factorization $K_{uu}(f) = K(f)K^*(f)$ such that $K^{-1}(f)$ is physically realizable, the overall transfer function of the realizable matched filter is

$$H(f) = H_1(f)H_2(f) = \frac{1}{K(f)}\left[\frac{R^*(f)e^{-j2\pi f t_0}}{K^*(f)}\right]^+ \tag{9.155}$$

Finite Memory Constraint

As a final comment on realizability conditions, we mention the constraint of finite memory time for filtering operations. Finite memory, in conjunction with physical realizability, implies that $h(t, s) = 0$ for $t < s$ and $t > s + T$, where T is called the *memory time* of the filter. The finite memory constraint is significant in situations where the linear estimation, rather than being implemented with a time-invariant filter, is performed by a multiplier-integrator device wherein the product of the incoming signal and a stored reference signal is integrated to form the estimate of the signal parameter. Practical limitations on the duration of the reference signal are equivalent to limitations on the memory time of the filter implementation. Similarly, in situations where the filtering operations are to be implemented with a digital computer, limitations in storage capability of the computer are equivalent to limitations in memory time. Incorporation of the finite memory constraint can be treated in the same manner as the constraint of physical realizability. In this case we prescribe the auxiliary condition

$$h(t, s) = w_1(t - s)h(t, s) \tag{9.156}$$

where

$$w_1(t) = 1 \qquad \text{for } 0 \leqslant t \leqslant T$$
$$= 0 \qquad \text{otherwise}$$

The mean-squared error under this constraint is given by (9.119) with w replaced by w_1 and the optimum filter must satisfy the condition

$$\int_{t-T}^{t} h(t, \sigma)k_{zz}(s, \sigma)\,d\sigma = k_{z\omega}(s, t) \qquad \text{for } s \leqslant t \leqslant s + T \tag{9.157}$$

The various techniques for solving the Wiener-Hopf equation are applicable to (9.157) with minor modifications. Several authors[12],[13],[17],[19] describe methods for solving equations of this type, particularly in the wide-sense stationary case where $K_{zz}(f)$ and $K_{\omega z}(f)$ are rational.

REFERENCES

1. A. Papoulis, *Probability, Random Variables and Stochastic Processes*, McGraw-Hill, 1965.
2. J. P. Costas, "Coding with Linear Systems," *Proc. IRE*, Vol. 40, pp. 1101–103 (September, 1952).
3. R. E. Maurer, "The Optimal Equalization of Random Channels," Communication Theory Group Report No. 9, Northeastern University, June, 1968.
4. J. H. Van Vleck and D. Middleton, "A Theoretical Comparison of the Visual, Aural, and Meter Reception of Pulsed Signals in the Presence of Noise, *Jour. Appl. Phys.*, Vol. 17, pp. 940–71 (November, 1946).
5. B. M. Dwork, "Detection of a Pulse Superimposed on Fluctuation Noise," *Proc. IRE*, Vol. 38 pp. 771–74 (July, 1950).
6. G. L. Turin, "An Introduction to Matched Filters," *IRE Trans. on Information Theory*, Vol. IT-6 pp. 311–29 (June, 1960).
7. D. W. Tufts, "Nyquist's Problem—The Joint Optimization of Transmitter and Receiver in Pulse Amplitude Modulation," *Proc. IEEE*, Vol. 53, pp. 248–59 (March, 1965).
8. M. R. Aaron and D. W. Tufts, "Intersymbol Interference and Error Probability," *IEEE Trans. on Information Theory*, Vol. IT-12, pp. 26–34 (January, 1966).
9. R. W. Lucky, "Automatic Equalization for Digital Communication," *Bell Sys. Tech. Jour.*, Vol. 44, pp. 547–88 (April, 1965).
10. R. W. Lucky, "Techniques for Adaptive Equalization of Digital Communication Systems," *Bell Sys. Tech. Jour.*, Vol. 45, pp. 255–86 (February, 1966).
11. Y. W. Lee, *Statistical Theory of Communication*, John Wiley & Sons, 1960.
12. J. H. Laning, Jr. and R. H. Battin, *Random Processes in Automatic Control*, McGraw-Hill, 1956.
13. W. B. Davenport, Jr. and W. L. Root, *An Introduction to the Theory of Random Signals and Noise*, McGraw-Hill, 1958.
14. D. Middleton, *An Introduction to Statistical Communication Theory*, McGraw-Hill, 1960.
15. H. L. Van Trees, *Detection, Estimation, and Modulation Theory*, Part I, John Wiley & Sons, 1968.
16. H. W. Bode and C. E. Shannon, "A Simplified Derivation of Linear Least Square Smoothing and Prediction Theory," *Proc. IRE*, Vol. 38, pp. 417–25 (April, 1950).
17. S. Darlington, "Linear Least-Squares Smoothing and Prediction, with Applications," *Bell Sys. Tech. Jour.*, Vol. 37, pp. 1221–94, September, 1958.
18. A. Papoulis, *The Fourier Integral and its Applications*, McGraw-Hill, 1962.
19. L. A. Zadeh and J. R. Ragazzini, "Optimum Filters for the Detection of Signals in Noise," *Proc. IRE*, Vol. 40, pp. 1223–31 (October, 1952).

SIGNAL DETECTION

10

10.1 Introduction

In this chapter we shall consider another type of signal processing operation where signal space concepts play an especially important role. Signal detection theory is also concerned with problems of estimation of signal parameters, but from a different viewpoint than described in Chapter 9. The optimum filtering problems in Chapter 9 were formulated with reference to a minimum mean-squared error criterion. In many applications, especially where the signal parameters form a finite set, a *probability of error* type of criterion may have a much greater practical significance. Consider, for example, an information transmission system which utilizes an m-ary alphabet.[1],[2] In this system, the transmitter produces only one of m possible signals from the set $\{s_i(t); i = 1, 2, \ldots, m\}$. Suppose that the signal is transmitted over a channel which introduces a random disturbance in the form of an additive noise process $\mathbf{u}$. The receiver must decide, on the basis of observations on *a particular realization* of the channel output $y(t) = s_i(t) + u(t)$, which of the m signals is most likely present. We can put this, and similar detection problems, into a parameter estimation context by expressing the signal as a function of two real variables $s(t, \theta)$. Then, in this example, $s_i(t) = s(t, \theta_i)$; $i = 1, 2, \ldots, m$ ($\theta_i \neq \theta_j$ for $i \neq j$). In response to the particular signal $y(t) = s_i(t) + u(t)$, the receiver produces an estimate $\hat{\theta}$ for the actual value θ of the parameter. If the noise statistics are known, the receiver can compute the m *a posteriori* probabilities [probabilities conditional on the received signal $y(t)$], $\Pr[\theta = \theta_i \mid y(t)]$; $i = 1, 2, \ldots, m$. The decision

$[\hat{\theta} = \theta_k]$ can be based on the relationship

$$\Pr[\theta = \theta_k \mid y(t)] \geqslant \Pr[\theta = \theta_i \mid y(t)]; \qquad i = 1, 2, \ldots, m \quad (10.1)$$

This receiver, besides choosing the most likely value of the parameter, is also able to evaluate the relative certainty of its decision; i.e., it can compute the probability of mistaking the true value of the parameter for any of the other $m - 1$ values of the parameter. These sharper results would be obscured by considering only the mean-squared error $E[(\boldsymbol{\theta} - \hat{\boldsymbol{\theta}})^2]$. It may very well turn out, however, that this receiver, (10.1), is also the one that minimizes mean-squared error.[2] In the optimum filtering problems considered previously, we needed only certain statistical averages, mean and correlation, for the signal and noise processes in order to arrive at a minimum mean-squared error receiver. The price we pay for designing a minimum error probability receiver is that more complete statistical information about the signal and noise processes is required. The receiver design will be based on probability density function representations for the processes.

In some situations, it is desirable to temper the decision strategy of the receiver according to known *a priori* probabilities $\Pr[\theta_i]$ for the transmitted signal or according to the assignment of unequal costs to different kinds of errors. In these situations, the receiver may not choose the most likely value of the signal parameter, but the decision strategy will still be based on the evaluation of the *a posteriori* probabilities $\Pr[\theta_i \mid y(t)]$. Standard methods for formulating the receiver strategy, taking these considerations into account, will be presented in the following two sections. In order to simplify the discussion, we shall consider only the binary case, where the signal parameter takes on one of two possible values. Extension of the methods to the multivalued case is straightforward. Following this, we shall apply these methods to some classical signal detection problems, obtaining easily interpretable results by assuming that signals are corrupted by an additive Gaussian noise process. Fortunately, these more restrictive results have a considerable practical significance because, in many physical systems, interference tends to be additive and strongly Gaussian in nature. Sections 10.2 and 10.3 give a brief summary of some statistical decision theory concepts in order to provide a background for the applications to follow. For a much more detailed treatment, from the standpoint of signal detection, the reader may consult References [2] and [5].

10.2 Likelihood Ratio Tests

In the binary signal detection problem, we want to decide, on the basis of receiver measurements, which one of two hypotheses H_0 or H_1 is true.

$$\begin{aligned} &H_0\text{: the signal } s_0(t) \text{ is present} \\ &H_1\text{: the signal } s_1(t) \text{ is present} \end{aligned} \quad (10.2)$$

In terms of parameter estimation, we can let

$$s(t, \theta) = (1 - \theta)s_0(t) + \theta s_1(t) \tag{10.3}$$

and the receiver must decide, on the basis of the received data, whether the true value of the parameter θ is equal to zero or equal to one. In many applications, e.g., radar, it is appropriate to make $s_0(t) = 0$, so that the hypothesis H_0 is actually "no signal present." Much of the notation employed obviously stems from this situation, although our generalization to two possible signals presents no difficulty. There are four possible outcomes for each decision. These are indicated in the table below.

Decision \ True state	H_0 is true	H_1 is true
Accept H_0	correct	type II error (miss)
Accept H_1	type I error (false alarm)	correct

Inasmuch as the penalties associated with committing either of the two types of errors might be substantially different, we assign *costs* C_f and C_m to the type I and type II errors, respectively. Now we can base the decision strategy on the values of the *a posteriori* probabilities Pr $[H_0 \mid \mathbf{y}]$ and Pr $[H_1 \mid \mathbf{y}]$ weighted by these costs.[1]

$$\begin{aligned} C(H_0 \mid \mathbf{y}) &= C_m \Pr [H_1 \mid \mathbf{y}] \\ C(H_1 \mid \mathbf{y}) &= C_f \Pr [H_0 \mid \mathbf{y}] \end{aligned} \tag{10.3}$$

The quantities $C(H_0 \mid \mathbf{y})$ and $C(H_1 \mid \mathbf{y})$ are the expected costs or *a posteriori risks* associated with a particular decision; i.e., $C(H_0 \mid \mathbf{y})$ is the risk associated with making the decision that H_0 is true on the basis of the received signal $\mathbf{y}$. A receiver which always selects the hypothesis giving the smaller *a posteriori* risk is called a *Bayes receiver*. We shall presently show that the Bayes receiver provides the minimum average risk [averaged over the ensemble of received signals (10.14)].

In order to evaluate the performance of the Bayes receiver, we must be able to relate *a posteriori* probabilities to the transmitter characteristics (*a priori* probabilities) and to the channel characteristics (probabilities associated with the random disturbances). To do this, we make use of signal space

[1] We use $\mathbf{y}$ to denote $y(t)$ as an element in the space of time functions ($L^2(T)$ in later applications). In this chapter, $\mathbf{y}$ should not be interpreted as a random process, but instead as a *particular realization* of a random process.

concepts. Consider a signal space S which contains all possible received signals and a partition [see (1.9)] of S into many small subsets $\{S_i; i = 1, 2, \ldots\}$. Now applying the *Bayes probability rule* for mutually exclusive events,[3] we can write

$$\Pr[H_0 \mid \mathbf{y} \in S_i] = \frac{q \Pr[\mathbf{y} \in S_i \mid H_0]}{p \Pr[\mathbf{y} \in S_i \mid H_1] + q \Pr[\mathbf{y} \in S_i \mid H_0]}$$

and

$$\Pr[H_1 \mid \mathbf{y} \in S_i] = \frac{p \Pr[\mathbf{y} \in S_i \mid H_1]}{p \Pr[\mathbf{y} \in S_i \mid H_1] + q \Pr[\mathbf{y} \in S_i \mid H_0]} \tag{10.4}$$

where

$$\begin{aligned} p &= \Pr[H_1] \\ q &= \Pr[H_0] = 1 - p \end{aligned} \tag{10.5}$$

are the *a priori* probabilities of the two hypotheses. We can express the conditional probabilities $\Pr[\mathbf{y} \in S_i \mid H_0]$ and $\Pr[\mathbf{y} \in S_i \mid H_1]$ symbolically in terms of two density functions $\ell(\boldsymbol{\sigma} \mid H_0)$ and $\ell(\boldsymbol{\sigma} \mid H_1)$ so that

$$\begin{aligned} \Pr[\mathbf{y} \in S_i \mid H_0] &= \int_{S_i} \ell(\boldsymbol{\sigma} \mid H_0)\, d\boldsymbol{\sigma} \\ \Pr[\mathbf{y} \in S_i \mid H_1] &= \int_{S_i} \ell(\boldsymbol{\sigma} \mid H_1)\, d\boldsymbol{\sigma} \end{aligned} \tag{10.6}$$

If S is an n-dimensional real space, the integrals in (10.6) represent conventional n-dimensional volume integrals and $\ell(\boldsymbol{\sigma} \mid H_0)$ and $\ell(\boldsymbol{\sigma} \mid H_1)$ are simply nth-order joint probability density functions. These functions are called *likelihood functions* and they provide the necessary characterization of the channel (they are independent of *a priori* probabilities).

If we partition S in such a way that the regions can all be made arbitrarily small, then the *a posteriori* probabilities and the likelihood functions can be considered essentially constant over each region and (10.4) can be rewritten in terms of individual points in S instead of subsets.

$$\Pr[H_0 \mid \mathbf{y}] = \frac{q\ell(\mathbf{y} \mid H_0)}{p\ell(\mathbf{y} \mid H_1) + q\ell(\mathbf{y} \mid H_0)} \tag{10.7}$$

$$\Pr[H_1 \mid \mathbf{y}] = \frac{p\ell(\mathbf{y} \mid H_1)}{p\ell(\mathbf{y} \mid H_1) + q\ell(\mathbf{y} \mid H_0)} \tag{10.8}$$

Combining (10.3) with (10.7) and (10.8), we see that the ratio of *a posteriori* risks is proportional to the ratio of likelihood functions.

$$\frac{C(H_0 \mid \mathbf{y})}{C(H_1 \mid \mathbf{y})} = \frac{pC_m\ell(\mathbf{y} \mid H_1)}{qC_f\ell(\mathbf{y} \mid H_0)} \triangleq \left[\frac{pC_m}{qC_f}\right]\lambda(\mathbf{y}) \tag{10.9}$$

where $\lambda(\mathbf{y})$ is referred to as the *likelihood ratio* for the channel. The Bayes receiver accepts H_0 whenever the overall ratio in (10.9) is less than unity.

(The ambiguity of the situation where the risks are equal is of no practical importance since in the usual applications this situation occurs with zero probability.) This strategy is equivalent to accepting H_0 whenever the likelihood ratio is below a threshold value λ_0 given by

$$\lambda_0 = \frac{qC_f}{pC_m} \tag{10.10}$$

The relation $\lambda(\mathbf{y}) = \lambda_0$ defines a surface in the signal space which partitions the space into two subsets R_0 and R_1 such that

$$\begin{aligned} \lambda(\mathbf{y}) &< \lambda_0 \qquad \text{for } \mathbf{y} \in R_0 \\ \lambda(\mathbf{y}) &> \lambda_0 \qquad \text{for } \mathbf{y} \in R_1 \end{aligned} \tag{10.11}$$

To summarize, the Bayes receiver computes the value of the likelihood ratio for the received signal and compares it with a threshold value determined

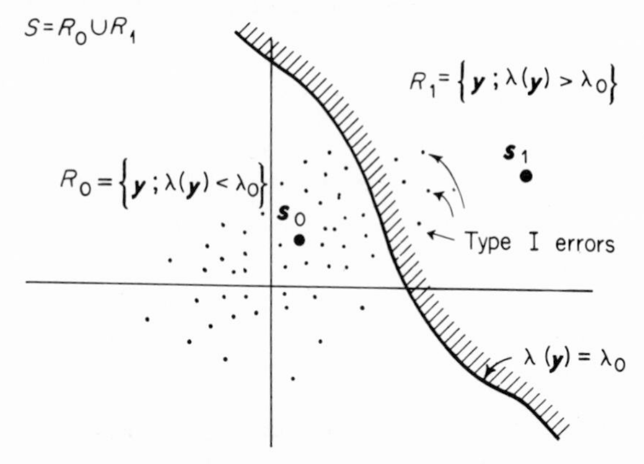

Figure 10.1. Signal space representation of likelihood ratio tests. The points clustered about s_0 represent various realizations of the received signal when s_0 is transmitted. Points to the right of the decision boundary result in type I errors.

by assigned costs and *a priori* probabilities. It then accepts H_0 if the likelihood ratio is below the threshold and accepts H_1 otherwise. In Figure 10.1, an attempt is made to show this pictorially.

If the channel disturbance is sufficient to cause the transmitted signal $s_0(t)$ to appear at the receiver as a signal in region R_1, then a type I error occurs. To characterize channel performance, we let P_f denote the probability of a type I error, given that H_0 is true. Using (10.6), P_f can be expressed in terms of one of the likelihood functions.

$$P_f = \int_{R_1} \ell(\boldsymbol{\sigma} \mid H_0)\, d\boldsymbol{\sigma} \tag{10.12}$$

Similarly, we let P_m denote the probability of a type II error, given that H_1 is true

$$P_m = \int_{R_0} \ell(\boldsymbol{\sigma} \mid H_1)\, d\boldsymbol{\sigma} \tag{10.13}$$

Taking into account the *a priori* probabilities and assigned costs, we can define an *average risk* $\bar{C}$ as

$$\bar{C} = pC_m P_m + qC_f P_f \tag{10.14}$$

For different choices of the decision regions R_0 and R_1, the average risk is different because of the dependence of P_m and P_f on R_0 and R_1. It seems intuitively evident that the Bayes receiver, which always decides on the basis of minimum *a posteriori* risk, should yield the minimum average risk. To prove this, we can choose an arbitrary strategy represented by decision regions R_0' and R_1' ($S = R_0' \cup R_1'$; $R_0' \cap R_1' = \varnothing$) and compare the resulting average risk $\bar{C}'$ with $\bar{C}_B$ for the Bayes regions given by (10.11).

$$\bar{C}' - \bar{C}_B = pC_m\left[\int_{R_0'} \ell(\boldsymbol{\sigma} \mid H_1)\, d\boldsymbol{\sigma} - \int_{R_0} \ell(\boldsymbol{\sigma} \mid H_1)\, d\boldsymbol{\sigma}\right] + qC_f\left[\int_{R_1'} \ell(\boldsymbol{\sigma} \mid H_0)\, d\boldsymbol{\sigma} - \int_{R_1} \ell(\boldsymbol{\sigma} \mid H_0)\, d\boldsymbol{\sigma}\right]$$

Subtracting out the common regions of integration (see Figure 10.2), we get

$$\bar{C}' - \bar{C}_B = pC_m\left[\int_{R_0' \cap R_1} \ell(\boldsymbol{\sigma} \mid H_1)\, d\boldsymbol{\sigma} - \int_{R_0 \cap R_1'} \ell(\boldsymbol{\sigma} \mid H_1)\, d\boldsymbol{\sigma}\right] + qC_f\left[\int_{R_1' \cap R_0} \ell(\boldsymbol{\sigma} \mid H_0)\, d\boldsymbol{\sigma} - \int_{R_1 \cap R_0'} \ell(\boldsymbol{\sigma} \mid H_0)\, d\boldsymbol{\sigma}\right] \tag{10.15}$$

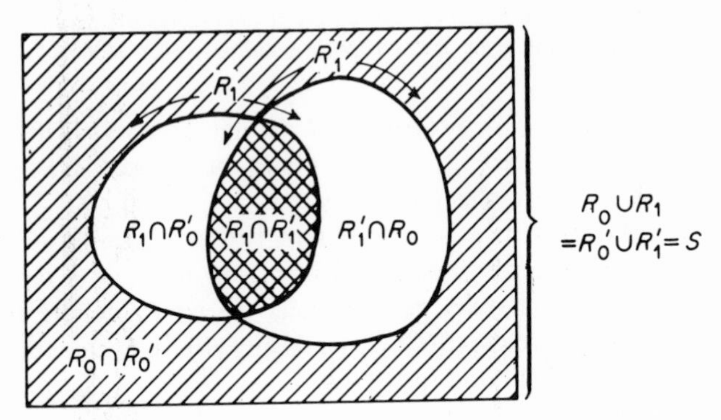

Figure 10.2. Identification of overlap of decision regions for two different decision strategies.

Rearranging the terms in (10.15), we have

$$\bar{C}' - \bar{C}_B = \int_{R_0' \cap R_1} \{pC_m \ell(\boldsymbol{\sigma} \mid H_1) - qC_f \ell(\boldsymbol{\sigma} \mid H_0)\} \, d\boldsymbol{\sigma} + \int_{R_1' \cap R_0} \{qC_f \ell(\boldsymbol{\sigma} \mid H_0) - pC_m \ell(\boldsymbol{\sigma} \mid H_1)\} \, d\boldsymbol{\sigma} \quad (10.16)$$

From the definition of R_0 and R_1, (10.11), the integrands in (10.16) are always positive over the region of integration; hence

$$\bar{C}' - \bar{C}_B \geqslant 0 \quad (10.17)$$

and $\bar{C}_B$ is the minimum average risk.

Some particular cases of the Bayes receiver are of special interest. If we assign equal costs so that $C_m = C_f = 1$, then $\bar{C}$, (10.14), is simply the overall error probability. In this case, $\lambda_0 = q/p$ and the receiver is referred to as the *ideal observer*. If, furthermore, the two hypotheses are assumed equiprobable, then $\lambda_0 = 1$ and the receiver is a *maximum-likelihood* receiver. A typical form of the dependence of the Bayes risk $\bar{C}_B$ on the *a priori* probability p is shown in Figure 10.3. For each value of p, a different λ_0 is required. It is of interest to examine the dependence of average risk on p when a fixed strategy (one particular value of λ_0) is employed. In this case, P_m and P_f are fixed and

$$\begin{aligned} \bar{C} &= pC_m P_m + (1 - p)C_f P_f \\ &= \{C_m P_m - C_f P_f\}p + C_f P_f \end{aligned} \quad (10.18)$$

Hence $\bar{C}$ is a straight line with slope $C_m P_m - C_f P_f$, as shown in Figure 10.3.

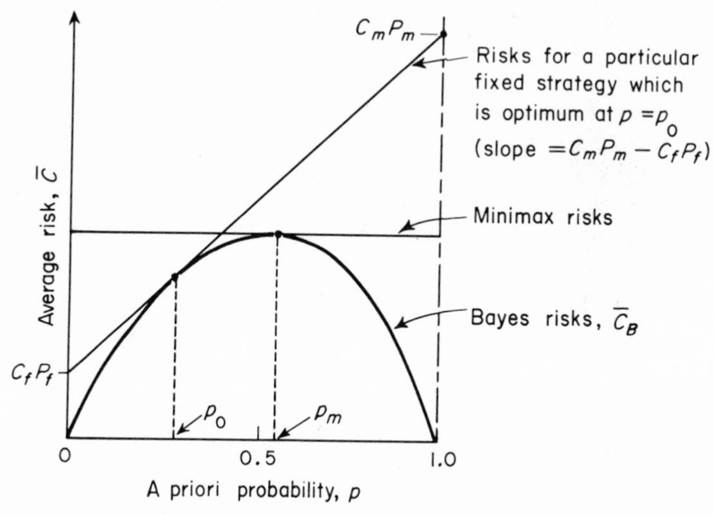

Figure 10.3. Typical dependence of average risk on *a priori* probabilities.

We see from Figure 10.3 that it is possible to encounter very large risks with a fixed strategy if the *a priori* probabilities change appreciably. When there is a high degree of uncertainty about the *a priori* probabilities, a favorite strategy is the *minimax criterion* which chooses a fixed strategy that minimizes the maximum average risk. This obviously corresponds to a fixed strategy line with zero slope; i.e., $C_m P_m = C_f P_f$, and the minimax risks are independent of p. The minimax strategy sets the threshold at

$$\lambda_0 = \frac{1 - p_m}{p_m} \cdot \frac{C_f}{C_m} \tag{10.19}$$

where p_m is the *a priori* probability for which $\bar{C}_B$ is a maximum.

10.3 Neyman-Pearson Theory

In choosing a decision strategy for a particular signal detection problem, it may turn out that we are more concerned with exercising direct control over the conditional error probabilities P_m and P_f rather than minimizing risks. This is the classical statistical approach to hypothesis testing problems. Some of the standard terminology is introduced below.

Null hypothesis: H_0
Alternate hypothesis: H_1
Critical region: R_1 = region of rejection of H_0
Size of critical region: P_f = probability of type I error
Power of a test: $1 - P_m$ = probability of *not* committing a type II error

A critical region of a given size P_f is said to be *most powerful* if P_m is a minimum.

The *Neyman-Pearson lemma* states that a likelihood ratio test is most powerful. In other words, for a preassigned false alarm probability P_f, the strategy which minimizes the miss probability P_m is the one which decides whether or not the likelihood ratio $\lambda(\mathbf{y})$ exceeds a threshold value λ_0. The proof is similar to that for showing that the Bayes risk is a minimum. Let $R_1 = \{\mathbf{y}; \lambda(\mathbf{y}) > \lambda_0\}$ define a critical region in the signal space S. Now let R_1' be any other critical region of the same size; i.e.,

$$P_f = \int_{R_1'} \ell(\boldsymbol{\sigma} \mid H_0)\, d\boldsymbol{\sigma} = \int_{R_1} \ell(\boldsymbol{\sigma} \mid H_0)\, d\boldsymbol{\sigma} \tag{10.20}$$

Then (see Figure 10.2) it follows that

$$\int_{R_1 \cap R_0'} \ell(\boldsymbol{\sigma} \mid H_0)\, d\boldsymbol{\sigma} = \int_{R_1' \cap R_0} \ell(\boldsymbol{\sigma} \mid H_0)\, d\boldsymbol{\sigma} \tag{10.21}$$

For the unprimed strategy,

$$1 - P_m = \int_{R_1} \ell(\boldsymbol{\sigma} \mid H_1)\, d\boldsymbol{\sigma}$$
$$= \int_{R_1 \cap R_1'} \ell(\boldsymbol{\sigma} \mid H_1)\, d\boldsymbol{\sigma} + \int_{R_1 \cap R_0'} \ell(\boldsymbol{\sigma} \mid H_1)\, d\boldsymbol{\sigma} \qquad (10.22)$$

but from the definition of R_1, the second term in (10.22) satisfies the inequality

$$\int_{R_1 \cap R_0'} \ell(\boldsymbol{\sigma} \mid H_1)\, d\boldsymbol{\sigma} > \lambda_0 \int_{R_1 \cap R_0'} \ell(\boldsymbol{\sigma} \mid H_0)\, d\boldsymbol{\sigma} \qquad (10.23)$$

Combining (10.22) and (10.23) and using the fact that $\ell(\mathbf{y} \mid H_1) < \lambda_0 \ell(\mathbf{y} \mid H_0)$ for $\mathbf{y} \in R_0$, we find

$$\int_{R_1 \cap R_0'} \ell(\boldsymbol{\sigma} \mid H_1)\, d\boldsymbol{\sigma} > \lambda_0 \int_{R_1 \cap R_0'} \ell(\boldsymbol{\sigma} \mid H_0)\, d\boldsymbol{\sigma}$$
$$= \lambda_0 \int_{R_1' \cap R_0} \ell(\boldsymbol{\sigma} \mid H_0)\, d\boldsymbol{\sigma} > \int_{R_1' \cap R_0} \ell(\boldsymbol{\sigma} \mid H_1)\, d\boldsymbol{\sigma} \qquad (10.24)$$

Hence

$$1 - P_m = \int_{R_1} \ell(\boldsymbol{\sigma} \mid H_1)\, d\boldsymbol{\sigma} > \int_{R_1'} \ell(\boldsymbol{\sigma} \mid H_1)\, d\boldsymbol{\sigma} \qquad (10.25)$$

and $1 - P_m$ is maximized by the critical region R_1.

A convenient device for displaying these results and also showing their relationship to Bayes and minimax solutions is the *receiver operating characteristic*. This is the graph for P_f and the maximum value of $1 - P_m$, shown in Figure 10.4. The graph depends only on the likelihood functions and not on *a priori* probabilities or assigned costs. A particular point on the graph is a particular Bayes solution, which does depend on *a priori* probabilities and assigned costs. Each point represents a different value of λ_0. Furthermore, the slope of the graph at any point is equal to λ_0 for the Bayes solution at that point. To show this, we note that the fixed strategy line in Figure 10.3 is tangent to the Bayes risk graph; hence

$$\frac{d\bar{C}_B}{dp} = pC_m \frac{dP_m}{dp} + qC_f \frac{dP_f}{dp} + C_m P_m - C_f P_f = C_m P_m - C_f P_f$$

and

$$\frac{d}{dP_f}(1 - P_m) = -\frac{dP_m}{dP_f} = \frac{qC_f}{pC_m} = \lambda_0 \qquad (10.26)$$

The minimax operating point is obtained by finding the intersection of the receiver operating characteristic with the straight line given by $C_m P_m = C_f P_f$.

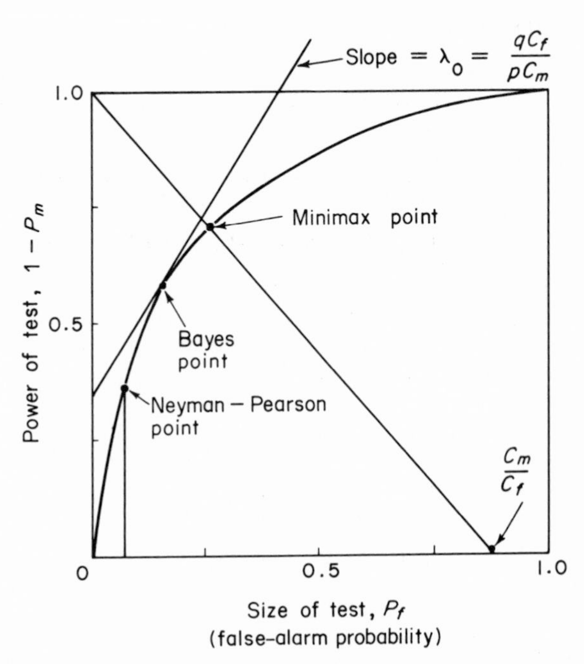

Figure 10.4. Receiver operating characteristic.

10.4 Detection of Binary Signals in White Gaussian Noise

The simplest practical application of the foregoing detection theory based on likelihood ratio tests is the design of a receiver to decide which of two known signals is present in a background of additive, zero-mean, white Gaussian noise. We shall assume that receiver observations can be made over a finite time interval T and that timing synchronization has been accomplished so that the interval can be selected for optimum detectability. We shall see later that the interval should be chosen to contain the maximum possible signal pulse energy.

$$\begin{aligned} H_0&: y(t) = s_0(t) + u(t); \qquad t \in T \\ H_1&: y(t) = s_1(t) + u(t); \qquad t \in T \end{aligned} \tag{10.27}$$

where $u(t)$ is a realization of a zero-mean random process with an autocorrelation function given by $k_{uu}(t, s) = m_{uu}(t, s) = N_0\,\delta(t - s)$.

Our approach will be to assume that the receiver is able to make an orthogonal projection of the received signal onto an n-dimensional subspace S of $L^2(T)$ with n arbitrary. Let $\{\boldsymbol{\varphi}_k; k = 1, 2, \ldots, n\}$ be a real orthonormal basis for S. The decision will be based on the value of the likelihood ratio for the real n-tuple $\{y_k\}$.

$$y(t) = \sum_{k=1}^{n} y_k \varphi_k(t); \qquad t \in T$$
$$= \sum_{k=1}^{n} (a_k + u_k)\varphi_k(t) \qquad \text{if } H_0 \text{ is true}$$
$$= \sum_{k=1}^{n} (b_k + u_k)\varphi_k(t) \qquad \text{if } H_1 \text{ is true} \tag{10.28}$$

where

$$y_k = \int_T y(t)\varphi_k(t)\,dt = (\mathbf{y}, \boldsymbol{\varphi}_k)$$

$$u_k = (\mathbf{u}, \boldsymbol{\varphi}_k); \qquad a_k = (\mathbf{s}_0, \boldsymbol{\varphi}_k); \qquad b_k = (\mathbf{s}_1, \boldsymbol{\varphi}_k) \tag{10.29}$$

For the case of additive noise, the likelihood functions, (10.6), are given by the joint probability density function $p_{u_1u_2\cdots u_n}(\sigma_1, \sigma_2, \ldots, \sigma_n)$ for the noise n-tuple $\{u_k\}$.

$$u(t) = y(t) - s_0(t) \Rightarrow \ell(y \mid H_0) = p_{u_1u_2\cdots u_n}(y_1 - a_1, y_2 - a_2, \ldots, y_n - a_n)$$
$$u(t) = y(t) - s_1(t) \Rightarrow \ell(y \mid H_1) = p_{u_1u_2\cdots u_n}(y_1 - b_1, y_2 - b_2, \ldots, y_n - b_n) \tag{10.30}$$

Probability Density Function for Jointly Gaussian Random Variables

For a zero-mean, Gaussian noise process, this joint probability density function is easily obtained. First, we define a *Gaussian random process.* If the random variables $\mathbf{x}_i = \mathbf{x}(t_i)$; $i = 1, 2, \ldots, r$ for any set $\{t_i\}$ and any r are jointly distributed, zero-mean, Gaussian random variables, then $\mathbf{x}(t)$ is a zero-mean, Gaussian random process. This means (by definition) that the joint characteristic function (7.11) is an exponential with a quadratic form as the exponent.[3]

$$Q_{x_1x_2\cdots x_r}(\nu_1, \nu_2, \ldots, \nu_r) = E\left[\exp\left(j\sum_{i=1}^{r} \mathbf{x}_i\nu_i\right)\right]$$
$$= \exp\left(-\tfrac{1}{2}\sum_{i=1}^{r}\sum_{j=1}^{r} m_{ij}\nu_i\nu_j\right) \tag{10.31}$$

where

$$m_{ij} = E[\mathbf{x}_i\mathbf{x}_j] = m_{xx}(t_i, t_j)$$

We note that such a process is completely characterized by its autocovariance function.

We now derive an expression for the joint characteristic function for the n random variables

$$\mathbf{a}_i = (\mathbf{x}, \boldsymbol{\varphi}_i) = \int_T \mathbf{x}(t)\varphi_i(t)\,dt; \qquad i = 1, 2, \ldots, n \tag{10.32}$$

where it is assumed the $\{\boldsymbol{\varphi}_i\}$ are real and $\{\mathbf{x}(t); t \in T\}$ is a zero-mean, Gaussian random process. A familiar statement is that a linear transformation on a Gaussian random process yields another Gaussian process. For our purposes, this can be demonstrated as follows. For the real n-tuple $\{\mu_i\}$ and the random n-tuple $\{\mathbf{a}_i\}$, we can write the random variable $(\mathbf{a}, \boldsymbol{\mu})$ as

$$(\mathbf{a}, \boldsymbol{\mu}) = \sum_{i=1}^{n} \mu_i \int_T \mathbf{x}(t)\varphi_i(t)\,dt = (\mathbf{x}, \boldsymbol{\xi}) \tag{10.33}$$

where

$$\xi(t) = \sum_{i=1}^{n} \mu_i \varphi_i(t); \qquad t \in T$$

For r sufficiently large, we can approximate $(\mathbf{x}, \boldsymbol{\xi})$ with a finite sum by subdividing T into r equal intervals. Then

$$(\mathbf{x}, \boldsymbol{\xi}) = \int_T \mathbf{x}(t)\xi(t)\,dt \cong \frac{T}{r}\sum_{k=1}^{r} \mathbf{x}(t_k)\xi(t_k); \qquad t_k \in T \tag{10.34}$$

Now using (10.34) in (10.31), we can write

$$E[e^{j(\mathbf{a}, \boldsymbol{\mu})}] = E[e^{j(\mathbf{x}, \boldsymbol{\xi})}] \cong \exp\left[-\frac{1}{2}\left(\frac{T}{r}\right)^2 \sum_{i=1}^{r}\sum_{j=1}^{r} m_{ij}\xi(t_i)\xi(t_j)\right] \tag{10.35}$$

But

$$\left(\frac{T}{r}\right)^2 \sum_{i=1}^{r}\sum_{j=1}^{r} m_{ij}\xi(t_i)\xi(t_j) \cong \iint_T m_{xx}(t, s)\xi(s)\xi(t)\,ds\,dt = (\mathbf{C}\boldsymbol{\mu}, \boldsymbol{\mu}) \tag{10.36}$$

where $\mathbf{C}$ is an $n \times n$ matrix with elements given by

$$c_{ij} = \iint_T m_{xx}(t, s)\varphi_j(s)\varphi_i(t)\,ds\,dt; \qquad i, j = 1, 2, \ldots, n \tag{10.37}$$

Hence, since r can be made arbitrarily large, we have shown that

$$Q_{a_1a_2\cdots a_n}(\mu_1, \mu_2, \ldots, \mu_n) = E[e^{j(\mathbf{a}, \boldsymbol{\mu})}] = e^{-\frac{1}{2}(\mathbf{C}\boldsymbol{\mu}, \boldsymbol{\mu})} \tag{10.38}$$

and $\mathbf{a} = \{\mathbf{a}_i\}$ are jointly Gaussian random variables with covariances $E[\mathbf{a}_i\mathbf{a}_j] = c_{ij}$ given by (10.37).

If it happens that the $\{\mathbf{a}_i\}$ are uncorrelated, then $\mathbf{C}$ is a diagonal matrix and

$$Q_{a_1a_2\cdots a_n}(\mu_1, \mu_2, \ldots, \mu_n) = \prod_{i=1}^{n} Q_{a_i}(\mu_i) \tag{10.39}$$

where

$$Q_{a_i}(\mu_i) \triangleq e^{-\frac{1}{2}\sigma_i^2\mu_i^2}; \qquad \sigma_i^2 \triangleq E[\mathbf{a}_i^2]$$

The joint probability density function can be obtained by taking the Fourier transform of each of the $Q_{a_i}(\mu_i)$.

$$p_{a_1a_2\cdots a_n}(\xi_1, \xi_2, \ldots, \xi_n) = \prod_{i=1}^{n} p_{a_i}(\xi_i)$$

where

$$p_{a_i}(\xi_i) = \frac{1}{\sqrt{2\pi}\,\sigma_i} e^{-\frac{1}{2}\left(\frac{\xi_i}{\sigma_i}\right)^2} \tag{10.40}$$

We note also that if the Gaussian random variables $\{\mathbf{a}_i\}$ are uncorrelated, then it follows that they are statistically independent, (7.17).

Likelihood Functions for Additive Gaussian Noise

Returning now to evaluation of the likelihood functions (10.30) for the white Gaussian noise process $\mathbf{u}$, we have, from (10.37),

$$E[\mathbf{u}_i\mathbf{u}_j] = \iint_T m_{uu}(t, s)\varphi_i(s)\varphi_j(t)\,ds\,dt \tag{10.41}$$

but since $m_{uu}(t, s) = N_0\,\delta(t - s)$, (10.41) becomes

$$E[\mathbf{u}_i\mathbf{u}_j] = N_0 \iint_T \delta(t - s)\varphi_i(s)\varphi_j(t)\,ds\,dt$$

$$= N_0 \int_T \varphi_i(t)\varphi_j(t)\,dt = N_0\,\delta_{ij} \tag{10.42}$$

Thus we see that, for any orthonormal basis for S, the representation for white Gaussian noise has uncorrelated coefficients with equal variance N_0. Substituting (10.40) into (10.30), we have

$$\ell(\mathbf{y} \mid H_0) = \frac{1}{(2\pi N_0)^{\frac{n}{2}}} \exp\left(-\frac{1}{2N_0}\sum_{k=1}^{n}(y_k - a_k)^2\right)$$

$$\ell(\mathbf{y} \mid H_1) = \frac{1}{(2\pi N_0)^{\frac{n}{2}}} \exp\left(-\frac{1}{2N_0}\sum_{k=1}^{n}(y_k - b_k)^2\right) \tag{10.43}$$

The likelihood ratio becomes

$$\lambda(\mathbf{y}) = \exp\left(-\frac{1}{2N_0}\sum_{k=1}^{n}(y_k - b_k)^2 - (y_k - a_k)^2\right) \tag{10.44}$$

Since the exponential function is monotonic in its argument, a threshold test

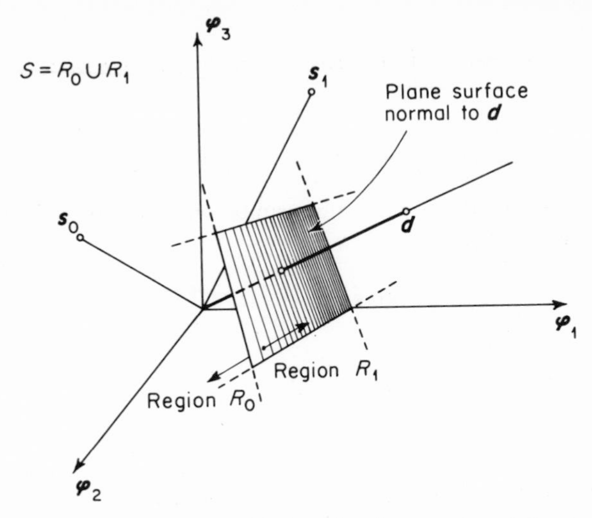

Figure 10.5. Decision surface for likelihood ratio test. R_0 and R_1 are separated by a plane surface normal to the difference signal vector **d**.

on $\lambda(\mathbf{y})$ is equivalent to a threshold test on the exponent and the decision rule becomes

$$\text{Accept } H_0 \text{ if } -\frac{1}{2N_0}\sum_{k=1}^{n}(y_k - b_k)^2 - (y_k - a_k)^2 < \ln \lambda_0$$

Equivalently,

$$\text{Accept } H_0 \text{ if } 2\sum_{k} y_k(b_k - a_k) < 2N_0 \ln \lambda_0 + \sum_{k=1}^{n} b_k^2 - \sum_{k=1}^{n} a_k^2 \qquad (10.45)$$

To simplify the expression, we define a *difference signal*

$$\begin{aligned} d(t) &= s_1(t) - s_0(t); \qquad t \in T \\ d_k &= b_k - a_k \end{aligned} \qquad (10.46)$$

Now the equation

$$\sum_{k=1}^{n} y_k\, d_k = \text{constant} \qquad (10.47)$$

represents a plane surface in S which is normal to **d** and which separates S into the two decision regions R_0 and R_1. This is shown pictorially in Figure 10.5.

If the $\{\boldsymbol{\varphi}_k;\, k = 1, 2, \ldots, n\}$ are chosen such that, for n sufficiently large, $s_0(t)$ and $s_1(t)$ are contained in S, then the inner products of n-tuples are equivalent to inner products in $L^2(T)$ [see (2.49)]. From this we obtain directly a "coordinate-free" decision rule corresponding to (10.45).

$$\text{Accept } H_0 \text{ if } (\mathbf{y}, \mathbf{d}) < N_0 \ln \lambda_0 + \tfrac{1}{2}(\mathbf{s}_1, \mathbf{s}_1) - \tfrac{1}{2}(\mathbf{s}_0, \mathbf{s}_0) \qquad (10.48)$$

Receiver Implementation

From (10.48), it is clear that the receiver is only required to form the inner product of the received signal with the difference of the transmitted signals

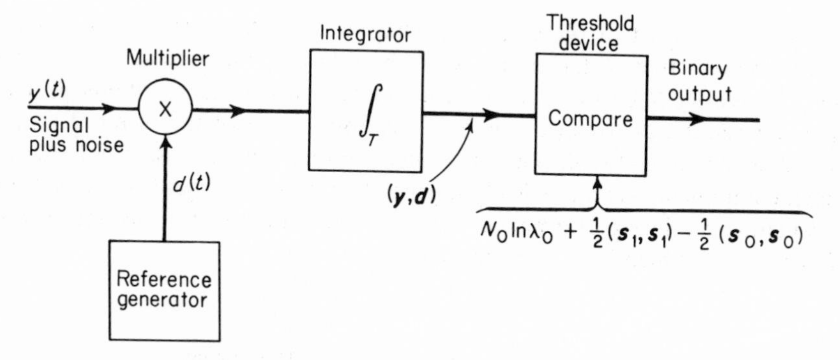

Figure 10.6. Multiplier-integrator implementation of likelihood ratio receiver for binary signals.

(hence, a reference signal is required at the receiver) and to compare this value with a preset threshold determined by λ_0 and the energy difference of the transmitted signals. Implementations of this operation can be accomplished by the methods suggested in Section 2.6. The inner product can be formed by the multiplier-integrator combination shown in Figure 10.6 or by the filter-sampler combination shown in Figure 10.7.

In the filter-sampler implementation, the impulse response of the filter is $h(t) = d(t_0 - t)$, so we see that the filter is actually the matched filter for the difference signal in white noise (see Section 9.4). This demonstrates the equivalence, in the white noise case, of minimum mean-squared error pulse amplitude estimation and likelihood ratio hypothesis testing. We can see that the hypothesis testing problem is equivalent to a pulse amplitude estimation problem by rewriting (10.3) in terms of the difference signal, so that the received signal is expressed by the random variables

$$\mathbf{y}(t) = \boldsymbol{\theta}\, d(t) + s_0(t) + \mathbf{u}(t); \qquad t \in T \tag{10.49}$$

and estimation of pulse amplitude $\boldsymbol{\theta}$ is equivalent to determining whether $s_0(t)$ or $s_1(t)$ is present.

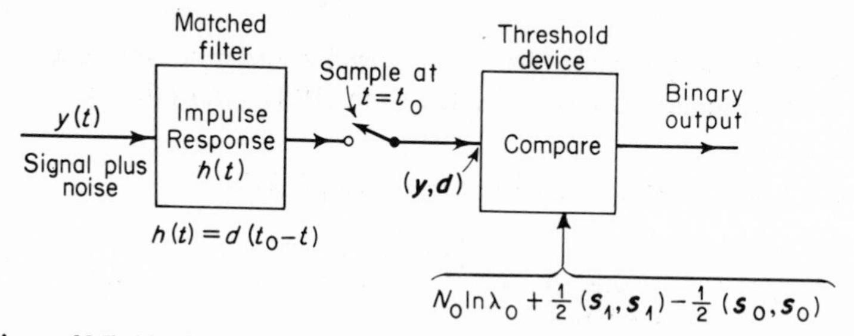

Figure 10.7. Matched filter-sampler implementation of likelihood ratio receiver. The reference signal can be considered as stored in the component values and component configuration of the filter.

Another aspect of these results, which is important from a conceptual standpoint, is that S does not have to be a large dimension subspace. We could have let S be the one-dimensional subspace which contains $\mathbf{d}$. The receiver rejects, as irrelevant data,[1] any component of the received signal which is orthogonal to $\mathbf{d}$. This is because, in the white Gaussian noise case, such components are statistically independent of which signal is present. In evaluating receiver performance, it is only the orthogonal projection of the noise onto S which is of importance.

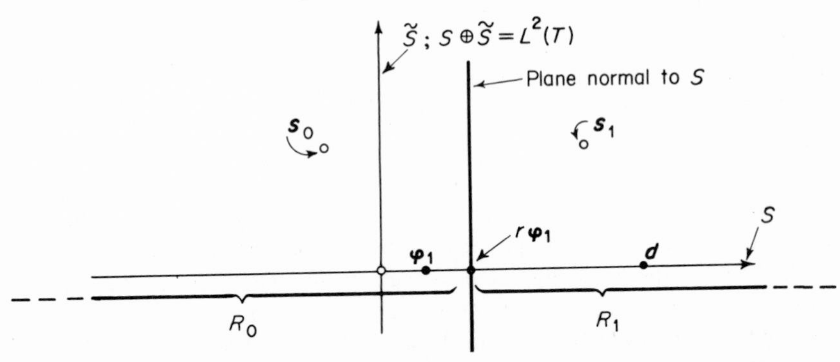

Figure 10.8. Decision regions in signal space for likelihood ratio tests.

Proceeding as suggested above, we let S be the subspace spanned by $\varphi_1(t) = \|\mathbf{d}\|^{-1} d(t)$ as indicated in Figure 10.8. Then the representation for the projection of the received signal on S is given by

$$y_1 = (\mathbf{y}, \boldsymbol{\varphi}_1) = \frac{1}{\|\mathbf{d}\|}(\mathbf{y}, \mathbf{d}) \tag{10.50}$$

and the decision rule (10.48) becomes

$$\text{Accept } H_0 \text{ if } y_1 < \frac{2N_0 \ln \lambda_0 + (\mathbf{s}_1, \mathbf{s}_1) - (\mathbf{s}_0, \mathbf{s}_0)}{2\,\|\mathbf{d}\|} \triangleq r \tag{10.51}$$

The likelihood functions (10.43) are now first-order density functions and receiver performance can be evaluated from (10.12) and (10.13).

$$\begin{aligned} P_m &= \Pr[y_1 < r \mid H_1] = \int_{R_0} \ell(\boldsymbol{\sigma} \mid H_1)\, d\boldsymbol{\sigma} \\ &= \frac{1}{\sqrt{2\pi N_0}} \int_{-\infty}^{r} e^{-\frac{1}{2N_0}(\sigma - b_1)^2}\, d\sigma; \qquad b_1 = (\mathbf{s}_1, \boldsymbol{\varphi}_1) \end{aligned} \tag{10.52}$$

By a change of integration variable in (10.52), we can express P_m in terms of the *cumulative distribution function* $\Phi(\alpha)$, for a zero-mean, unit-variance,

Gaussian random variable. The values of $\Phi(\alpha)$ are readily available in mathematical tables.[2]

$$P_m = \frac{1}{\sqrt{2\pi}} \int_{-\infty}^{\alpha_1} e^{-\frac{\omega^2}{2}} \, d\omega \triangleq \Phi(\alpha_1) \tag{10.53}$$

where we have let

$$\omega = \frac{\sigma - b_1}{\sqrt{N_0}}; \qquad \alpha_1 = \frac{r - b_1}{\sqrt{N_0}} = \frac{\ln \lambda_0}{2\rho} - \rho \tag{10.54}$$

and

$$\rho \triangleq \frac{\|\mathbf{d}\|}{2\sqrt{N_0}}$$

Similarly,

$$P_f = \Pr\,[y_1 > r \mid H_0] = \int_{R_1} \ell(\boldsymbol{\sigma} \mid H_0) \, d\boldsymbol{\sigma}$$

$$= \frac{1}{\sqrt{2\pi N_0}} \int_r^{\infty} e^{-\frac{1}{2N_0}(\sigma - a_1)^2} \, d\sigma; \qquad a_1 = (\mathbf{s}_0, \boldsymbol{\varphi}_1)$$

and we let

$$\omega = \frac{\sigma - a_1}{\sqrt{N_0}}; \qquad \alpha_2 = \frac{\ln \lambda_0}{2\rho} + \rho \tag{10.55}$$

Then we have

$$P_f = \frac{1}{\sqrt{2\pi}} \int_{\alpha_2}^{\infty} e^{-\frac{\omega^2}{2}} \, d\omega = 1 - \Phi(\alpha_2) = \Phi(-\alpha_2) \tag{10.56}$$

Using (10.53) and (10.56) and tabulated values for $\Phi(\alpha)$, the receiver operating characteristic can be constructed for various values of the parameter ρ, as shown in Figure 10.9. It is clear that performance improves as ρ increases. This parameter is referred to as the *signal-to-noise ratio*. Since ρ is proportional to $\|\mathbf{d}\|$, receiver performance depends on how much energy in the difference signal can be concentrated into the observation interval T. In some applications, transmitted pulse energies $(\mathbf{s}_0, \mathbf{s}_0)$ and $(\mathbf{s}_1, \mathbf{s}_1)$ are fixed at some maximum level. In this situation, $(\mathbf{d}, \mathbf{d}) = (\mathbf{s}_1, \mathbf{s}_1) + (\mathbf{s}_0, \mathbf{s}_0) - 2(\mathbf{s}_1, \mathbf{s}_0)$, which is maximized by making $s_0(t) = -s_1(t)$ for $t \in T$, whereupon $(\mathbf{d}, \mathbf{d}) = 4(\mathbf{s}_1, \mathbf{s}_1)$ and

$$\rho^2 = \frac{(\mathbf{s}_1, \mathbf{s}_1)}{N_0} \tag{10.57}$$

[2] A convenient approximation for low error probability situations is given by [1] $\Phi(-\alpha) \cong \frac{1}{\sqrt{2\pi}\alpha} e^{-\frac{\alpha^2}{2}}$. The approximation error is less than 10 percent for α greater than 3.

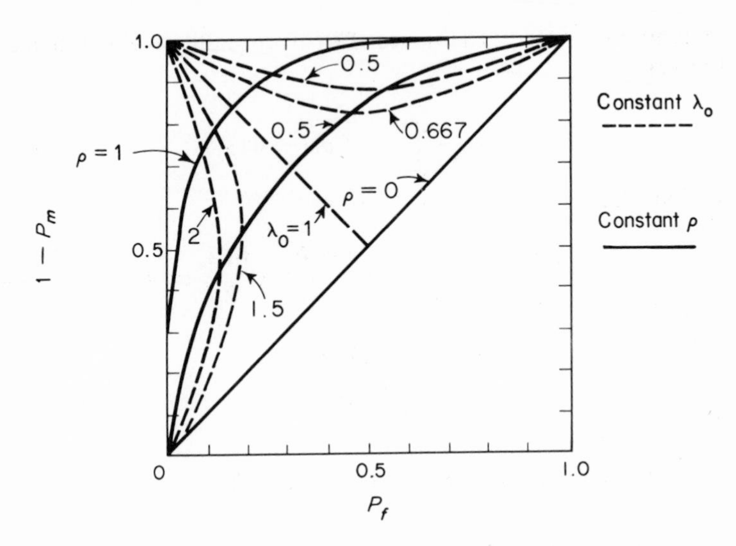

Figure 10.9. Receiver operating characteristic for binary signal detection.

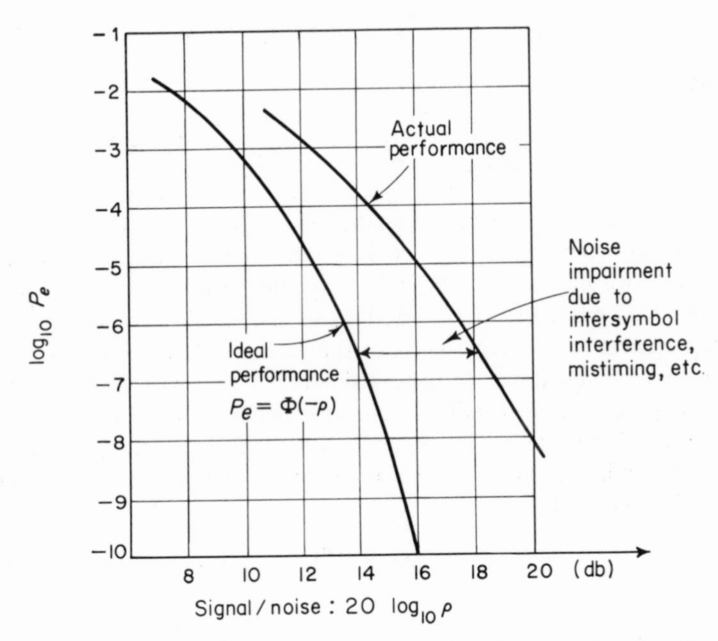

Figure 10.10. Error performance of maximum-likelihood receiver for binary signals in Gaussian noise.

In a binary data communication system, data pulses might be transmitted every T seconds. Signal power would be $(1/T)(\mathbf{s}_1, \mathbf{s}_1)$. To avoid appreciable intersymbol interference, the channel bandwidth must be at least $1/2T$. The noise power in this bandwidth would be N_0/T (N_0 is a double-sided spectral density). By this argument, we see that ρ^2 has a sensible interpretation as a signal-to-noise power ratio. In a typical data communication system, equal costs are assigned to the two types of errors and it might be assumed that the two hypothesis are equiprobable. In this case, the receiver is a maximum-likelihood receiver; $\lambda_0 = 1 \Rightarrow \alpha_2 = -\alpha_1 = \rho$ and the overall error probability P_e is simply

$$P_e = \tfrac{1}{2}(P_m + P_f) = \Phi(-\rho) \tag{10.58}$$

For high performance data systems, $P_e \leqslant 10^{-6}$ say, the receiver operating characteristic is a poor mechanism for displaying performance since the operating points will all be crowded into the upper left-hand corner. The conventional means for showing performance of these systems is a graph of the logarithms of error probability and signal-to-noise ratio[4] as shown in Figure 10.10. The graph corresponding to (10.58) represents idealized performance of the data system. Actual performance will be somewhat poorer due to intersymbol interference, timing error, etc. These additional impairments are conveniently characterized on the performance graph as an equivalent *noise impairment*, i.e., the increase in signal-to-noise required to maintain a given error probability on the idealized performance graph.

Exercise 10.1. A receiver makes orthogonal projections onto a five-dimensional subspace of $L^2(T)$. The components of $s_0(t)$ and $s_1(t)$, relative to an orthonormal basis, are

$$s_0(t)\colon \{0.5, 1.0, 1.0, 1.0, 0.5\}$$

$$s_1(t)\colon \{-0.5, -1.0, -1.0, -1.0, -0.5\}$$

A particular received signal $y(t)$ in additive white Gaussian noise has the components

$$y(t)\colon \{0.3, 0.8, -0.6, -1.5, 0.2\}$$

Suppose $s_1(t)$ has an *a priori* probability of 0.6 and that the cost of mistaking $s_0(t)$ for $s_1(t)$ is twice that of mistaking $s_1(t)$ for $s_0(t)$. Assuming that the noise coefficients have a variance of 0.6, what decision will be made by a Bayes decision rule? For this particular $y(t)$, what is the probability that the decision will be wrong?

Exercise 10.2. A Neyman-Pearson detector is designed to detect the presence or absence of a signal $s(t)$ in white Gaussian noise with $N_0 = 1$ watt/Hz. Let $s(t)$ be a rectangular pulse with height $4T^{-\frac{1}{2}}$ and width T. Suppose that the decision threshold is set to yield $P_f = 0.01$. Evaluate P_m.

10.5 Detection of Binary Signals in Colored Gaussian Noise

In generalizing the binary detection problem to include noise which does not have a flat power spectral density (colored noise), we immediately encounter the difficulty that components of the received signal orthogonal to **d** are correlated with components which are proportional to **d**. In the white noise case, the noise coefficients $\{\mathbf{u}_k\}$ were uncorrelated for *any* orthonormal basis and we could let a normalized version of **d** be one of the basis vectors. Recalling the development in Section 7.5, we note that there is a *particular* orthonormal set which makes the expansion coefficients of an arbitrary zero-mean process uncorrelated. This is the basis for the Karhunen-Loève expansion, where the basis functions are the eigenfunctions of the integral operator whose kernel is the autocovariance function of the process; i.e.,

$$\int_T k_{uu}(t, s)\varphi_i(s)\, ds = \lambda_i \varphi_i(t); \qquad t \in T, \quad i = 1, 2, \ldots \tag{10.59}$$

From (7.51), with $\mathbf{u}_i = (\mathbf{u}, \boldsymbol{\varphi}_i)$, we have

$$E[\mathbf{u}_i \mathbf{u}_j^*] = \lambda_i\, \delta_{ij} \tag{10.60}$$

so that the components are uncorrelated and the variance of $\mathbf{u}_i$ is λ_i. Using this basis, we shall be able to derive a likelihood ratio receiver by the same approach used in the previous section. Note that we do not require that the noise be a stationary process in this derivation. The joint probability density function for the $\mathbf{u}_k$; $k = 1, 2, \ldots, n$ is given by (10.40) with $\sigma_k^2 = \lambda_k$; hence the likelihood functions for the received signal projected onto an n-dimensional subspace of $L^2(T)$ are given by

$$\begin{aligned} \ell(\mathbf{y} \mid H_0) &= \frac{1}{\left(\sqrt{2\pi}\right)^n \prod_{k=1}^{n} \lambda_k^{\frac{1}{2}}} \exp\left(-\frac{1}{2}\sum_{k=1}^{n} \frac{(y_k - a_k)^2}{\lambda_k}\right) \\ \ell(\mathbf{y} \mid H_1) &= \frac{1}{\left(\sqrt{2\pi}\right)^n \prod_{k=1}^{n} \lambda_k^{\frac{1}{2}}} \exp\left(-\frac{1}{2}\sum_{k=1}^{n} \frac{(y_k - b_k)^2}{\lambda_k}\right) \end{aligned} \tag{10.61}$$

where $\{y_k\}$, $\{a_k\}$, and $\{b_k\}$ are the n-tuples representing received and transmitted signals as in (10.28). The likelihood ratio is

$$\begin{aligned} \lambda(\mathbf{y}) &= \frac{\ell(\mathbf{y} \mid H_1)}{\ell(\mathbf{y} \mid H_0)} \\ &= \exp\left(-\frac{1}{2}\sum_{k=1}^{n} \frac{1}{\lambda_k}\{(y_k - b_k)^2 - (y_k - a_k)^2\}\right) \end{aligned} \tag{10.62}$$

Because of the monotonicity of the exponential function, a threshold decision on $\lambda(\mathbf{y})$ is equivalent to a threshold decision on the exponent and the decision rule becomes

$$\text{Accept } H_0 \text{ if } -\frac{1}{2}\sum_{k=1}^{n}\frac{1}{\lambda_k}\{(y_k - b_k)^2 - (y_k - a_k)^2\} < \ln \lambda_0$$

or, equivalently,

$$\text{Accept } H_0 \text{ if } \sum_{k=1}^{n} y_k \frac{(b_k - a_k)}{\lambda_k} = \sum_{k=1}^{n} y_k \frac{d_k}{\lambda_k} < \ln \lambda_0 + \sum_{k=1}^{n}\frac{b_k^2}{2\lambda_k} - \sum_{k=1}^{n}\frac{a_k^2}{2\lambda_k} \tag{10.63}$$

The n-dimensional space is separated into decision regions R_0 and R_1 by the hyperplane defined by

$$\sum_{k=1}^{n} y_k \frac{d_k}{\lambda_k} = \sum_{k=1}^{n} y_k f_k = \text{constant} \tag{10.64}$$

Since the $\{\boldsymbol{\varphi}_k\}$ are given by (10.59), the sum in (10.64), for n sufficiently large, can be expressed as $(\mathbf{y}, \mathbf{f})$, where $f(t)$ is the solution of a Fredholm integral equation of the first kind.[8]

$$\int_T k_{uu}(t, s) f(s)\, ds = d(t); \qquad t \in T \tag{10.65}$$

Now the coordinate-free description of the decision rule is

$$\text{Accept } H_0 \text{ if } (\mathbf{y}, \mathbf{f}) < \ln \lambda_0 + \tfrac{1}{2}(\mathbf{s}_1, \mathbf{g}_1) - \tfrac{1}{2}(\mathbf{s}_0, \mathbf{g}_0) = r \tag{10.66}$$

where we have let $\mathbf{g}_1 - \mathbf{g}_0 = \mathbf{f}$, and

$$\begin{aligned} \int_T k_{uu}(t, s) g_1(s)\, ds &= s_1(t); \qquad t \in T \\ \int_T k_{uu}(t, s) g_0(s)\, ds &= s_0(t); \qquad t \in T \end{aligned} \tag{10.67}$$

Equation (10.65) presents some difficulties since the existence of a solution is questionable unless $d(t)$ or $k_{uu}(t, s)$ have certain types of singular behavior or unless the integration interval T becomes infinite. One practical approach to the solution of (10.65) is to assume $\mathbf{d}$ is approximated by a linear combination of eigenfunctions; then $\mathbf{f}$ is a linear combination of these same eigenfunctions with coefficients $f_k = d_k/\lambda_k$. Another approach[5] is to introduce a singularity into the kernel by assuming that the noise has a white component; i.e.,

$$k_{uu}(t, s) = k'_{uu}(t, s) + N_0\, \delta(t - s) \tag{10.68}$$

where $k'_{uu}(t, s)$ is the autocovariance of a finite-variance process. In this case, (10.65) becomes a Fredholm equation of the second kind,[8]

$$\int_T k'_{uu}(t, s) f(s)\, ds + N_0 f(t) = d(t); \qquad t \in T \tag{10.69}$$

and a solution will generally exist. Another possibility which should not be overlooked, in situations where transmitted signals can be modified somewhat, is to first select the receiver characteristic $f(t)$, then choose $d(t)$ according to (10.65).

Once $f(t)$ is determined, the receiver implementation can be accomplished as shown in Figure 10.6 or 10.7 with $f(t)$ replacing $d(t)$. An interesting correspondence with the earlier solution results when the noise process is stationary and the observation interval is long. In (10.65), we let the autocovariance function be $k_{uu}(t - s)$ and assume that the finite integration interval can be replaced by $(-\infty, \infty)$ without appreciably modifying the solution. Then, taking Fourier transforms,

$$F(f) \cong \frac{D(f)}{K_{uu}(f)} \qquad \text{for large } T \tag{10.70}$$

In the filter-sampler implementation, the filter transfer function in this case is

$$H(f) \cong \frac{D^*(f)}{K_{uu}(f)} e^{-j2\pi f t_0} \qquad \text{for large } T \tag{10.71}$$

which is the matched filter for the signal $\mathbf{d}$ in noise with power spectral density $K_{uu}(f)$ (see Section 9.4).

Receiver Performance

The Karhunen-Loève expansion was very helpful for determining the likelihood ratio receiver, but evaluation of receiver performance is difficult with a signal space defined in this manner because of the complicated nature of the integration limits in (10.12) and (10.13). It is possible to find a one-dimensional subspace such that components of the received signal outside this space can be disregarded. Thus the detection problem becomes one-dimensional as in the white noise case.[6] By analogy with the white noise problem, we are tempted at first to let this be the space spanned by $f(t)$. It is conceptually more revealing, however, to let S be the one-dimensional space spanned by $\chi_1(t) = \|\mathbf{d}\|^{-1} d(t)$ as we did previously. In the colored noise case, we use a different kind of projection of $\mathbf{y}$ onto S (not an orthogonal projection). Let $\tilde{S}$ be the subspace of all vectors orthogonal to $f(t)$. Now the entire space can be expressed as the direct sum of S and $\tilde{S}$, i.e., any signal can be expressed uniquely as the sum of a signal in S and a signal in $\tilde{S}$ (see Exercise 3.2).

$$\mathbf{y} = (\mathbf{y}, \boldsymbol{\psi}_1)\boldsymbol{\chi}_1 + \tilde{\mathbf{y}}; \qquad \tilde{\mathbf{y}} \in \tilde{S}, \quad (\tilde{\mathbf{y}}, \boldsymbol{\psi}_1) = 0$$

where[3]

$$(\boldsymbol{\chi}_1, \boldsymbol{\psi}_1) = 1 \Rightarrow \boldsymbol{\psi}_1 = \frac{\|\mathbf{d}\|}{(\mathbf{d}, \mathbf{f})} \mathbf{f} \tag{10.72}$$

The projections of the transmitted signals onto S are given by

$$\begin{aligned} a_1 &= (\mathbf{s}_0, \boldsymbol{\psi}_1) = \frac{\|\mathbf{d}\|}{(\mathbf{d}, \mathbf{f})} (\mathbf{s}_0, \mathbf{f}) \\ b_1 &= (\mathbf{s}_1, \boldsymbol{\psi}_1) = \frac{\|\mathbf{d}\|}{(\mathbf{d}, \mathbf{f})} (\mathbf{s}_1, \mathbf{f}) \end{aligned} \tag{10.73}$$

as shown in Figure 10.11. The threshold hyperplane normal to $\mathbf{f}$, (10.64),

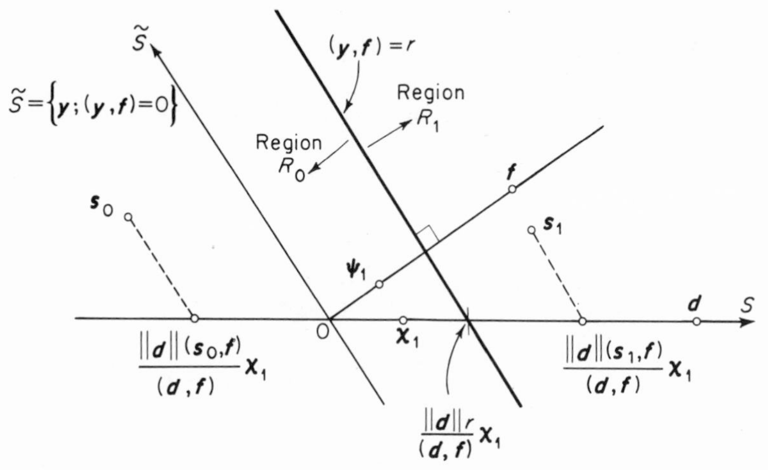

Figure 10.11. Signal space representation for likelihood ratio test with additive colored noise. Note that $\tilde{S}$ is not orthogonal to S.

intersects S at

$$\frac{\|\mathbf{d}\| \, r}{(\mathbf{d}, \mathbf{f})} \boldsymbol{\chi}_1$$

Components of noise in S are uncorrelated with (hence statistically independent of) the corresponding components in $\tilde{S}$ (Exercise 10.3). The variance of the coefficient of the noise projection is

$$\begin{aligned} E[\mathbf{u}_1^2] &= \iint\limits_T k_{uu}(t, s)\psi_1(s)\psi_1^*(t) \, ds \, dt \\ &= \frac{\|\mathbf{d}\|^2}{(\mathbf{d}, \mathbf{f})^2} \iint\limits_T k_{uu}(t, s) f(s) f^*(t) \, ds \, dt = \frac{\|\mathbf{d}\|^2}{(\mathbf{d}, \mathbf{f})} \end{aligned} \tag{10.74}$$

[3] It is helpful to think of $\boldsymbol{\psi}_1$ as the first vector in the basis reciprocal to $\{\boldsymbol{\chi}_1, \boldsymbol{\chi}_2, \boldsymbol{\chi}_3, \ldots\}$ spanning $S \oplus \tilde{S}$.

The likelihood functions for $y_1 = (\mathbf{y}, \boldsymbol{\psi}_1)$ are thus given by

$$\ell(y_1 \mid H_0) = \frac{(\mathbf{d}, \mathbf{f})^{\frac{1}{2}}}{\sqrt{2\pi}\, \|\mathbf{d}\|} e^{-\frac{(\mathbf{d},\mathbf{f})}{2(\mathbf{d},\mathbf{d})}(y_1 - a_1)^2}$$
$$\ell(y_1 \mid H_1) = \frac{(\mathbf{d}, \mathbf{f})^{\frac{1}{2}}}{\sqrt{2\pi}\, \|\mathbf{d}\|} e^{-\frac{(\mathbf{d},\mathbf{f})}{2(\mathbf{d},\mathbf{d})}(y_1 - b_1)^2} \tag{10.75}$$

and the receiver performance, from (10.12) and (10.13), is

$$P_m = \int_{R_0} \ell(\sigma \mid H_1)\, d\sigma = \frac{(\mathbf{d}, \mathbf{f})^{\frac{1}{2}}}{\sqrt{2\pi}\, \|\mathbf{d}\|} \int_{-\infty}^{\frac{r\|\mathbf{d}\|}{(\mathbf{d},\mathbf{f})}} e^{-\frac{(\mathbf{d},\mathbf{f})}{2(\mathbf{d},\mathbf{d})}(\sigma - b_1)^2}\, d\sigma$$
$$= \frac{1}{\sqrt{2\pi}} \int_{-\infty}^{\alpha_1} e^{-\frac{\omega^2}{2}}\, d\omega = \Phi(\alpha_1)$$

where, using (10.66) and the hint in Exercise 10.4,

$$\alpha_1 = \frac{\ln \lambda_0}{2\rho} - \rho \tag{10.76}$$

Similarly,

$$P_f = \int_{R_1} \ell(\sigma \mid H_0)\, d\sigma = \frac{(\mathbf{d}, \mathbf{f})^{\frac{1}{2}}}{\sqrt{2\pi}\, \|\mathbf{d}\|} \int_{\frac{r\|\mathbf{d}\|}{(\mathbf{d},\mathbf{f})}}^{\infty} e^{-\frac{(\mathbf{d},\mathbf{f})}{2(\mathbf{d},\mathbf{d})}(\sigma - a_1)^2}\, d\sigma$$
$$= \frac{1}{\sqrt{2\pi}} \int_{\alpha_2}^{\infty} e^{-\frac{\omega^2}{2}}\, d\omega = 1 - \Phi(\alpha_2) = \Phi(-\alpha_2)$$

where

$$\alpha_2 = \frac{\ln \lambda_0}{2\rho} + \rho \tag{10.77}$$

and we have defined

$$\rho \triangleq \tfrac{1}{2}(\mathbf{d}, \mathbf{f})^{\frac{1}{2}} \tag{10.78}$$

From these relations, we see that characterization of receiver performance is exactly the same as in the white noise case except that the signal-to-noise ratio parameter ρ is given by $\frac{1}{2}(\mathbf{d}, \mathbf{f})^{\frac{1}{2}}$ instead of $\|\mathbf{d}\|/2\sqrt{N_0}$. In the white noise case, receiver performance did not depend on the shape of the signal waveforms, only on the amount of energy in the interval T. In the colored noise case, performance does depend on the shape of signal waveforms. Rather than working out the solution for particular examples, we can more effectively demonstrate this in the general case by choosing $d(t)$ equal to one of the eigenfunctions in (10.59). If $d(t) = \varphi_k(t)$, then, from (10.65), $f(t) = (1/\lambda_k)\varphi_k(t)$. In this case, $\mathbf{f} \in S$ and the projections onto S are orthogonal projections (see Figure 10.11). The signal-to-noise ratio parameter is simply

$$\rho = \frac{1}{2}\left(\frac{1}{\lambda_k}\right)^{\frac{1}{2}} \tag{10.79}$$

Clearly, the performance is better when $d(t)$ is an eigenfunction with a smaller eigenvalue. This is intuitively appealing, since it says that the best signals are the ones that least resemble the additive noise.

Exercise 10.3. For noise with autocorrelation function $k_{uu}(t, s)$, let $\mathbf{u}_1$ be the projection of $\mathbf{u}$ onto S along $\tilde{S}$ and $\mathbf{u}_2$ be the projection of $\mathbf{u}$ onto $\tilde{S}$ along S as shown in Figure 10.11. Show that

$$E[\mathbf{u}_1(t)\mathbf{u}_2(s)] = 0; \qquad t, s \in T$$

Exercise 10.4. Derive equations (10.76) and (10.77). *Hint:* Show that $(\mathbf{s}_0, \mathbf{g}_1) = (\mathbf{s}_1, \mathbf{g}_0)$.

Exercise 10.5. Let $s_0(t)$ and $s_1(t)$, with $s_0(t) = -s_1(t)$, be equiprobable, unit-energy, rectangular pulses in Gaussian noise with spectral density $K_{uu}(f)$ of a lowpass nature. Show the implementation for the maximum-likelihood receiver in the limiting case as the pulse width becomes very small compared to the reciprocal of the noise bandwidth. Express error performance for this case.

10.6 Bandpass-type Signals—Non-coherent Detection

As noted in Chapter 4, many physical systems employ signals of a bandpass nature. These signals might result, for example, from amplitude modulation of a sinusoidal carrier with frequency f_0 by signal waveforms whose highest frequency components are considerably less than f_0 (see Figure 10.12).

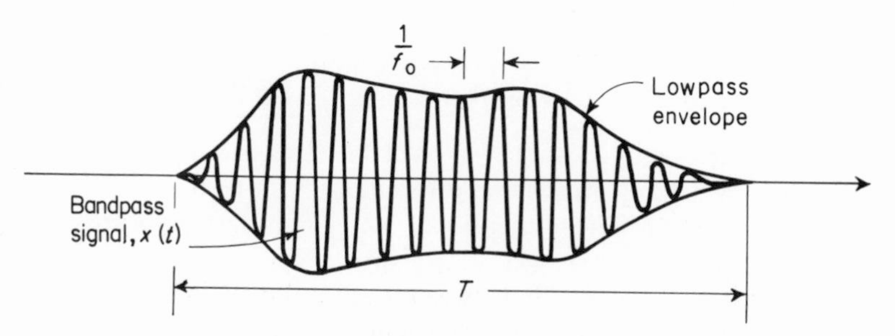

Figure 10.12. Typical bandpass-type signal pulse.

The foregoing signal detection methods are, of course, applicable to this type of signal but a special problem exists which makes a separate treatment worthwhile. This is the problem of carrier phase uncertainty and it can arise in the following way. The likelihood ratio receiver forms the inner product $(\mathbf{y}, \mathbf{d})$ of the received signal $y(t)$ and a reference signal $d(t)$. When $y(t)$ and $d(t)$ have a highly oscillatory character as illustrated in Figure 10.12, then signal arrival time must be very accurately known if $(\mathbf{y}, \mathbf{d})$ is to have any significance in hypothesis testing. We can demonstrate this as follows (see Exercise 4.8).

$$(\mathbf{y}, \mathbf{d}) - (\mathbf{y}', \mathbf{d}) = \tfrac{1}{2} \operatorname{Re} [(1 - e^{j\theta})(\boldsymbol{\omega}, \boldsymbol{\gamma})] \tag{10.80}$$

where $\omega(t)$ and $\gamma(t)$ are the complex envelopes of $y(t)$ and $d(t)$, and $y'(t)$ is a phase-shifted (by an amount θ) version of $y(t)$. Now if $y(t)$ (or $d(t)$) is time-shifted by a small amount Δ compared to T, the factor $(\boldsymbol{\omega}, \boldsymbol{\gamma})$ is not significantly different but the factor $1 - e^{j\theta}$ may be greatly changed. A time shift $\Delta = 1/2f_0$ completely reverses the polarity of the receiver output. It is clear that timing must be accurate to a small fraction of $1/2f_0$ rather than a small fraction of T as would be required in the lowpass case. As a practical matter, it may not be feasible to establish such precise timing and an alternate form of detection must be employed. A receiver which does not make use of any carrier phase information is called a *non-coherent* detector. The development of such a receiver based on likelihood ratio tests is the topic of this section.

Suppose that the transmitted signals $s_0(t)$ and $s_1(t)$ are subjected to a phase shift θ and additive, zero-mean, white Gaussian noise having a spectral density of N_0 watts/Hz. Then on the basis of the received signal $y(t)$ we want to decide between H_0 and H_1 where

$$\begin{aligned} H_0&: y(t) = s_0'(t) + u(t); \qquad t \in T \\ H_1&: y(t) = s_1'(t) + u(t); \qquad t \in T \end{aligned} \tag{10.81}$$

and where $s_0'(t)$ and $s_1'(t)$ are the phase-shifted versions of the transmitted signals. In terms of complex envelopes relative to f_0, (10.81) becomes

$$\begin{aligned} H_0&: \omega(t) = e^{j\theta}\alpha(t) + \eta(t); \qquad t \in T \\ H_1&: \omega(t) = e^{j\theta}\beta(t) + \eta(t); \qquad t \in T \end{aligned} \tag{10.82}$$

We proceed in exactly the same way as in the previous white noise problem (Section 10.4) by making orthogonal projections of the complex envelopes of received signals onto a finite-dimensional (complex) subspace S spanned by the orthonormal set $\{\varphi_k(t); k = 1, 2, \ldots, n\}$.

There is a simple relationship between the real and imaginary parts of the complex inner products and the corresponding inner products of the real bandpass signals. This relationship will be used frequently so we write it here for reference.

$$\begin{aligned} (\mathbf{x}_1, \mathbf{x}_2) &= \tfrac{1}{2} \operatorname{Re} (\boldsymbol{\gamma}_1, \boldsymbol{\gamma}_2) \\ (\mathbf{x}_1, \hat{\mathbf{x}}_2) &= -\tfrac{1}{2} \operatorname{Im} (\boldsymbol{\gamma}_1, \boldsymbol{\gamma}_2) \end{aligned} \tag{10.83}$$

where $x_1(t)$ and $x_2(t)$ are real bandpass signals with complex envelopes $\gamma_1(t)$ and $\gamma_2(t)$ and $\hat{x}_2(t)$ is a $90°$ phase-shifted version of $x_2(t)$; i.e., $\hat{x}_2(t) = \operatorname{Re} [\gamma_2(t)e^{-j\frac{\pi}{2}}e^{j2\pi f_0 t}]$ (see Exercise 4.8).

Indicating real and imaginary components by subscripts R and I, we let

$$\begin{aligned}(\boldsymbol{\omega}, \boldsymbol{\varphi}_k) &= w_k = w_{R_k} + jw_{I_k} \\ (\boldsymbol{\alpha}, \boldsymbol{\varphi}_k) &= a_k = a_{R_k} + ja_{I_k} \\ (\boldsymbol{\beta}, \boldsymbol{\varphi}_k) &= b_k = b_{R_k} + jb_{I_k} \\ (\boldsymbol{\eta}, \boldsymbol{\varphi}_k) &= n_k = n_{R_k} + jn_{I_k}\end{aligned} \tag{10.84}$$

Using (10.83) and $k_{uu}(t, s) = N_0\,\delta(t - s)$ for the autocovariance of the bandpass noise, it is easy to show that the real and imaginary noise components are all uncorrelated and each component has a variance $2N_0$. It follows, similar to (10.30), that the likelihood function for $\boldsymbol{\omega}$, conditional on H_0 and for a given θ, is

$$\ell(\boldsymbol{\omega} \mid H_0 \text{ and } \theta) = \frac{1}{(4\pi N_0)^n} \exp\left(-\frac{1}{4N_0}\left\{\sum_{k=1}^{n}(w_{R_k} - a_{R_k}\cos\theta + a_{I_k}\sin\theta)^2 + (w_{I_k} - a_{R_k}\sin\theta - a_{I_k}\cos\theta)^2\right\}\right) \tag{10.85}$$

Expanding the exponent in (10.85), rearranging terms, and writing the terms involving θ in polar form, we obtain

$$\ell(\boldsymbol{\omega} \mid H_0 \text{ and } \theta) = \frac{1}{(4\pi N_0)^n} \exp\left(-\frac{1}{4N_0}\{A_0 + B_0 - 2C_0\cos(\theta + \xi)\}\right)$$

where

$$\begin{aligned}A_0 &= \sum_{k=1}^{n} w_{R_k}^2 + w_{I_k}^2 \\ B_0 &= \sum_{k=1}^{n} a_{R_k}^2 + a_{I_k}^2 \\ C_0^2 &= \left(\sum_{k=1}^{n} w_{R_k}a_{R_k} + w_{I_k}a_{I_k}\right)^2 + \left(\sum_{k=1}^{n} w_{R_k}a_{I_k} - w_{I_k}a_{R_k}\right)^2\end{aligned} \tag{10.86}$$

Now we want to interpret the phase shift as a random disturbance introduced by the channel. It could also be interpreted as a random parameter associated with the transmitter or as an uncertainty in the timing of reference signals at the receiver. Interpreting $\boldsymbol{\theta}$ as a random channel parameter, we characterize its effect on the detection problem by modifying the likelihood functions for the channel. This is done by assuming a probability density function for $\boldsymbol{\theta}$, whereupon

$$\ell(\boldsymbol{\omega} \mid H_0) = \int_0^{2\pi} \ell(\boldsymbol{\omega} \mid H_0 \text{ and } \theta = \sigma)p_\theta(\sigma)\,d\sigma \tag{10.87}$$

Expressions for the likelihood functions have been worked out for a variety of density functions for $\boldsymbol{\theta}$.[2] The case which is of most frequent practical interest is the noncoherent case with $\boldsymbol{\theta}$ distributed uniformly over 0 to 2π. Other density functions lead to receivers which are called *partially coherent.*

Using the uniform density, (10.87) becomes

$$\ell(\boldsymbol{\omega} \mid H_0) = \frac{1}{(4\pi N_0)^n} e^{-\frac{1}{4N_0}(A_0+B_0)} \int_0^{2\pi} e^{\frac{C_0}{2N_0}\cos(\sigma+\xi)} \frac{d\sigma}{2\pi} \tag{10.88}$$

The integral in (10.88) can be expressed as a known function of the argument $C_0/2N_0$. This function is $I_0(x)$, called the *modified Bessel function of the first kind, order zero.* Repeating the same steps for the likelihood function conditional on H_1, we can write the likelihood ratio.

$$\lambda(\boldsymbol{\omega}) = \frac{\ell(\boldsymbol{\omega} \mid H_1)}{\ell(\boldsymbol{\omega} \mid H_0)} = e^{\frac{1}{4N_0}(B_0-B_1)} \cdot \frac{I_0(C_1/2N_0)}{I_0(C_0/2N_0)} \tag{10.89}$$

where B_1 and C_1 are the quantities corresponding to B_0 and C_0 in (10.86) with b_{R_k} and b_{I_k} replacing a_{R_k} and a_{I_k}. These quantities can be expressed as inner products in $L^2(T)$ provided n is large enough so that S contains the transmitted signals $\alpha(t)$ and $\beta(t)$. (If the basis functions were chosen properly, $n = 2$ would suffice.)

$$B_0 = \sum_{k=1}^{n} a_{R_k}^2 + a_{I_k}^2 = \sum_{k=1}^{n} |a_k|^2 = (\boldsymbol{\alpha}, \boldsymbol{\alpha})$$

$$B_1 = (\boldsymbol{\beta}, \boldsymbol{\beta})$$

$$\sum_{k=1}^{n} w_{R_k} a_{R_k} + w_{I_k} a_{I_k} = \operatorname{Re}(\boldsymbol{\omega}, \boldsymbol{\alpha})$$

$$\sum_{k=1}^{n} w_{R_k} a_{I_k} - w_{I_k} a_{R_k} = -\operatorname{Im}(\boldsymbol{\omega}, \boldsymbol{\alpha})$$

Hence,

$$C_0^2 = |(\boldsymbol{\omega}, \boldsymbol{\alpha})|^2; \qquad C_1^2 = |(\boldsymbol{\omega}, \boldsymbol{\beta})|^2 \tag{10.90}$$

Using these relations, we see that the non-coherent likelihood ratio receiver must evaluate the magnitude of inner products of the received signals with reference versions of the transmitted signals. The decision rule, from (10.89) and (10.90), is

$$\text{Accept } H_0 \text{ if } I_0\left(\frac{|(\boldsymbol{\omega}, \boldsymbol{\beta})|}{2N_0}\right) < \lambda_0 e^{\frac{1}{4N_0}[(\boldsymbol{\beta},\boldsymbol{\beta})-(\boldsymbol{\alpha},\boldsymbol{\alpha})]} I_0\left(\frac{|(\boldsymbol{\omega}, \boldsymbol{\alpha})|}{2N_0}\right) \tag{10.91}$$

The decision rule is considerably simplified in the case apt to be of most interest in binary data communication. In this case, we would probably use signals of equal energy and assign equal costs and *a priori* probabilities. Then $\lambda_0 = 1$; $(\boldsymbol{\beta}, \boldsymbol{\beta}) - (\boldsymbol{\alpha}, \boldsymbol{\alpha}) = 0$; and since $I_0(x)$ is a monotonic function of its argument, the decision rule is

$$\text{Accept } H_0 \text{ if } |(\boldsymbol{\omega}, \boldsymbol{\beta})| - |(\boldsymbol{\omega}, \boldsymbol{\alpha})| < 0 \tag{10.92}$$

A conceptual form of the receiver, using complex envelopes, is shown in Figure 10.13. To show an actual receiver implementation, we use the

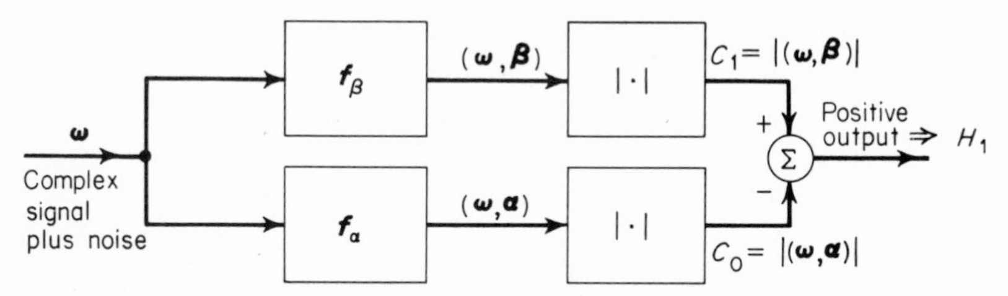

Figure 10.13. Complex envelope representation of non-coherent maximum-likelihood receiver for equal energy binary signals in white noise.

relations in (10.83).

$$\begin{aligned} \text{Re}\,(\boldsymbol{\omega}, \boldsymbol{\alpha}) &= 2(\mathbf{y}, \mathbf{s}_0) \\ \text{Im}\,(\boldsymbol{\omega}, \boldsymbol{\alpha}) &= -2(\mathbf{y}, \hat{\mathbf{s}}_0) \end{aligned} \tag{10.93}$$

Hence

$$C_0^2 = |(\boldsymbol{\omega}, \boldsymbol{\alpha})|^2 = 4[(\mathbf{y}, \mathbf{s}_0)^2 + (\mathbf{y}, \hat{\mathbf{s}}_0)^2]$$

and

$$C_1^2 = |(\boldsymbol{\omega}, \boldsymbol{\beta})|^2 = 4[(\mathbf{y}, \mathbf{s}_1)^2 + (\mathbf{y}, \hat{\mathbf{s}}_1)^2] \tag{10.94}$$

A multiplier-integrator implementation, along with squarers and 90° phase shifters, follows directly from (10.94). This is shown in Figure 10.14.

The complex envelope form in Figure 10.13 suggests a filter-sampler implementation which might be preferable from a practical standpoint because envelope detectors can be quite simply implemented. Since the

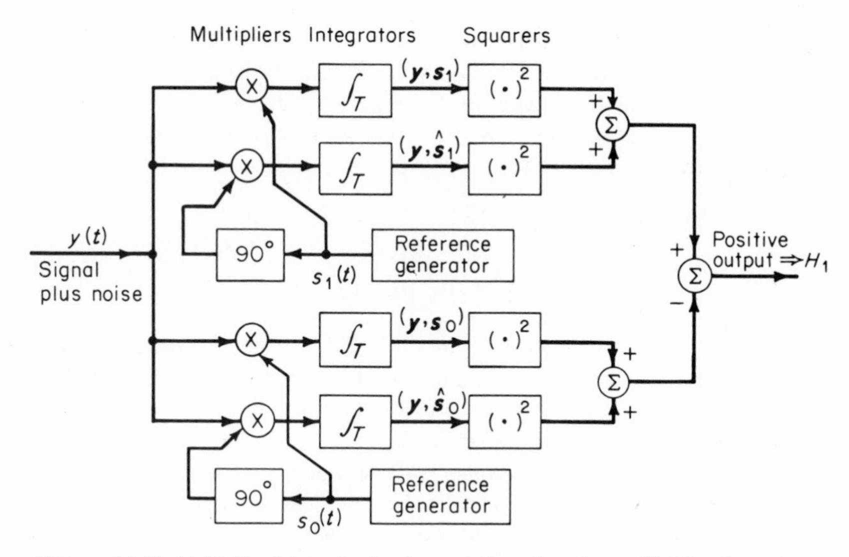

Figure 10.14. Multiplier-integrator implementation of maximum-likelihood receiver.

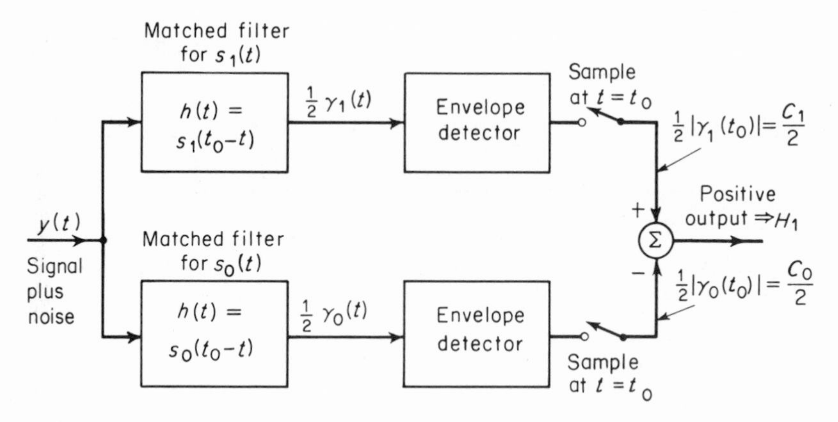

Figure 10.15. Filter-sampler implementation of maximum-likelihood receiver.

magnitude of a complex envelope is the envelope of the corresponding bandpass signal [see (4.40) and (4.46)], the desired quantities C_1 and C_0 can be obtained by sampling the outputs of the envelope detectors in the receiver shown in Figure 10.15. From (4.57) we can write

$$\begin{aligned}\gamma_1(t_0) &= \tfrac{1}{2}\int_{-\infty}^{\infty}\omega(\tau)\lambda(t_0-\tau)\,d\tau \\ &= \tfrac{1}{2}(\boldsymbol{\omega}, \boldsymbol{\beta}) \qquad \text{if } \lambda(t) = \beta(t_0 - t)\end{aligned} \tag{10.95}$$

where $\lambda(t)$ is the complex envelope of the impulse response of a real bandpass filter. The impulse response of the bandpass filter in the top path is $s_1(t_0 - t)$; hence it is the matched filter for one of the transmitted signals.

In selecting fixed signals for binary communication, it seems reasonable that we would want the maximum difference in receiver output for the two signals in the noise-free case. With $\eta(t) = 0$, the output difference of the configuration in Figure 10.13 is

$$\begin{aligned}(C_1 - C_0 \mid H_1, \eta = 0) &- (C_1 - C_0 \mid H_0, \eta = 0) \\ &= |(\boldsymbol{\beta}, \boldsymbol{\beta})| - |(\boldsymbol{\beta}, \boldsymbol{\alpha})| - |(\boldsymbol{\alpha}, \boldsymbol{\beta})| + |(\boldsymbol{\alpha}, \boldsymbol{\alpha})| \\ &= 2\,\|\boldsymbol{\alpha}\|^2 - 2\,|(\boldsymbol{\alpha}, \boldsymbol{\beta})|\end{aligned} \tag{10.96}$$

The quantity in (10.96) is clearly maximized by using orthogonal complex envelopes, $(\boldsymbol{\alpha}, \boldsymbol{\beta}) = 0$. This is in contrast to the completely coherent case in Section 10.4 where output difference was maximized by $s_0(t) = -s_1(t)$. This polarity reversal signaling scheme is obviously inappropriate for the noncoherent receiver since a 180° phase shift on one signal would make it look exactly like the other signal. It should also be noted that orthogonality of the bandpass signals is not sufficient since $(\mathbf{s}_0, \mathbf{s}_1) = 0$ does not imply that $(\boldsymbol{\alpha}, \boldsymbol{\beta}) = 0$, although the converse is true. Orthogonality of the complex envelopes can be accomplished in many ways. In particular, it is possible to obtain orthogonal complex envelopes while maintaining constant envelopes of the bandpass signal, i.e., by pure phase modulation. This is of practical significance in situations where signal energy constraints are imposed in terms of peak signal value. Also, in phase modulation systems, some degree of immunity against impulsive type interference can be obtained with peak signal limiters at the receiver.

Example 10.1 *Frequency-shift keying:* A simple way to obtain orthogonal complex envelopes with bandpass signals having constant envelope is to let

$$\begin{aligned} \alpha(t) &= e^{-j2\pi\nu t}; \qquad 0 \leqslant t < T \\ \beta(t) &= e^{+j2\pi\nu t}; \qquad 0 \leqslant t < T \end{aligned} \tag{10.97}$$

In this case,

$$(\boldsymbol{\alpha}, \boldsymbol{\beta}) = \int_0^T e^{-j2\pi(2\nu)t}\, dt = e^{-j2\pi\nu t} \cdot \frac{\sin 2\pi\nu T}{2\pi\nu} \tag{10.98}$$

These complex envelopes simply introduce positive and negative frequency shifts from f_0 and the signaling scheme is referred to as *frequency-shift keying.* The smallest value of ν which makes the complex envelopes orthogonal, from (10.98), is $\nu = 1/2T$. The magnitude of the transfer functions in the matched filter-sampler implementation are sketched in Figure 10.16. It can be seen that the receiver functions as a crude form of a frequency discriminator. ▌

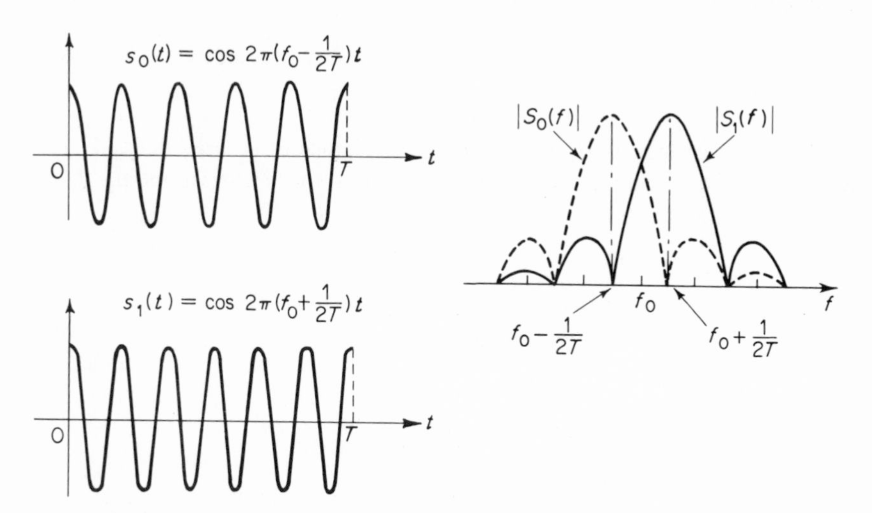

Figure 10.16. Signaling waveforms and bandpass filter functions for frequency-shift keying.

Receiver Performance

In order to evaluate receiver performance for the case of equal-energy, orthogonal complex envelope signals, it is convenient to use normalized versions of the signals as basis functions for S. Hence $\varphi_1(t) = \|\boldsymbol{\alpha}\|^{-1}\alpha(t)$ and $\varphi_2(t) = \|\boldsymbol{\beta}\|^{-1}\beta(t)$. Then $a_1 = (\boldsymbol{\alpha}, \boldsymbol{\varphi}_1) = \|\boldsymbol{\alpha}\|$ and $a_2 = (\boldsymbol{\alpha}, \boldsymbol{\varphi}_2) = 0$ and, from (10.88), the likelihood function conditional on H_0 for the orthogonal projections of received signals onto S is given by

$$\ell(w_{R_1}, w_{I_1}, w_{R_2}, w_{I_2} \mid H_0) = \frac{e^{-\frac{\|\boldsymbol{\alpha}\|^2}{N4_0}}}{(4\pi N_0)^2} e^{-\frac{1}{4N_0}[(w_{R_1}{}^2 + w_{I_1}{}^2) + (w_{R_2}{}^2 + w_{I_2}{}^2)]} I_0\left(\frac{\|\boldsymbol{\alpha}\| \sqrt{w_{R_1}^2 + w_{I_1}^2}}{2N_0}\right) \tag{10.99}$$

Changing to a polar coordinate description of the variables

$$\begin{aligned} w_{R_1} &= z_0 \cos \mu;\ w_{I_1} = z_0 \sin \mu \Rightarrow dw_{R_1}\, dw_{I_1} = z_0\, dz_0\, d\mu \\ w_{R_2} &= z_1 \cos \nu;\ w_{I_2} = z_1 \sin \nu \Rightarrow dw_{R_2}\, dw_{I_2} = z_1\, dz_1\, d\nu \end{aligned} \tag{10.100}$$

we can now write a new likelihood function with two of the variables, z_0 and z_1, proportional to receiver outputs on each of the two paths in the configuration shown in Figure 10.13. This likelihood function, from (10.99) and (10.100), is

$$\ell(z_0, z_1, \mu, \nu \mid H_0) = \frac{1}{(4\pi N_0)^2} e^{-\frac{\|\boldsymbol{\alpha}\|^2}{4N_0}} e^{-\frac{1}{4N_0}(z_0{}^2 + z_1{}^2)} I_0\left(\frac{\|\boldsymbol{\alpha}\|\, z_0}{2N_0}\right) \tag{10.101}$$

The probability of a type I error (accepting H_1 when H_0 is true) is given by integrating the likelihood function over the critical region R_1, (10.12).

$$P_f = \underbrace{\iiiint}_{R_1} \ell(z_0, z_1, \mu, \nu \mid H_0) z_0 z_1\, dz_0\, dz_1\, d\mu\, d\nu \tag{10.102}$$

The receiver accepts H_1 whenever $z_1 > z_0$; hence R_1 is easily defined. The portion of the two-dimensional receiver output space corresponding to R_1 is shown in Figure 10.17.

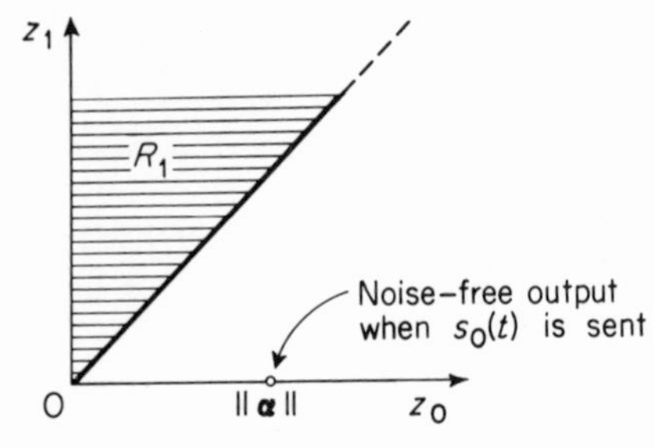

Figure 10.17. Critical region in receiver output space for non-coherent maximum-likelihood receiver.

Substituting (10.101) into (10.102), we have

$$P_f = \Pr[z_1 > z_0 \mid H_0]$$
$$= \frac{1}{(4\pi N_0)^2} e^{-\frac{\|\boldsymbol{\alpha}\|^2}{4N_0}} \int_0^{2\pi} d\mu \int_0^{2\pi} d\nu \int_0^{\infty} z_0 \, dz_0 e^{-\frac{z_0^2}{4N_0}} I_0\left(\frac{z_0 \|\boldsymbol{\alpha}\|}{2N_0}\right) \int_{z_0}^{\infty} z_1 e^{-\frac{z_1^2}{4N_0}} dz_1 \tag{10.103}$$

The integration over μ, ν, and z_1 in (10.103) is easily evaluated. When this is done, we find

$$P_f = \frac{1}{2N_0} e^{-\frac{\|\boldsymbol{\alpha}\|^2}{4N_0}} \int_0^{\infty} z_0 e^{-\frac{z_0^2}{2N_0}} I_0\left(\frac{z_0 \|\boldsymbol{\alpha}\|}{2N_0}\right) dz_0 \tag{10.104}$$

Making the change of variables, $z^2 = z_0^2/N_0$ and $a^2 = \|\boldsymbol{\alpha}\|^2/4N_0$, (10.104) can be written as

$$P_f = \tfrac{1}{2} e^{-\frac{\|\boldsymbol{\alpha}\|^2}{8N_0}} \int_0^{\infty} z e^{-\frac{1}{2}(z^2+a^2)} I_0(az) \, dz$$
$$= \tfrac{1}{2} e^{-\frac{\|\boldsymbol{\alpha}\|^2}{8N_0}} Q(a, 0) = \tfrac{1}{2} e^{-\frac{\|\boldsymbol{\alpha}\|^2}{8N_0}} \tag{10.105}$$

The integral in (10.105), with a variable lower limit, appears frequently in signal detection studies.[2],[5] It is often referred to as the *Marcum Q-function* and tabulated values are available.[7]

$$Q(a, b) \triangleq \int_b^{\infty} z e^{-\frac{1}{2}(z^2+a^2)} I_0(az) \, dz$$

and

$$Q(a, 0) = 1; \qquad Q(0, b) = e^{-\frac{b^2}{2}} \tag{10.106}$$

Because of symmetry in this problem, the likelihood function conditional on H_1 has the same form as (10.101) with z_0 and z_1 interchanged. Hence $P_m = P_f = P_e$ in the case of equiprobable signals. Noting that $\|\boldsymbol{\alpha}\|^2 = 2(\mathbf{s}_0, \mathbf{s}_0) = 2(\mathbf{s}_1, \mathbf{s}_1)$, the overall error probability has the simple expression

$$P_e = \tfrac{1}{2} e^{-\frac{1}{4}\rho^2}; \qquad \rho^2 = \frac{(\mathbf{s}_0, \mathbf{s}_0)}{N_0} \tag{10.107}$$

where ρ, the signal-to-noise ratio parameter, is the ratio of signal pulse energy to noise power spectral density as in (10.57). From (10.107), the error performance graph can be constructed and comparisons with the phase-coherent case can be made (Figure 10.10). One way to ascribe noise impairment values to the random phase channel might be to compare performance with a coherent case using orthogonal signals since the non-coherent case is evaluated for orthogonal complex envelopes. For orthogonal signals, $(\mathbf{d}, \mathbf{d}) = (\mathbf{s}_0, \mathbf{s}_0) + (\mathbf{s}_1, \mathbf{s}_1) = 2(\mathbf{s}_0, \mathbf{s}_0)$ and $P_e = \Phi(-\rho/\sqrt{2})$ in the coherent case, where ρ is defined as in (10.107). These error performances are shown

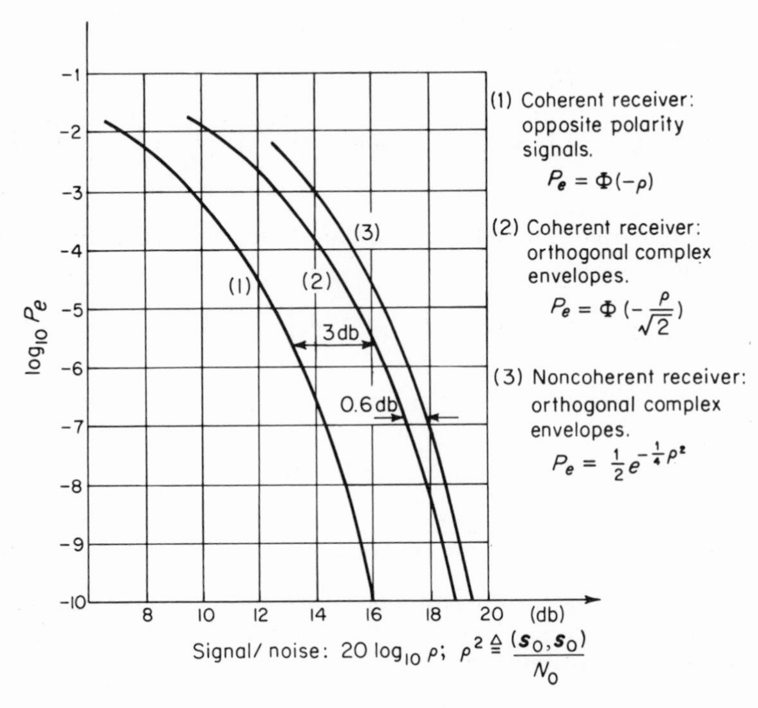

Figure 10.18. Error performance for coherent and non-coherent receivers. (Equal energy signals.)

in Figure 10.18. At large signal-to-noise ratio, the noise impairment introduced by the random phase channel becomes negligible. For error probabilities in the neighborhood of 10^{-7}, the impairment is only about 0.6 db. In these high signal-to-noise ratio situations, we conclude that the main penalty of non-coherent operation is in the 3-db loss of signal-to-noise ratio resulting from the necessity of using orthogonal signals rather than opposite polarity signals.

For radar and sonar detection problems, the situation where $s_0(t) = 0$ is of considerable importance. In this case, the likelihood ratio, from (10.89), is

$$\lambda(\boldsymbol{\omega}) = e^{-\frac{\|\boldsymbol{\beta}\|^2}{4N_0}} I_0\left(\frac{|(\boldsymbol{\omega}, \boldsymbol{\beta})|}{2N_0}\right) \tag{10.108}$$

Because of the monotonicity of $I_0(x)$, a likelihood ratio test is equivalent to a threshold test on $|(\boldsymbol{\omega}, \boldsymbol{\beta})|$ and the decision rule is

$$\text{Accept } H_0 \text{ if } |(\boldsymbol{\omega}, \boldsymbol{\beta})| < r;$$

where r is given implicitly by

$$I_0\left(\frac{r}{2N_0}\right) = \lambda_0 e^{\frac{\|\boldsymbol{\beta}\|^2}{4N_0}} \tag{10.109}$$

The receiver is the same as shown previously (Figures 10.13–10.15), with the lower half missing and at the output a threshold device set at the value r, determined either from assigned costs and *a priori* probabilities or from a prescribed false alarm probability P_f. Performance expressions are not as simple as for the equal energy, two-signal case (see Exercise 10.7). The receiver operating characteristic, obtained by numerical evaluation of the Q-functions, is presented in References [2] and [5].

Exercise 10.6. Show that the real and imaginary parts of the complex noise coefficients in (10.84) are uncorrelated; i.e.,

$$E[\mathbf{n}_{R_k}\mathbf{n}_{R_j}] = E[\mathbf{n}_{I_k}\mathbf{n}_{I_j}] = 0 \qquad \text{for } k \neq j$$

and

$$E[\mathbf{n}_{R_k}\mathbf{n}_{I_j}] = 0 \qquad \text{for all } k \text{ and } j$$

and that each component has variance $2N_0$.

Exercise 10.7. For the detection problem with $s_0(t) = 0$ and a decision threshold r given by (10.109), show that

$$P_f = Q\left(0, \frac{r}{\sqrt{2N_0}\,\|\boldsymbol{\beta}\|}\right) = e^{-\frac{r^2}{4N_0\|\boldsymbol{\beta}\|^2}}$$

$$1 - P_m = Q\left(\frac{\|\boldsymbol{\beta}\|}{\sqrt{2N_0}}, \frac{r}{\sqrt{2N_0}\,\|\boldsymbol{\beta}\|}\right)$$

where $Q(a, b)$ is defined in (10.106).

REFERENCES

1. J. M. Wozencraft and I. M. Jacobs, *Principles of Communication Engineering*, John Wiley & Sons, 1965.
2. H. L. Van Trees, *Detection, Estimation, and Modulation Theory*, Part I, John Wiley & Sons, 1968.
3. A. Papoulis, *Probability, Random Variables and Stochastic Processes*, McGraw-Hill, 1965.
4. W. R. Bennett and J. R. Davey, *Data Transmission*, McGraw-Hill, 1965.
5. C. W. Helstrom, *Statistical Theory of Signal Detection*, Pergamon Press, 1960.
6. T. Kailath, "A Projection Method for Signal Detection in Colored Gaussian Noise," *Trans. IEEE*, Vol. IT-13, No. 3, pp. 441–47 (July, 1967).
7. J. I. Marcum, "Table of Q-functions," Rand Corporation Rpt. RM-339, January, 1950.
8. R. Courant and D. Hilbert, *Methods of Mathematical Physics*, Vol. I, Interscience, 1953.

INDEX